Advances in

ECOLOGICAL RESEARCH

VOLUME 19

Advances in
ECOLOGICAL RESEARCH

Edited by

M. BEGON

Department of Zoology, University of Liverpool, Liverpool, L69 3BX, UK

A. H. FITTER

Department of Biology, University of York, York, YO1 5DD, UK

E. D. FORD

Center for Quantitative Science, University of Washington, 3737 15th Avenue, Seattle, WA 98195, USA

A. MACFADYEN

23 Mountsandel Road, Coleraine, Northern Ireland

VOLUME 19

1989

ACADEMIC PRESS

Harcourt Brace Jovanovich, Publishers
London San Diego New York Berkeley
Boston Sydney Tokyo Toronto

This book is printed on acid-free paper ♾

ACADEMIC PRESS LTD.
24/28 Oval Road
London NW1

United States Edition published by
ACADEMIC PRESS INC.
San Diego, CA 92101

British Library Cataloguing in Publication Data
Advances in ecological research.
Vol. 19
1. Ecology
I. Begon, Michael
574.5

ISBN 0–12–013919–7

Typeset by Latimer Trend & Company Ltd, Plymouth
Printed in Great Britain by T. J. Press (Padstow) Ltd., Padstow, Cornwall

Contributors to Volume 19

F. ATHIAS-BINCHE, *Laboratoire Arago, 66650 Banyuls-Sur-Mer, France.*

E. A. BERNAYS, *Department of Entomology, University of Arizona, Tucson, Arizona 85721, USA.*

M. BILGENER, *Biological Sciences Department, Boston University, Boston, Mass. 02215, USA.*

G. COOPER DRIVER, *Biological Sciences Department, Boston University, Boston, Mass. 02215, USA.*

D. EAMUS, *Institute of Terrestrial Ecology, NERC, Bush Estate, Penicuik, Midlothian, EH26 0QB.*

*T. M. FARRELL, *Department of Zoology, Oregon State University, Corvallis, Oregon 97331, USA.*

H. C. FRITTS, *Laboratory of Tree-Ring Research, University of Arizona, Tuscon, Arizona 85721, USA.*

P. G. JARVIS, *Department of Forestry and Natural Resources, University of Edinburgh, The King's Buildings, Mayfield Road, Edinburgh EH9 3JU.*

B. A. MENGE, *Department of Zoology, Oregon State University, Corvallis, Oregon 97331, USA.*

R. K. MONSON, *Department of Environmental, Population and Organismic Biology, Campus Box 334, University of Colorado, Boulder, Colorado 80309, USA.*

T. W. SWETNAM, *Laboratory of Tree-Ring Research, University of Arizona, Tuscon, Arizona 85721, USA.*

*Present address: *Hopkins Marine Station, Stanford University, Pacific Grove, California 93950, USA*

Preface

The new team of editors are pleased to see the production of this, their second volume in the series Advances in Ecological Research. The six papers in this volume cover, in the tradition of the series, a wide variety of topics: population and community ecology, evolutionary and physiological ecology, animals, plants and plant–animal interactions, and marine and terrestrial habitats. We are certain that this breadth will be maintained in future volumes and would particularly like to encourage potential authors to review, in a readable but rigorous fashion, areas of ecology which they feel should be brought to the attention of a wider ecological audience. Ecology is a science which is forever in danger of falling apart into separate, virtually independent sub-disciplines. Advances in Ecological Research will be an important vehicle in maintaining the essential unity of the subject.

The increase in global mean concentration of carbon dioxide, from the mid-eighteenth century to the present day is well established, and the concentration is still rising rapidly. Popular attention has focused recently on the global warming that has in the past and is likely in the future to result from this – the so-called 'greenhouse effect'. Drs Eamus and Jarvis, by contrast, concentrate in their paper on the direct effects of this increase on natural and commercial temperate trees and forests. A simplistic view might be that, since CO_2 is the substrate for photosynthesis, elevated CO_2 concentrations should lead to enhanced rates of photosynthesis and thus to enhanced plant growth and yield. Eamus and Jarvis show, on the contrary, that it is necessary to take into account the complex interaction between CO_2 concentration, photosynthesis and other environmental variables, as well as the relationship between carbon assimilation and plant respiration, growth and yield. Their article shows, too, how difficult it can be to use traditional experimental approaches in some areas of ecology: trees are simply too large and long-lived for it to be possible to measure the effects of CO_2 directly, so that inferences must be made from smaller and briefer samples.

Plants with C_4 photosynthesis have a net photosynthetic rate which continues to increase with increasing radiation flux rather than levelling off above a saturation level (as it does in C_3 plants). Plants with Crassulacean acid metabolism (CAM) conserve water by maintaining their stomata open at night (and absorbing CO_2) but closed during the daytime. They can do so because they can dissociate the process of CO_2 fixation from that of energy capture. Both C_4 and CAM plants have clearly arisen from C_3 ancestors, and in his paper, Russell Monson discusses the evolution of the two strategies by tracing the evolutionary pathways through C_3–C_4 intermediate photosynthesis on the one hand, and the various modes of CAM on the other.

'Dendrochronology' is the discipline concerned with dating structures (including archaeological ruins and early Dutch paintings) from yearly tree-ring width sequences. 'Dendroclimatology' and 'dendrohydrology' are commonly used to refer to dendrochronological studies of climatic and hydrologic phenomena. 'Dendroecology', by contrast, is a relatively young science (the first conference devoted to it was held in August 1986). Harold Fritts and Thomas Swetnam introduce the science of Dendroecology and discuss its applications in forest environments of the past and present.

A major goal of community ecology is to determine the causes of spatial and temporal variation in community structure. Species diversity, species composition, relative abundance, trophic complexity, size structure and spatial structure are all components of community structure. Recent years have witnessed a rapid increase in the number of studies devoted to the quantitative dynamics of simple ecological communities, and for no habitat has this been truer than for the marine littoral. Thus, it is entirely appropriate that Bruce Menge and Terrence Farrell, in their paper, should use these marine studies in community regulation to evaluate relevant community theory.

Drs Bernays, Cooper Driver and Bilgener discuss the relationships between herbivores – both vertebrates and arthropods – and plant tannins. It has usually been assumed in the past that tannin-production has evolved in plants because of the defensive activity of tannins against herbivores, and indeed that the mode of action of tannins is anti-digestive. In this paper, accounts of tannin chemistry, potential modes of action, and the vexed question of the quantitative evaluation of tannins are followed by reviews of the biological activity of tannins on herbivores. It seems that as yet, the evidence for both the accepted wisdoms – 'defensive', 'anti-digestive' – is not truly conclusive.

Finally, Dr Athias-Binche examines the ecology of a little-studied group, the Uropodid mites. These exist near the end of food chains in the soil subsystem: they feed on living organic matter and thus depend on habitats rich in decomposing organic matter and decomposer micro-organisms. They are found mainly in forest soils. Here, demographic studies carried out in the Massane Mediterranean beech forest (Eastern Pyrenees mountains, southern France) are reviewed, and the results related to general ecological principles.

Contents

The Direct Effects of Increase in the Global Atmospheric CO_2 Concentration on Natural and Commercial Temperate Trees and Forests

D. EAMUS and P. G. JARVIS

On the Evolutionary Pathways Resulting in C_4 Photosynthesis and Crassulacean Acid Metabolism (CAM)

RUSSELL K. MONSON

Dendroecology: A Tool for Evaluating Variations in Past and Present Forest Environments

H. C. FRITTS and T. W. SWETNAM

Community Structure and Interaction Webs in Shallow Marine Hard-Bottom Communities: Tests of an Environmental Stress Model

B. A. MENGE and T. M. FARRELL

Herbivores and Plant Tannins

E. A. BERNAYS, G. COOPER DRIVER and M. BILGENER

General Ecological Principles Which are Illustrated by Population Studies of Uropodid Mites

F. ATHIAS-BINCHE

The Direct Effects of Increase in the Global Atmospheric CO_2 Concentration on Natural and Commercial Temperate Trees and Forests

D. EAMUS and P. G. JARVIS

ADVANCES IN ECOLOGICAL RESEARCH Vol. 19
ISBN 0-12-013919-7

I. INTRODUCTION

A. The Problem

There is an extensive literature documenting the evidence for an increase in global mean CO_2 concentrations from the mid 18th century through to the present day. The evidence is based upon ice core data (Neftel *et al.*, 1985; Pearman *et al.*, 1986; Fifield, 1988); inferences from tree ring data (La Marche *et al.*, 1984; Hamburg and Cogbill, 1988; Hari and Arovaara, 1988); climate modelling based on fossil fuel consumption and the CO_2 airborne fraction (see Gifford, 1982; Keepin, Mintzer and Kristoferson, 1986); and most recently measurements of global atmospheric CO_2 concentrations at different sites around the world (see Crane, 1985). Much of this evidence has been presented and evaluated elsewhere (e.g. Gifford, 1982; Ausubel and Nordhaus, 1983; Gates, 1983; Bolin, 1986; Keepin *et al.*, 1986; Wigley, Jones and Kelly, 1986). There is a consensus that prior to the industrial revolution in northern Europe about 130 years ago, the global atmospheric CO_2 concentration was about 270 to 280 $\mu mol\ mol^{-1}$. Today (1989) the concentration is *ca* 350 $\mu mol\ mol^{-1}$ and is increasing at *ca* 1.2 $\mu mol\ mol^{-1}$ per year (Conway *et al.*, 1988). Throughout this review, it will be presumed that atmospheric CO_2 concentrations will double to approximately 700 $\mu mol\ mol^{-1}$ by the mid to late 21st century and we shall not discuss here the evidence for past or future increases in the atmospheric CO_2 concentration.

Other reviews of this topic include Kramer (1981), Strain and Bazzaz (1983), Sionit and Kramer (1986), Kramer and Sionit (1987).

B. Aims of this Review

Because CO_2 is the substrate for photosynthesis, one would expect that elevated CO_2 concentrations must result in enhanced rates of photosynthesis which in turn must lead to enhanced plant growth and yield. This simplistic view does not take into account the complex interactions between CO_2 concentration, photosynthesis and other environmental variables and the more complex relationships between carbon assimilation and plant respiration, growth and yield. The major aim of this review is to assess what we know of these relationships in trees and to predict the consequences of an increase in CO_2 upon temperate zone forests.

Information concerning the reaction of trees and forests to increase in the atmospheric CO_2 concentration is particularly important because forests cover about one-third of the land area of the world and carry on about two-thirds of the global photosynthesis (Kramer, 1981). The total amount of carbon stored in terrestrial ecosystems has diminished over recent centuries as a result of anthropogenic actions, especially forestry clearance (Bolin, 1977; Woodwell *et al.*, 1978; Houghton *et al.*, 1983). Recently, about 2.5×10^6 m^3 of wood have been consumed annually (FAO, 1982). FAO predicts an increasing global rate of forest clearance for the remainder of this century. If this occurs, the present area of forest may be further reduced by as much as 20% by the year 2000. On a global scale, this further reduction in area of forest will both exacerbate the rise of CO_2 in the atmosphere through oxidation of wood and wood products, and reduce the sink strength for CO_2 (Houghton *et al.*, 1987; Marland, 1988; Jarvis, 1989). By contrast, the area of temperate and boreal forest has been increasing in Europe, North America and some parts of Asia: in the UK, for example, the area of afforested land is projected to continue to increase at a rate of about 30 000 ha per year. However, the increase in area of temperate and boreal forest is too small to compensate for the amount of CO_2 released into the global atmosphere by clearance of tropical forests (Houghton *et al.*, 1987).

Although the database is small, there has been a sufficient number of studies of the effects of CO_2 on northern temperate forest and woodland species to demonstrate that the primary effects are on photosynthesis and stomatal action, although direct effects on other processes such as leaf initiation and microbial action may yet be convincingly demonstrated. Because of the complexity of forest ecosystems, there may be many consequences of long-term changes in the rates of carbon gain and water loss by trees and stands. We identify four main reasons for being concerned about the rise in CO_2 and its effect on trees and forests:

(*a*) the enhancement of biological knowledge about the functioning of tree species of major ecological and economic importance,
(*b*) the impact on the productivity and value of the economic product,
(*c*) the impact on the ecology and environment of woods and forests, and
(*d*) the downstream, socio-economic consequences.

C. Spatial and Temporal Scales

The effects of a major environmental variable on plants and ecological systems can be examined at spatial scales ranging from sub-cellular through to the biome or geographical region and with timescales ranging from seconds or parts of seconds for rapid biophysical processes, to centuries for evolutionary changes affecting land-use systems on the larger, spatial scales. The consequences of a doubling in the global, atmospheric CO_2 concentration may be seen over these ranges of both spatial and temporal scales (Figure 1). For example, an increase in CO_2 may affect the primary photosynthetic carboxylation at the spatial scale of the chloroplast over the timescale of seconds or minutes, whereas assimilation of CO_2 and transpiration of water by leaves of trees is influenced by CO_2 at a timescale of hours, or of weeks if allowance is made for acclimation of the plants to a high CO_2 environment. The growth of seedlings and individual trees is influenced by an increase in CO_2 concentrations over periods of weeks or years, although particular growth processes such as carbon allocation or nutrient uptake may have shorter timescales. The functioning of stands may be affected at timescales of the order of hours, but the main impact of the effects of a rise in CO_2 will be seen in changing properties of stands with timescales of a year or more. Effects on forest ecosystem processes are likely to be seen over decades and effects on regional land-use over centuries (Strain, 1985).

This review examines the likely effects of the projected increase in global atmosphere CO_2 concentration on these spatial and temporal scales. However, the information we have to draw upon is severely restricted to certain spatial and temporal scales, determined largely by the convenience and feasibility of experimental methods with the current levels of technology and resources.

D. Physiological Action of CO_2

The main plant physiological roles of CO_2 are as a substrate and activator for photosynthetic carbon assimilation, with concomitant effects upon photorespiration, and as an environmental variable determining stomatal aperture. Processes of dark fixation of CO_2, predominantly in roots, may also be concentration-dependent, but the CO_2 concentration in soil environments dominated by respiration is far in excess of the range of atmospheric concentrations under consideration. All species of major ecological and economic importance in northern temperate forests, both overstorey and understorey, have photosynthesis of the typical C_3 pattern, and, therefore, we may expect the main effects of rising CO_2 concentrations to be similar to effects seen on other C_3 species. However, the amount of data for temperate and boreal forest species is small, especially in comparison with comparable data for agricultural crops (cf. Warrick *et al.* (1986) with Shugart *et al.* (1986)).

It has been inferred from several experiments on the growth response of herbaceous plants to elevated CO_2 concentrations that CO_2 has a direct stimulating effect on leaf growth, not driven through increased availability of substrate (e.g. Morison and Gifford, 1983) and the same inference has been

	Temporal scale				
Spatial scale	Seconds/Minutes	Hours/Days	Weeks/Months	Years/Decades	Centuries
Cell	enzyme activation; fluorescence; carboxylation; transport/ partitioning	enzyme kinetics; organelle structure	acclimation of cellular processes		
Leaf/Shoot		assimilation; transpiration; stomatal action	acclimation of assimilation; senescence		
Seedling/Tree			growth and carbon allocation; nutrient uptake; root/shoot dynamics	crown properties; branching	
Plantation/ Woodland		canopy properties of photosynthesis, transpiration, light interception	nutrient uptake; WUE; productivity	canopy structure; competition; harvest index	
Forest/ Ecosystem				yield; rotation length; WUE	natural and artificial selection; land use; species composition

Fig. 1. The range of temporal and spatial scales available in the study of the effects of elevated CO_2 upon trees. WUE = water use efficiency.

drawn for tree seedlings (Tolley and Strain, 1984a). However, these inferences are unsubstantiated at present, although possible mechanisms have been proposed (Sionit *et al.*, 1981; Sasek *et al.*, 1985).

The allocation of carbon within plants is clearly influenced by the supply of CO_2 and, as will be discussed later, this appears to result from change in the balance between rates of uptake of CO_2 and inorganic nutrients, especially nitrogen.

There are suggestions, too, that increase in atmospheric CO_2 concentration will lead to changes in mycorrhizal development and in root exudations, thus affecting rhizosphere activities (Strain and Bazzaz, 1983). Such effects remain to be clearly demonstrated and are most probably the result of greater availability of carbohydrates. There is, however, a stronger supposition, also still awaiting clear demonstration, that changes in the composition (especially the C:N ratio) of leaf and other litter, will lead to significant changes in rates of mineralisation and hence recycling of nutrients through the soil (Luxmore, 1981).

E. The Data Base

1. The leaf and shoot scale

There has now been an appreciable number of short-term experiments with a timescale of hours or days on single leaves, shoots or young trees of temperate forest origin in assimilation chambers, where the response of CO_2 uptake and stomatal action has been characterised in relation to a range of ambient CO_2 concentrations, and many more on agricultural crops and other herbaceous species (see Strain and Cure, 1986). However, the majority of these experiments has been on unacclimated plants that have been grown in air containing the current, atmospheric CO_2 concentration. There have been only a few experiments on woody plants that have been grown for a substantial period in higher CO_2 concentrations, so that the plants were at least partially acclimated anatomically and physiologically to an appropriate CO_2 environment. Much of this work and the growth studies considered below, has been carried out in North America and only a few studies of physiological processes have been made at the leaf scale on species important in other forests.

2. The seedling, sapling and individual tree scale

Prior to 1983, the motivation for experiments in which young trees were grown at elevated CO_2 concentrations was derived from horticultural practice and the aims of the experiments generally were to produce improved planting stock. For this purpose, young trees were often grown in unusual conditions with the CO_2 increased by a factor of $\times 4$ or more (Hårdh, 1967; Funsch *et al.*, 1970; Yeatman, 1970; Krizek *et al.*, 1971; Siren and Alden,

1972; Tinus, 1972; Laiche, 1978; Canham and MacCavish, 1981; Lin and Molnar, 1982; Mortensen, 1983). These experiments covered a range of broadleaved and coniferous species, mostly ornamentals of horticultural interest or forest trees of importance in North America. The evidence from these experiments indicates that growth in height, leaf area, stem diameter and dry weight of temperate trees is generally increased by large increases in the ambient CO_2 concentration but growth was extremely variable and in some species no growth response was elicited. Because of the unusual environmental conditions and the high concentrations of ambient CO_2 used, the results are of marginal relevance to the situation we are addressing.

It is only since the realisation of an appropriate scenario for the projected rise in the atmospheric CO_2 concentration that relevant experiments have been done. Since 1983 some nine species of conifers and 15 species of broadleaves have been subjected to elevated ambient CO_2 in the range $\times 2$ to $\times 2.5$ and the results published in each case in 15 papers. Regarding each species in each paper as a separate experiment, three experiments were done in glasshouses, five in open-topped chambers and 23 in controlled environment growth cabinets, chambers or rooms. In 32 cases plants were grown in pots or boxes and in only one case were plants grown directly into soil, and this was, necessarily, in an open-topped chamber. Two experiments lasted for three years, another for two and a half years and another for one year. The remainder of the experiments lasted 32–287 days with a median of around 120 days. Nutrients were added in 22 experiments but there was a series of nine in which no nutrients were added. In one experiment nitrogen supply was a variable and in another phosphorus supply. In five experiments water supply was a variable and light was a variable in three others. In one experiment, competition was studied; in all others it was excluded.

Most of these experiments were done with native North American species, of little or no economic or ecological importance in Europe or elsewhere. Only three were with species grown extensively in Europe, for example (Higginbotham *et al.*, 1985 with *Pinus contorta*; Mortensen and Sandvik, 1987 with *Picea abies*; and Hollinger, 1987 with *Pseudotsuga menziesii*) whilst only two others were with Australasian species. Few of the experiments were with native species from elsewhere or with other commercial forest species grown widely outside North America, such as *Picea sitchensis*. These experiments are, thus, limited in their direct relevance to temperate and boreal forests worldwide, although substantial information about the effects of CO_2 on young trees can be adduced from them.

3. *The plantation and woodland scale*

There have been no observations of *stand* physiology on any timescale in relation to ambient CO_2 concentration. The nearest approach to this is a single study of the assimilation and transpiration of an individual tree of

Eucalyptus, growing on a lysimeter and enclosed in a tall, open-topped, plastic chamber in south-east Australia (Wong and Dunin, 1987). Essentially, measurement was made of the CO_2 and water vapour exchange of an unacclimated tree in a giant cuvette. This experiment was, therefore, a physiological study of an individual tree and is not of direct relevance either to stand functioning or, because of the short timescale of the experiment, to the rise in atmospheric CO_2. It did, however, demonstrate that a single tree behaves much as one would anticipate from cuvette studies on individual leaves.

Although open-air exposure to CO_2 is now being practised in agricultural crops in several places, it is hardly surprising that uncontained stands of trees, or areas of woodland, have not been exposed to high CO_2 concentrations in experiments. The sheer scale of the problem renders this a phenomenally expensive approach: Allen *et al.* (1985), have estimated the annual cost of CO_2 alone for this to exceed $US4 $\times 10^6$ in a 40 m tall stand of tropical forest! In the absence of any directly determined data and with the likelihood that none will be forthcoming, recourse must be made at this and larger scales to modelling the functioning of the system with respect to the atmospheric CO_2 concentration. We would emphasise, however, that inability to address the problem directly at the scale of concern imposes a severe constraint on our capability to make adequate assessment of the consequences of the rise in atmospheric CO_2 concentration for stand-scale processes.

F. Problems with the Data

It is evident that there is a serious dilemma with respect to the acquisition of data regarding the response of trees and forest to increased levels of CO_2. For technical reasons experiments have been confined to small numbers of comparatively small and young trees *enclosed* in relatively small volumes of space. There are significant problems with this approach and these problems are necessarily exacerbated with respect to trees and forest, because of the large size of mature trees and the areal extent of woodlands and forest.

Juvenile trees are known to be more sensitive to environmental change (Higginbotham *et al.*, 1985), and consequently the response elicited may be larger in juvenile trees than in mature trees. Conversely, the long-term totally acclimated response to elevated CO_2 may be larger than anticipated, *if* there are accumulative effects on tree morphology and ecosystem processes. In either case, the present experiments may introduce bias in the estimation of the predicted response for mature stands of trees unless such possibilities are taken into account.

The majority of experiments so far have been carried out in controlled environment chambers or rooms, and these have the advantage that the foliage of the plants is generally well-coupled to the atmosphere prescribed by the operator. In comparison, in glasshouses and open-topped chambers, ventilation can be very much poorer and there may be substantial feedback between the exchanges of mass and energy at the plant surfaces and the local, leaf environment. Consequently in glasshouses and open-topped chambers the plants are usually very poorly coupled to the atmosphere and substantial practical problems may result, particularly with respect to overheating (e.g. Sürano *et al.*, 1986). More importantly, however, the poor coupling to the atmosphere may result in a substantial shift in the importance of the driving variables for gas exchange by the plants and, hence, lead to fluxes of water vapour and carbon dioxide that are substantially different from the fluxes that would be expected in response to the weather from well-coupled vegetation such as trees and forests in the field (Jarvis and McNaughton, 1986).

Controlled environment chambers and rooms do, however, have other problems, not least that the plants must be grown in pots and that the volume of space available is usually rather small. This puts a major constraint on the length of the experiment and this is very evident in the relatively short periods over which young trees have been exposed to elevated CO_2 concentrations. Whilst large responses to CO_2 have been elicited in many experiments during the initial few weeks of exposure, the responses have diminished with time and after more than a year have virtually disappeared. It is not yet clear whether this is an artefact of the restricted rooting conditions or a physiological response of the plants. Possible interaction between the response of young trees to an increase in CO_2 and the availability of nutrients is an important question (Kramer, 1981), but it is unlikely that it can be elucidated by growing plants with their roots constrained in a fixed volume of soil.

Even with plants growing in open-topped chambers, the opportunity of rooting the plants into the soil directly has only been taken in one experiment, and then only as a secondary aspect (Surano *et al.*, 1986).

G. Acclimating, Acclimated and Adapted Trees

Most of the early studies of the effects of CO_2 on trees used non-acclimated plants and leaves. Trees were grown at the current global CO_2 concentration (*ca* 325 to 340 µmol mol^{-1}) and leaf or shoot responses to elevated CO_2 were measured in the short-term over periods of minutes or hours. Such a protocol is no longer acceptable, since we do not know whether the response observed after a few hours will be maintained by tissue grown in elevated CO_2 concentrations over a period of years. There is evidence that the responses of photosynthesis, transpiration and growth to increase in the ambient CO_2 concentration change with time. However, since much of the information available has been obtained on non-acclimated trees, the data will be briefly reviewed.

The process of acclimation is multi-faceted and may occur over a timescale of hours to generations, depending upon the process studied. Photosynthesis and stomatal conductance *may* acclimate over a period of hours or days; changes in stomatal density in response to elevated CO_2 concentrations have been recorded over periods of weeks (Woodward and Bazzaz, 1988) and months (Oberbauer *et al.*, 1985), but also, it has been claimed, over generations (Woodward, 1987). Whether the latter is the result of acclimation or of natural selection and genetic adaptation is open to question. In trees and forests both complete acclimation and genetic adaptations are likely to require centuries.

The use of the term 'acclimated' may perhaps be better replaced by 'acclimating'. This is particularly pertinent to many studies where trees have been grown for months or years at the current global CO_2 concentration and then transferred to an elevated ambient CO_2 atmosphere. Such a sudden, single-step perturbation may induce oscillations in many plant processes and it is likely that subsequent measurements are made on *acclimating* trees. In some studies (Tolley and Strain, 1984a; Higginbotham *et al.*, 1985; Oberbauer *et al.*, 1985) seeds have been sown directly into an elevated ambient CO_2 concentration and the resulting plants may resemble more closely the acclimated trees of the next century.

Two further points concerning the experimental data must be made. Trees growing today with a lifespan of several decades are subject to a *gradually* increasing CO_2 concentration. Long-term physiological responses to a slow increase in CO_2 concentration (*ca* 1.2 μmol mol^{-1} a^{-1} presently) are likely to differ somewhat from the responses generated by large, immediate, single-step increase in CO_2 concentration, even though plants are accustomed to wide short-term local, diurnal and seasonal variations (usually < 100 μmol mol^{-1}), depending on location and latitude (Jarvis, 1989). This problem of acclimation was highlighted recently for crop species, for which it was shown that the weighted average, short-term (non-acclimated) CO_2 assimilation rate was increased by 52% for a doubling of the ambient CO_2 concentration, whilst the acclimated assimilation rate was enhanced by only 29% (Cure and Acock, 1986). This result was ascribed to the lack of *sustained*, active sinks for photosynthate in the acclimated plants.

It is evident that the data available at present are severely limited in their applicability by the conditions in which the experiments have been done. Not only must the results be interpreted with great caution, but also considerable care must be taken in using information obtained in the experiments to parameterize models used for predicting the consequences of increased CO_2 levels at larger, spatial and temporal scales.

The following sections review the results obtained so far about the influence of an increase in CO_2 concentration on seedlings, trees and forests over the range of spatial scales shown on the left-hand side of Figure 1 and at the range of temporal scales along the top of Figure 1. We shall, necessarily, concentrate upon the entries in the top, left-hand area of Figure 1 for which substantial information is available, even though this information is of limited relevance to the likely responses of stands of trees over one or more generations.

II. THE CELLULAR SCALE

Measurements at the cellular scale include the measurement of fluorescence induction phenomena, enzyme activity, carbon partitioning within cellular pools and cellular transport properties. The latter two processes, although directly relevant to the understanding of growth and differentiation with respect to the influence of environmental perturbations (such as changes in CO_2 concentration), have not received any attention to date, and we can only highlight this deficiency. It will be shown later that source–sink relationships (and hence, carbon partitioning and cellular transport) directly influence photosynthesis and assimilate allocation and this gap in our knowledge is critical to our poor understanding and central to our relative inability to extrapolate with confidence from leaf and shoot scale studies to a stand or regional scale.

Enzyme activities have been better studied, generally as part of wider programmes investigating leaf/shoot responses. The major emphasis by far has fallen on the primary carboxylating enzyme. In addition to acting as a substrate for ribulose bisphosphate carboxylase-oxygenase (rubisco), CO_2 combines with this enzyme as the first step in a two-step chemical activation process. The inactive enzyme can bind with a molecule of CO_2 in a slow, reversible reaction. When this enzyme–CO_2 complex subsequently reacts with Mg^{2+} in a rapid, reversible reaction, the active enzyme–CO_2–Mg^{2+} complex is formed. Carboxylation requires a second CO_2 molecule to bind at a third site. The concentration of CO_2, therefore, influences enzyme activity *via* its influence upon the activation state as well as through its role as a substrate. Changes in activation state alter V_{max} and K_m (Edwards and Walker, 1983). Growth at elevated CO_2 concentration directly influences this activation process (Sage *et al.*, 1987, 1988). What little is known about trees in these connections will be taken up in the following sections that treat acclimation and adaptation at the leaf scale.

III. THE LEAF AND SHOOT SCALE

A. Stomatal Conductance

1. Non-acclimated tree studies

With some exceptions, stomatal conductance decreases in response to increases in CO_2 concentration. The mechanism by which CO_2 affects stomatal aperture and conductance (g_s) remains speculative (see Eamus, 1986a) and will not be dealt with in this review. However, it is clear that stomata respond to the intercellular space CO_2 concentration (C_i) rather than to the ambient CO_2 concentration (C_a) (Mott, 1988). Responses of g_s to CO_2 concentration follow one of the curves shown in Figure 2. Broadly, we may classify trees into those with stomata that are quite sensitive to CO_2 (− −) and those which have little sensitivity (+, 0, −).

Table 1 summarizes some of the available data for tree species using non-acclimated shoots. There are two points to note from Table 1. Firstly, there is a paucity of data, and secondly, interactions between other environmental variables and CO_2 concentration on g_s has been poorly studied. The latter, in particular, requires investigation, since in one study (Beadle *et al.*, 1979), the direction of response of g_s was reversed by imposing water stress. Modification of the response of g_s to elevated CO_2 concentrations by the environment is likely, and without this information, extrapolation from single shoot studies in optimal conditions in cuvettes in the laboratory to the field remains impossible.

2. Acclimating tree studies

Table 2 summarizes recent data concerning the response of g_s to elevated CO_2 levels in acclimating trees. Several points can be noted from this table. The length of time allowed for acclimation is short; in all but two cases, it is less than one growing season. It is unknown whether acclimation of g_s to elevated CO_2 levels occurs over a longer time period (Morison, 1985). Oberbauer *et al.* (1985) observed changes in stomatal density over 123 days for *Pentaclethra macroloba*, but no change over 60 days for *Ochroma lagopus*. Decreases in stomatal density for eight temperate, arboreal species in the past 200 years have been found (Woodward, 1987). Long-term studies of acclimation of g_s and stomatal index [(100 × no. of stomatal pores)/(no. of stomatal pores + total number of epidermal cells)] to elevated CO_2 concentration are required.

The general response of g_s to elevated CO_2 levels is a decrease of 10–60% (Oberbauer *et al.*, 1985; Tolley and Strain, 1985; Hollinger, 1987). However, notable exceptions to this are apparent for well-watered and water-stressed *Pinus radiata* (Conroy *et al.*, 1986); *Pinus taeda* grown under high light levels (Tolley and Strain, 1985) and *Pseudotsuga menziesii* (Hollinger, 1987). The

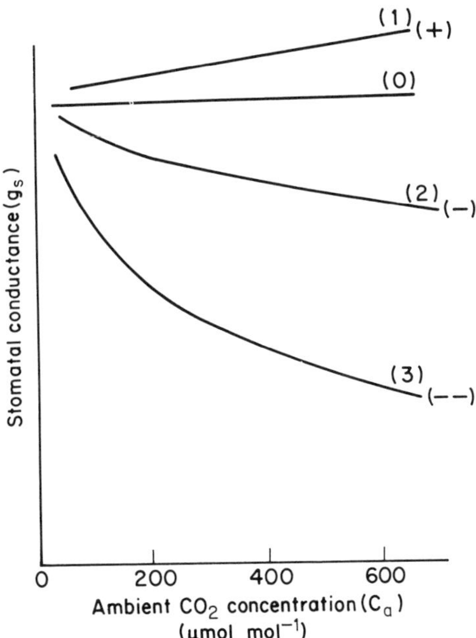

Fig. 2. Diagram to show the major responses of stomatal conductance (g_s) to change in the ambient CO_2 concentration (C_a).

Table 1

Response of stomatal conductance to changing CO_2 concentration in a range of unacclimated tree species. Refer to Figure 2 for curve form. C_a: ambient CO_2 concentration. C_i: mean intercellular space CO_2 concentration.

Species	Response curve	Max C_a or C_i (µmol mol^{-1})	Source
Picea sitchensis	—	600 C_a	Ludlow and Jarvis (1971)
Picea sitchensis			
unstressed	—	600 C_a	Beadle *et al.* (1979)
water stressed	+	600 C_a	
Picea rubra	—	600 C_i	D. Eamus (unpub)
Pinus taeda	+	400 C_i	Teskey *et al.* (1986)
Fraxinus pennsylvanica	+	400 C_i	Davis *et al.* (1987)
Populus deltoides	—	1000 C_a	Regehr *et al.* (1975)
Eucalyptus pauciflora	—	300 C_i	Wong *et al.* (1978)
Malus pumila	—	400 C_i	Warrit *et al.* (1980)

Table 2
Response of stomatal conductance (g_s) (mmol m^{-2} s^{-1}) to elevated CO_2 concentrations (µmol mol^{-1}) in a range of acclimating tree species

Species	g_s at current C_a	g_s at elevated C_a	Elevated C_a	Period of study	Source
Pinus radiata					
well watered	20	20	660	22 weeks	Conroy *et al.* (1986b)
stressed	20	20	660		
nutrient stressed	22	12	660		
Pinus taeda					
high light	41–74	41–74	675, 1000	56 days[a]	Tolley and Strain (1985)
low light	82–131	41–82			
Pinus taeda	48	30	500	15 months	Fetcher *et al.* (1988)
Liquidambar styraciflua					
high light	74–131	41–90	675, 1000	56 days[a]	Tolley and Strain (1985)
low light	49–115	20–61			
Liquidambar styraciflua	700	350	500	15 months	Fetcher *et al.* (1988)
Ochroma lagopus	299	127	675	60 days[a]	Oberbauer *et al.* (1986)
Pentaclethra macroloba	156	45	675	123 days	
Pinus radiata	105	68	640	120 days	Hollinger (1987)
Nothofagus fusca	78	66	640	120 days	
Pseudotsuga menziesii	160	166	640	120 days	
Pinus ponderosa					
07.00 h (time of day)	47	63	*ca* 500	1 year	Surano *et al.* (1986)
11.00 h (time of day)	98	55	500	1 year	
13.00 h (time of day)	47	63	500	1 year	
18.00 h (time of day)	63	29	500	1 year	

[a] started from seed.

response of g_s to elevated CO_2 concentrations varied according to the time of day in *Pinus ponderosa* (Surano *et al.*, 1986), possibly the result of variations in needle temperature. It is not possible to state whether the apparent variation in response amongst species is a true reflection of inter-species differences, differences in tree age, duration of the experiment, CO_2 concentrations used, degree of acclimation or other additional variables.

Stomatal sensitivity to CO_2 increased in *P. radiata* grown at elevated CO_2 concentrations (Hollinger, 1987) but decreased in *Betula pendula* and *Picea sitchensis* (P. G. Jarvis, A. P. Sandford and A. Brenner, unpublished). Stomatal sensitivity to CO_2 also varies with photon flux density: in response to an increase, stomatal sensitivity to CO_2 may increase (Morison and Gifford, 1983) or decrease (Beadle *et al.*, 1979). Abscisic acid both reduces g_s and increases stomatal sensitivity to CO_2 in crop plants but whether this is true for trees is unknown. In view of the long-term influence of drought upon subsequent stomatal behaviour in crop plants (Eamus, 1986b) and trees (Schulte *et al.*, 1987), this area requires further study in trees.

As the difference in vapour pressure between leaf and ambient air (D_l) increases, g_s decreases (e.g. Watts and Neilson, 1978; Johnson and Ferrell, 1983; Roberts 1983) and stomatal sensitivity to D_l varies between species and with plant water status (Ludlow and Jarvis, 1971; Johnson and Ferrell, 1983; Schulte *et al.*, 1987). Growth at elevated CO_2 significantly influences stomatal response to D_l. The relative closure induced by a stepwise increase in D_l was decreased by growth at elevated CO_2 for *P. radiata* and *P. menziesii*, but not for *Nothofagus fusca*, so that the stomata of trees grown in an elevated CO_2 concentration were less sensitive to dry air (Hollinger, 1987). Temperature can also modify the response of g_s to D_l and possibly also to CO_2 (Johnson and Ferrell, 1983).

It is clear from the above that a detailed study of the interactions amongst temperature, CO_2 concentration, D_l, photon flux density and plant water status is required to determine more realistically the influence of elevated CO_2 upon g_s. Stomatal sensitivity to CO_2 may well be reduced under high photon flux densities in plants of high relative water content (that is, not water stressed). These are the conditions frequently used in experiments in which stomatal sensitivity to CO_2 is assessed and this may explain the observation that in many species of trees stomata show a low sensitivity to CO_2 concentration. In crop plants a role for indole-3-acetic acid and calcium in the control of stomata has been postulated. We are not aware of any studies investigating these regulating factors with respect to stomatal response to elevated CO_2 concentrations in trees.

B. Photosynthesis

1. Non-acclimated tree studies

The relationship between CO_2 concentration and rate of photosynthesis has the form shown in Figure 3. The exact form of the relationship varies somewhat amongst species and with other environmental variables but is essentially the same for all C_3 species (von Caemmerer and Farquhar 1981, 1984). Table 3 summarizes some of the available data for non-acclimated trees. From this table, it can be seen that doubling the ambient CO_2 concentration invariably increases photosynthetic rate. However, the percentage increase in carbon dioxide assimilation (A) varies widely from 20% to 300%. Furthermore, although the absolute increase in assimilation at elevated CO_2 concentration decreases with the imposition of additional stress variables, the percentage increase is frequently larger, so that an increase in CO_2 concentration may ameliorate the effects of stresses of various kinds (see below).

2. Acclimating tree studies

Table 4 summarizes recent data on the influence of elevated CO_2 concentrations upon CO_2 assimilation in acclimating trees. Although there are some exceptions, in the majority of these experiments a doubling of the CO_2 concentration resulted in enhanced rates of assimilation, which sometimes more than doubled. The large variability (e.g. Tolley and Strain, 1984a) may reflect true inter-species differences, differences in the CO_2 concentration

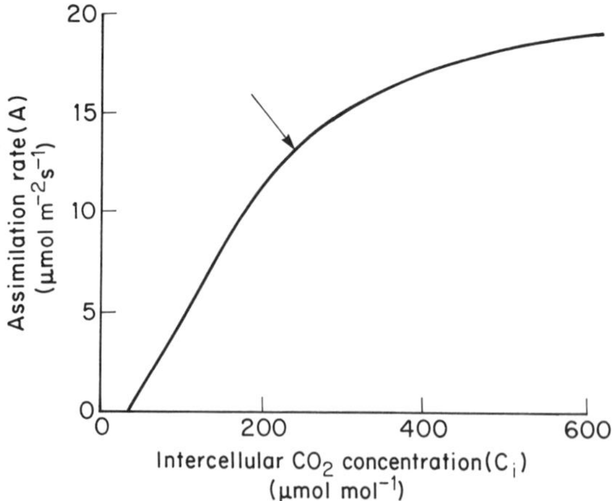

Fig. 3. The relationship between rate of assimilation of CO_2 in photosynthesis (A) and mean intercellular space CO_2 concentration (C_i) in *Picea sitchensis*. (Adapted from Jarvis and Sandford, 1986.)

Table 3

Influence of a doubling of CO_2 concentration on assimilation rate (A) ($\mu mol\ m^{-2}\ s^{-1}$) determined for a range of non-acclimated tree species. All figures are approximate having been read from graphs.

Species	A at present C_a	A at doubled C_a	Percentage increase	Source
Picea engelmanii				
warm	3·5	4·8	37	Delucia (1987)
chilled	0·5	0·6	20	
Picea rubra	10·0	12·0	20	D. Eamus (unpubl)
Picea sitchensis				
unstressed	9·1	11·4	25	Beadle *et al.* (1981)
water stressed	1·1	3·4	209	
Picea sitchensis	12·6	20·2	60	Ludlow and Jarvis (1971)
Pinus sylvestris				
warm	6·0	7·5	25	Strand and Öquist (1985)
chilled	3·0	5·0	66	
Pinus taeda	6·0	8·0	33	Teskey *et al.* (1986)
Populus deltoides	15·8	18·9	20	Regehr *et al.* (1975)
Pseudotsuga menziesii				
high light	1·6	2·6	60·5	Brix (1968)
low light	0·2	0·3	50	
Quercus suber	2·0	4·0	100	Tenhunen *et al.* (1984)
Malus pumila				
unstressed	24·0	45·0	87	Jones and Fanjul (1983)
water stressed	4·0	10·0	150	
Malus pumila				
extension shoot	18·2	27·3	50	Watson *et al.* (1978)
spurs + fruit	9·1	15·9	75	
spurs − fruit	8·2	15·9	94	

Table 4

Influence of elevated CO_2 concentrations (μmol mol^{-1}) upon assimilation rate (A) for a range of tree species acclimating for various periods to elevated CO_2.

Species	% change in A	Elevated C_a	Duration of experiment	Source
Liquidambar styraciflua				
well watered	ns	675, 1000	56 days[a]	Tolley and Strain (1984a)
water stressed	ns			
Liquidambar styraciflua				
high light	17–100	675, 1000	56 days[a]	Tolley and Strain (1986)
low light	20–100			
Liquidambar styraciflua (high light)				
1– 4 weeks	41, 44	675, 1000	112 days	Tolley and Strain (1984b)
4– 8 weeks	33, 43			
8–12 weeks	22, 10			
12–16 weeks	−15, −23			
Liquidambar styraciflua	*ca* −10	500	15 months	Fetcher *et al.* (1988)
Nothofagus fusca	+45	620	120 days[c]	Hollinger (1987)
Ochroma lagopus	−49	675	60 days[a]	Oberbauer *et al.* (1985)
Pentaclethra macroloba	−, ns	675		
Carya ovata	*ca* 80	700	90 days[d]	Williams *et al.* (1986)
Liriodendron tulipifera	*ca* 80	700		
Acer saccharinum	*ca* 80	700		
Fraxinus lanceolata	*ca* 30	700		
Platanus occidentalis	0	700		
Quercus rubra	*ca* 80	700		
Populus euramericana	−29	660	15 days[e]	Gaudillère and Mousseau (1989)
Picea abies				
24 h day	67	1650	75 days	Mortensen (1983)
10 h day	68	1650	75 days	

Pinus contorta				
saturating light	96	1000	*ca* 150 days[a]	Higginbotham *et al.* (1985)
	43	2000		
saturating CO_2	−5	1000		
	−55	2000		
Pinus radiata	55	620	120 days[c]	Hollinger (1987)
Pinus radiata				
adequate P	200	660	22 weeks[b]	Conroy *et al.* (1986b)
low P	−35	660		
Pinus taeda				
well watered	207	675	56 days[a]	Tolley and Strain (1984a)
water stressed	83	675		
well watered	−ns	1000		
water stressed	−53	1000		
Pinus taeda				
high light	0–38	675, 1000	56 days[a]	Tolley and Strain (1986)
low light	10–38			
Pinus taeda (high light)				
1– 4 weeks	ns, ns	675, 1000	84 days	Tolley and Strain (1984b)
4– 8 weeks	ns, 47			
8–12 weeks	207, ns			
12–16 weeks	−45, −34			
Pinus taeda	12·4	500	15 months	Fetcher *et al.* (1988)
Pseudotsuga menziesii	32	620	120 days[c]	Hollinger (1987)

[a] started from seed
[b] started with 8-week-old plants
[c] started with 10-month-old plants
[d] started with 1-year-old plants
[e] started with cuttings

used, duration of the experiment, stress induced by other variables, attack by pathogens, mites and insects, or other aspects of the cultural conditions.

The increase in assimilation rate of leaves grown in elevated CO_2 is the result of both increased substrate concentration for rubisco and increased activation of the rubisco, since CO_2 activates this enzyme. In addition the quantum yield (ϕ) is increased and photorespiration is reduced (Figure 4) with increasing CO_2 concentration, as a result of enhanced competition by CO_2 relative to O_2 for active sites on rubisco (Pearcy and Björkman, 1983; Ehleringer and Björkman, 1981). Elevated CO_2 can also cause a reduction in the compensation photon flux density and this may possibly also delay leaf senescence and lead to a larger population of more effective shade leaves in the tree crown.

Tolley and Strain (1984a,b) showed for *Liqidambar styraciflua* and *P. taeda* that the rate of net assimilation declined as the duration of the experiment increased, particularly at high photon flux densities. From analysis of the relation between A and C_i, it could be shown that the observed reduction in photosynthesis was not the result of stomatal closure, since C_i increased. After 16 weeks, *L. styraciflua* grown at a CO_2 concentration of 1000 μmol mol^{-1} had a rate of CO_2 assimilation 23% *less* than the control trees. For low-light grown trees, a peak in photosynthetic rate was observed, followed by a decline in the control values. A similar reduction has been observed in crop plants (Raper and Peedin, 1978; Sionit *et al.*, 1981; Wulff

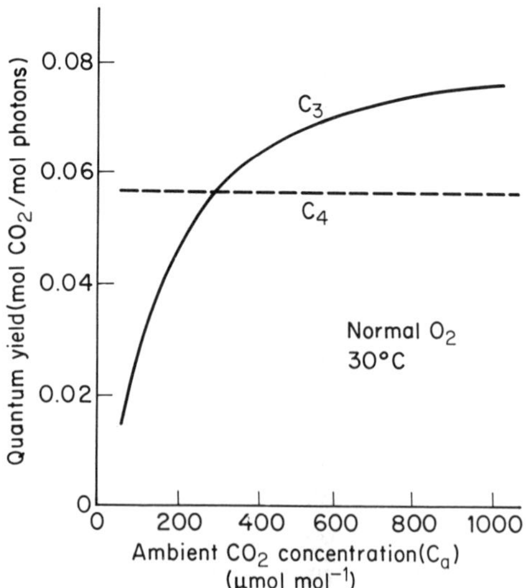

Fig. 4. Relationship between quantum yield (ϕ, mole of CO_2 assimilated per mole of photons absorbed) and ambient CO_2 concentration (C_a) in C_3 plants. (Adapted from Pearcy and Björkman, 1983.)

and Strain, 1982; DeLucia *et al.*, 1985; Ehret *et al.*, 1985) as well as in some woody species (Oberbauer *et al.*, 1985; Williams *et al.*, 1986).

Several possible causes of this reduction in photosynthetic capacity have been proposed. Starch accumulation in the leaves of plants grown in elevated CO_2 concentration has been noted (De Lucia *et al.*, 1985; Ehret *et al.*, 1985). An inverse relationship between leaf starch levels and photosynthesis has been observed in annual crop species (Mauney *et al.*, 1979) where a decrease in quantum yield has been associated with starch accumulation. Such a decline indicates inhibition of the production and/or the consumption of NADPH and ATP (De Lucia *et al.*, 1985) which may in extreme cases be the result of chloroplast disruption by starch accumulation (Wulff and Strain, 1982).

Where starch accumulation is observed, this clearly indicates that photosynthate production exceeds demand. There is accumulating evidence that the maintenance of high rates of photosynthesis in response to elevated CO_2 concentrations requires the existence of *sustained*, active carbon sinks (Koch *et al.*, 1986; Downton *et al.*, 1987). In the absence of sustained sinks, assimilation declines as a result of a series of feedback processes. This point is particularly pertinent to experiments with pot-grown seedlings, in which continuing growth of both root and shoot is constrained by a fixed (generally small) volume of soil and amount of nutrients.

Carboxylation efficiency has been observed to decline following growth at elevated CO_2 in a number of species (see Berry and Downton, 1982). A decline in photosynthetic capacity of tree species as a result of growth at high CO_2 has been noted on several occasions by comparing the rates of photosynthesis of trees grown at elevated CO_2 with the rates of control trees, both *measured* at the control concentration of CO_2, or through comparison of the entire A/C_i response function (Oberbauer *et al.*, 1985; Williams *et al.*, 1986; Hollinger, 1987; Fetcher *et al.*, 1988; P. G. Jarvis, A. P. Sandford and A. Brenner, unpublished).

Such reductions in efficiency may be the result of a decrease in the amount or activity of rubisco as has been observed in crop plants (Pearcy and Björkman, 1983; Porter and Grodzinski, 1984; Peet *et al.*, 1985; Sage *et al.*, 1987). A reduction in the amount or activity of this enzyme to match the rate of production of ATP and NADPH set by the rate of electron transport through the photosystems (Pearcy and Björkman, 1983), would release nitrogen to downstream enzyme systems concerned with the utilization of assimilate. Field and Mooney (1986) have argued that plants maximize efficiency of resource use by the reallocation of resources, principally nitrogen, to maintain a balance between all the components of the photosynthetic system. Thus, changes in the growth CO_2 concentration influence the orthophosphate limitation of photosynthesis and can cause changes in the amount and activity of rubisco, *via* the reallocation of nitrogen. Conroy *et al.*

(1986b) have shown that electron flow was not enhanced by elevated CO_2 concentrations in *P. radiata*. However, in a study of fruiting *Citrus* trees, Koch *et al.* (1986) found that the activity of rubisco was increased in one species and not affected in another, and postulated that it was the rate of turnover of the enzyme that was influenced by CO_2 concentrations. Sage (1989), however, has proposed that the activation state of rubisco is an effective indicator of the acclimation potential of a species to elevated CO_2. This proposal deserves further attention.

Additional hypotheses have been proposed to explain the reduction in assimilation observed in crop plants after extended periods of growth at elevated CO_2. These include increase in the amount or activity of carbonic anhydrase (Chang, 1975); sequestration of phosphate as sugar phosphates (Pearcy and Björkman, 1983); and inability of the Calvin cycle to regenerate ribulose bisphosphate or orthophosphate (von Caemmerer and Farquhar, 1981, 1984). Indeed, in a study of five crop species, Sage (1989) showed that a period of growth at elevated CO_2 increased the extent to which assimilation was stimulated by a decrease in O_2 concentration or an increase in CO_2 concentration. This indicates that the limitation imposed upon assimilation rate by the capacity to regenerate orthophosphate (Pi) was decreased by growth in elevated CO_2. This result could be attributed to one or all of (*a*) an absolute increase in Pi regeneration capacity, resulting from an increase in the enzymes synthesizing sucrose and starch, (*b*) an increase in Pi regeneration relative to RuBP regeneration capacity, or (*c*) a decline in rubisco and RuBP-regeneration capacity. This last possibility may be a stress response. Little is known about the mechanisms by which photosynthesis of herbaceous plants respond to long-term CO_2 enhancement (Sage, 1989) and even less is known about these mechanisms within trees. It seems clear, however, that elevated CO_2 will influence photosynthesis through the nitrogen and phosphorus economies both of the entire tree and the tree stand (Evans, 1988).

3. *Compensation for stress*

Elevated CO_2 has been shown to ameliorate the effects of water stress (Tolley and Strain, 1984b; 1985; Conroy *et al.*, 1986), such that the percentage increase in biomass accumulation following a period of drought is larger in elevated CO_2 than in ambient CO_2. This amelioration has several possible causes. The onset of drought is often delayed, because plants grown in elevated CO_2 have a larger root mass and are able to acquire water from a larger soil volume. CO_2-induced stomatal closure may also delay the onset of drought, although this may well be offset by increase in leaf area. A delay in the onset of drought is associated with a delay in drought-induced stomatal closure, so that photosynthesis is maintained. The rate of assimilation after the period of drought is generally higher than in control plants and also recovers faster (Tolley and Strain, 1984; Sionit *et al.*, 1985; Hollinger, 1987). A possible effect of elevated CO_2 upon osmoregulatory processes and maintenance of turgor remains untested, although it has been noted that the solute potential of trees grown in elevated CO_2 is more negative than in controls (Tolley and Strain, 1985).

Electron flow, subsequent to PS II, was affected in drought-stressed *P. radiata* grown at a CO_2 concentration of 330 μmol mol^{-1}, but was not affected in plants grown at 660 μmol mol^{-1} (Conroy *et al.*, 1986b), indicating one mechanism whereby elevated CO_2 may ameliorate the effect of drought stress.

The ameliorating effect of elevated CO_2 is apparently species-specific, since under identical growth conditions, elevated CO_2 moderated the effects of a drying soil on *L. styraciflua*, but not on *P. taeda* (Tolley and Strain, 1985).

In C_4 weeds the thermal stability of phospho-enol-pyruvate carboxylase has been shown to increase in response to high CO_2 concentrations (Simon *et al.*, 1984) and the temperature optimum for photosynthesis in C_3 *Larrea divaricata* also increased in elevated CO_2 concentrations (Pearcy and Björkman, 1983). Similarly, the thermal tolerance of *P. ponderosa* increased after growth at a CO_2 concentration of 500 μmol mol^{-1} but decreased after growth at 650 μmol mol^{-1} (Surano *et al.*, 1986). It is unclear whether this thermal effect in *P. ponderosa* resulted directly from the higher CO_2 concentrations or whether it was influenced by the thermal environment within the open-topped chambers, since needle abscission and chlorosis developed during the experiment. This point requires clarification since it is likely that global mean temperatures will also increase as CO_2 levels increase, and this increase may be larger in temperate regions than nearer the equator.

Nutrition can also influence the photosynthetic response to CO_2. In well-watered *P. radiata* adequately supplied with P, CO_2 assimilation was enhanced by 221% by doubling the CO_2 concentration (Conroy *et al.*, 1986a, 1986b). With insufficient P, however, photosynthesis was decreased by approximately 35% (Conroy *et al.*, 1986b). Significantly, acclimation of the trees to low P supply occurred after 21 weeks growth in a CO_2 concentration of 330 μmol mol^{-1} but not in plants grown in 660 μmol mol^{-1} CO_2. The decrease in photosynthesis at elevated CO_2 concentrations in P-deficient trees was attributed to structural changes in the chloroplast thylakoid membranes and a concomitant decrease in the ability of the photon-harvesting proteins to trap and transfer energy (Conroy *et al.*, 1986b).

C. Photorespiration

CO_2 and O_2 compete for active sites on rubisco. When CO_2 is the substrate, carbon enters the photosynthetic carbon reduction cycle (the PCR or Calvin cycle): when O_2 is the substrate, phosphoglycolate and 3-phosphoglycerate are formed and carbon enters the photorespiratory carbon oxidation cycle (the PCO cycle), in which O_2 is consumed and CO_2 released (hence photorespiration). The release of CO_2 represents a significant loss of previously fixed carbon but this can be inhibited either by reducing the O_2 concentration or by raising the CO_2 concentration, since it is the ratio of O_2:CO_2 concentrations that determines the relative rates of the PCR and PCO cycles (Figure 5). A reduction in photorespiration also results in an enhanced availability of NADPH and ATP (which are consumed in photorespiration) and this may have significant, positive feedback effects on photosynthesis.

Whilst a significant reduction in photorespiration as a result of increasing the ambient CO_2 concentration has scarcely been demonstrated in trees, there is every reason to believe that the observed increases in net CO_2 assimilation, resulting from elevated CO_2 concentration, are partially attributable to reduction in photorespiration. Mortensen (1983) has shown that elevated CO_2 concentrations caused a significant reduction in photorespiration of *Picea abies*.

IV. THE SEEDLING, SAPLING AND INDIVIDUAL TREE SCALE

A. Growth over Periods of Weeks and Months

In almost all experiments done to date, growth of both broadleaves and conifers was increased by an approximate doubling of the ambient CO_2 concentration from between 20 to 120% with a median of *ca* 40%. An exception was *Pseudotsuga menziesii* which showed no increase in growth in response to CO_2, in contrast to *Pinus radiata*, in a 120-day experiment (Hollinger, 1987). Some of this variability is the result of experiments of

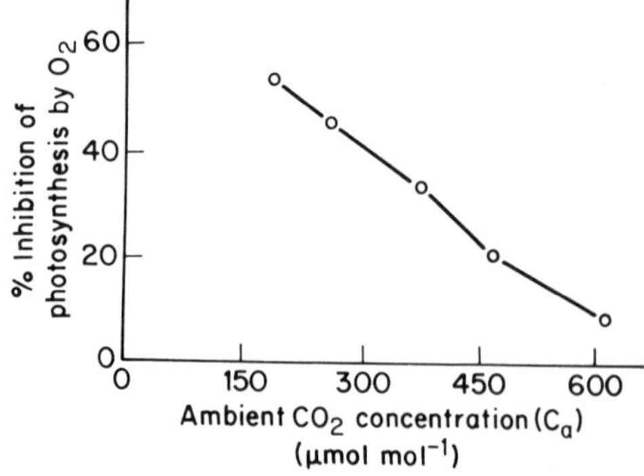

Fig. 5. Influence of ambient CO_2 concentration (C_a) on inhibition of CO_2 assimilation by oxygen in *Solanum tuberosum*. (Adapted from Ku, Edwards and Tanner, 1977.)

different length and some is related to stress associated with the experimental treatments, but much of it is not readily related to species and is largely inexplicable and can only be attributed to differences in experimental procedure and environments. Koch *et al.* (1986) suggested that effects of increased CO_2 on growth are likely to be large and persistent when there is a large sink capacity for carbon in the plant. In the longer experiments reviewed here, lack of sink capacity may well have attenuated the response and caused much of the variability (see Tables 5 and 6).

In most of the experiments, relative growth rates (1/W.dW/dt) were less than 5% per day and in many of the experiments, less than 2% per day. These rates are very low in comparison with reported rates of up to 10% per day for seedlings of conifers and 20% per day for seedlings of broadleaves (Jarvis and Jarvis, 1964; Ingestad, 1982; Ingestad and Kahr, 1985). In experiments in which nutrients were not supplied, the relative growth rates were lower still, less than 1% per day. The relative growth rate of tree seedlings does decline rapidly with age (Rutter, 1957) and this is particularly evident in the experiments by Brown and Higginbotham (1986) on interactions between CO_2 enrichment and nitrogen supply with *Populus tremuloides* and *Picea glauca*. None the less, the suspicion remains that very low relative growth rates indicate that the experiments were, in many cases, being done in far from optimum growth conditions. This may well, of course, reflect conditions in the real world, as in the experiments in which nutrients were not added, but the reasons for low relative growth rates in the experiments are often not clear.

B. Growth Partitioning

In a number of the experiments, an increase in CO_2 concentration resulted in an increase in leaf number, leaf area and leaf weight per plant (e.g. Tolley and Strain, 1984a; Sionit *et al.*, 1985; Conroy *et al.*, 1986a; Koch *et al.*, 1986; Brown and Higginbotham, 1986), leaf thickness (Rogers *et al.*, 1983) and leaf weight per unit area (e.g. Conroy *et al.*, 1986a; Brown and Higginbotham, 1986). Increases in leaf thickness and in weight per unit area may be associated with both an increase in cell size (e.g. Conroy *et al.*, 1986a) and an increase in the number of layers of cells in the mesophyll (Thomas and Harvey, 1983). It is, however, still an open question as to whether these effects result from direct action of CO_2 on leaf initiation, as suggested by Tolley and Strain (1984a), for example, or whether they result from an enhanced supply of substrate. Whilst a direct effect of CO_2 upon root initiation and growth has been observed (J.F. Farrar, pers. comm.), such an effect upon leaf primordial activity remains to be shown.

In most of the experiments too, an increase in CO_2 concentration led to an increase in the weight of both coarse and fine roots and this was especially

Table 5
Available data for the influence of elevated CO_2 concentrations on the growth of conifers.

Species	C_a	Growth location	Duration	Nutrients added	Other variables	Growth response	Author
Picea abies	×3	GR; P	118 days	x	light	+	Mortensen and Sandvik (1987)
Picea glauca	×2	GR; P	100 days		nitrogen	+	Brown and Higginbotham (1987)
Pinus contorta	×3, ×5	GR; P	150 days	x		+	Higginbotham *et al* (1985)
Pinus echinata	×2	GC; P	168 days			+	O'Neil *et al.* (1987)
Pinus echinata	×2	GC; P	287 days			+	Norby *et al.* (1987)
Pinus ponderosa	<×2	OTC; soil; P	2·5 years	x		+	Surano *et al.* (1986)
Pinus radiata	×2	GR; P	120 days	x		+	Hollinger *et al.* (1987)
Pinus radiata	×2	GC; P	254 days	x	water; phosphorus	+	Conroy *et al.* (1986a)
Pinus taeda	<×2½	OTC; P	*ca* 200 days	x			Thomas and Harvey (1983)
Pinus taeda	×2½	GC; P	84 days	x	light	+	Tolley and Strain (1984a)
Pinus taeda	×2½	GC; P	84 days	x	water	+, −	Tolley and Strain (1984b)
Pinus taeda	×2	GH; P	3 years	x		+	Sionet *et al.* (1985); Telewski and Strain (1987); Fetcher *et al.* (1988)
Pinus taeda	×2½	OTC; P	90 days	x		+	Rogers *et al.* (1983)
Pinus virginiana	<×3	OTC; P	122 days			+	Luxmore *et al.* (1986)
Pseudotsuga menziesii	×2	GR; P	120 days	x		+	Hollinger (1987)

[a] OTC—open-topped chamber; GC—growth chamber; GR—growth room; GH—glasshouse; P—pot

Table 6
Available data for the influence of elevated CO_2 concentrations on the growth of broadleaves.

Species	C_a	Growth location	Duration	Nutrients added	Other variables	Growth response	Author
Populus tremuloides	×2	GR; P	100 days	x		+	Brown and Higginbotham (1987)
Nothofagus fusca	×2	GR; P	120 days	x		+	Hollinger (1987)
Carya ovata					light and competition	No effect on community biomass	Williams *et al.* (1986)
Liriodendron tulipifera		GC;	90 days				Williams *et al.* (1986)
Quercus rubra		GC;	90 days				Williams *et al.* (1986)
Platanus occidentalis		GC;	90 days				Williams *et al.* (1986)
Acer saccharinum		GC;	90 days				Williams *et al.* (1986)
Fraxinus lanceolata		GC;	90 days				Williams *et al.* (1986)
Quercus alba	×2	GC; P	280 days			+	Norby *et al.* (1986a)
Quercus alba	×2	GC; P	210 days			+	O'Neil *et al.* (1987)
Quercus alba	×2	GC; P	280 days			+	Norby *et al.* (1986b)
Liriodendron tulipifera	×2	GC; Box	32 days			+	O'Neil *et al.* (1987)
Liquidambar styraciflua	< ×2½	OTC; P	90 days	x		+	Rogers *et al.* (1983)
Liquidambar styraciflua	<2½	OTC; P	*ca* 200 days	x			Thomas and Harvey (1983)
Liquidambar styraciflua	<2½	OTC; P	*ca* 200 days	x	light	+	Tolley and Strain (1984a)
Liquidambar styraciflua	<2½	GC; P	113 days	x	water	+	Tolley and Strain (1984b)
Liquidambar styraciflua	×2	GH; P	3 years	x		+	Sionit *et al.* (1985); Telewski and Strain (1987); Fetcher *et al.* (1988)
Citrus sinensis	×2	GC; P	305 days	x		+	Downton *et al.* (1987)
Robinia pseudoacacia	×2	GC; P	150 days				Norby (1987)
Alnus glutinosa	×2	GC; P	90 days				Norby (1987)

OTC—open-topped chamber; GC—growth chamber; GR—growth room; GH—glasshouse; P—pot.

large in *Pinus contorta* (Higginbotham *et al.*, 1985). Fruit production was increased by 70% in Valencia oranges (Downton *et al.*, 1987) and there were also concomitant increases in soluble solids. Stem diameter, weight or volume also generally increased (Tolley and Strain, 1984a; Conroy *et al.*, 1986a; Surano *et al.*, 1986; Mortensen and Sandvik, 1987) as well as height. In the longer experiments, it was noticeable that ring width was larger but there was no change in the average density of the wood (Telewski and Strain, 1987; Kienast and Luxmore, 1988). However, such observations on the juvenile wood of very young trees provide little guide to eventual changes in mature wood.

In addition to the changes in the amounts of leaves and roots, changes in response to an increase in CO_2 concentration were also observed in the dynamics of the leaf population and have been inferred in the dynamics of the population of fine roots. In experiments with *Quercus alba* the leaf area duration increased as a result of delayed leaf abscission (Norby *et al.*, 1986a, b), whereas in the relatively long-term experiment with *Pinus ponderosa* (Surano *et al.*, 1986) the sapling exposed to the highest CO_2 concentration had by the end of the study lost most two-year-old and many one-year-old needles. This last result may be an experimental artefact caused by very poor coupling between the needles and the atmosphere in the open-topped chambers and the excessively high needle temperatures that resulted.

The allocation of carbon within the plant is usually expressed as the root:shoot ratio, although this is not explicit in terms of whether there is, for example, an increase in the weight of roots or a decrease in the weight of leaves. In most of the experiments reviewed here, when nutrients have been supplied in adequate amounts sufficiently often, increase in CO_2 concentration resulted in a decrease or in no change in the root:shoot ratio (Tolley and Strain, 1984a; Sionit *et al.*, 1985; Koch *et al.*, 1986; Conroy *et al.*, 1986a; Brown and Higginbotham, 1987; Hollinger, 1987; Mortensen and Sandvik, 1987). One marked exception to this was noted with *P. contorta* in which root weight increased by $\times 15$ in response to a $\times 2$ increase in CO_2 concentration, although regularly fertilized (Higginbotham *et al.*, 1985). A similar result was obtained with *Castanea sativa* in which root weight not only increased substantially but shoot weight also decreased in response to a doubling of ambient CO_2 concentration, although apparently well-fertilized (Mousseau and Enoch, 1989).

In contrast, in the experiments in which no nutrients were added and trees were growing on a fixed capital of nutrients in low fertility soil, an increase in the atmospheric CO_2 concentration led to a substantial increase in the root:shoot ratio (Norby *et al.*, 1986a, b; Luxmore *et al.*, 1986; Norby *et al.*, 1987; O'Neill *et al.*, 1987a). This was shown in some experiments to be made up particularly of an increase in the amount of fine roots (Luxmore *et al.*, 1986; Norby *et al.*, 1986).

These results are in general accordance with the hypothesis that allocation of carbohydrates within the plant depends on the balance between the rate of supply of carbon to the leaves and the rate of supply of nutrients, particularly nitrogen, to the roots (Reynolds and Thornley, 1982; MacMurtrie and Wolf, 1983; Makela, 1986; Ågren and Ingestad, 1987). Essentially, the response of a tree to a low rate of supply of nitrogen is to increase the amount of roots, whereas the response to a low rate of supply of carbon is to increase the area of leaves. When the rate of supply of carbon is increased, as in these experiments, the proportion of leaf mass declines somewhat and the proportion of root mass increases, since relatively, the rate of supply of nitrogen to the roots is insufficient in relation to the supply of carbon.

A general conclusion from these measurements of growth is that increase in CO_2 concentration primarily leads to plants getting larger more quickly and that the majority of the changes observed are normal ontogenetic changes associated with growth and development.

C. Compensation for Stress

Increase in CO_2 concentration can compensate for stress-induced reduction in growth. Compensation in plants grown in low photon flux densities and double the current atmospheric CO_2 concentration was complete for *Liquidambar styraciflua* but only partial for *Pinus taeda* (Tolley and Strain, 1984a), but a tripling in CO_2 concentration led to over-compensation for *Picea abies* (Mortensen and Sandvik, 1987).

Full compensation for water stress by a doubling of the CO_2 concentration was also shown in *L. styraciflua* and partial compensation in *P. taeda* by Tolley and Strain (1984b), although Sionit *et al.* (1985) demonstrated only partial compensation in older seedlings of *L. styraciflua* growing in glasshouse sections. Compensation for water stress was also found with *P. radiata* (Conroy *et al.*, 1986b).

D. Water Use Efficiency

An increase in atmospheric CO_2 concentration is likely to result in a decreased stomatal conductance, and this will reduce the rate of transpiration per unit leaf area. Conversely, rates of photosynthesis are generally increased in elevated CO_2 environments. Consequently, an increase in water use efficiency (WUE), defined as (mol CO_2 assimilated)/(mol H_2O transpired), would be expected as a result of growing plants in elevated ambient CO_2 concentrations. Only a few reports on trees actually quantify WUE (e.g. Oberbauer *et al.*, 1985; Norby *et al.*, 1986; Hollinger, 1987), but an increase in WUE can be inferred from many reports. Where measured, water use efficiency has been shown to increase in response to the increase in CO_2

concentration, largely because water use did not increase in proportion to the increase in plant size (Rogers *et al.*, 1983; Norby *et al.*, 1986a). Figure 6 shows the positive linear relationship between CO_2 concentration and WUE for *L. styraciflua* observed by Rogers *et al.* (1983).

A possible benefit from increased WUE as a result of stomatal closure is a reduced rate of water consumption per unit leaf area, and hence a decrease in the likelihood of drought developing. However, since elevated CO_2 concentrations frequently result in an increase in total leaf area, this increase in WUE may be offset by increase in leaf area per tree. Increased WUE may also result in increased growth during drought in comparison to droughted control trees, since drought is delayed and less severe in elevated CO_2 environments (see Acclimating Tree Studies).

E. Nutrient Stress—a Limiting Factor?

It has been suggested that the nutrient stress commonly experienced by trees growing in woods and forests could completely negate any benefits to growth that might result from the increase in CO_2 concentration. In the majority of experiments referred to so far, attempts have been made to eliminate nutrient stress from consideration by supplying nutrients regularly to the plant-pots. In most cases this seems to have been successful and nutrient stress has been avoided, although in some cases the growth responses do suggest that some

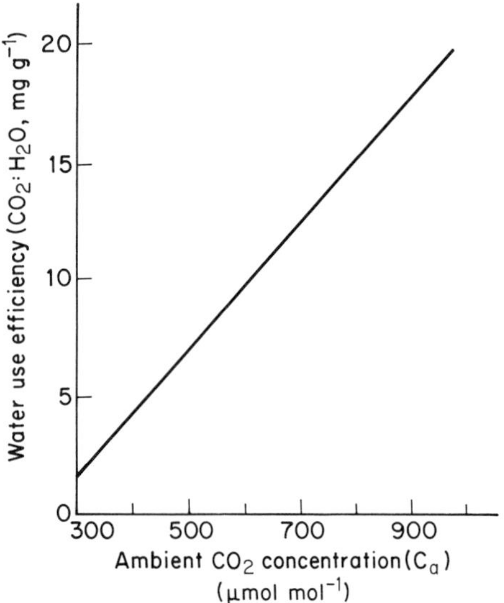

Fig. 6. Relationship between water use efficiency (WUE) and ambient CO_2 concentration (C_a) for *Liquidambar styraciflua*. (Adapted from Rogers *et al.*, 1983 and Tolley and Strain, 1985.)

nutrient stress did occur. In two experiments the limiting factor hypothesis has been tested by investigating the response to increase in CO_2 concentration by seedlings provided with a range of nitrogen concentrations in sand culture with *Populus tremuloides* and *Picea glauca* (Brown and Higginbotham, 1986) and additions of different amounts of phosphorus to a phosphorus-deficient soil with *Pinus radiata* (Conroy *et al.*, 1986a,b). Unfortunately, in both experiments the *amount* of nutrients added was controlled and no attempt was made to follow Ingestad principles and to control the *rate of supply* of the nutrients in relation to the endogenous capacity for growth (Ingestad, 1982; Ingestad and Lund, 1986).

In both these experiments growth increased in response to an increase in CO_2 concentration at each amount of added nutrient. Plant weight, leaf weight and plant height were increased particularly at the higher additions of nutrients and root weight at the lower additions. In both experiments the increase in growth was proportionately larger in the treatments with the larger additions of nutrients, but substantial growth increases (13% with P and 31% with N) were observed at the lowest additions of nutrients. The response of *P. tremuloides* was, however, only temporary and did not persist after 50 days.

In a series of experiments carried out at Oak Ridge by R. J. Luxmore and his colleagues, seedlings of *Q. alba* (Norby *et al.*, 1986a,b; O'Neill *et al.*, 1987b), *Liriodendron tulipifera* (O'Neill *et al.*, 1987a), *Pinus virginiana* (Luxmore *et al.*, 1986) and *Pinus echinata* (O'Neill *et al.*, 1987b) have been grown on soils of low fertility with a fixed, small capital of nutrients and their growth and uptake of a range of nutrients investigated in relation to root and rhizosphere properties. The four species investigated at Oak Ridge showed large (37–99%) increases in growth in response to a doubling of the ambient CO_2 concentration: the corresponding increases in relative growth rates ranged from 23 to 40%. These increases were largely the result of an increase in root weight which was consistently larger than the increase in total dry weight. Analysis of the plant nutrient contents showed four patterns of nutrient uptake as follows:

(*a*) a larger increase in uptake of the nutrient than in the assimilation of carbon, leading to an increase in the nutrient concentration in the plant;

(*b*) an increase in uptake of the nutrient proportional to the increase in carbon assimilation, leading to no change in the nutrient concentration in the plant;

(*c*) an increase in uptake of the nutrient less than the increase in carbon assimilation, leading to a decline in nutrient concentration in the plant; and

(*d*) no increase in uptake of the nutrient, so that the nutrient was progressively diluted as carbon was assimilated, leading to a large, progressive decline in the concentration of the nutrient in the plant.

The two broadleaves behaved similarly, showing no increase in uptake of N and S at the high CO_2 concentration (pattern *d*), but with all other nutrients (P, K, Ca, Mg, Al, Fe, Zn) following pattern (*b*) or (*c*). In contrast, in *P. virginiana* N and Ca showed no change in concentration (i.e. pattern *b*) and P, Mg and K declined in concentration (pattern *c*, *c* and *d*, respectively) whilst the trace elements increased in concentration in the plant (pattern *a*). Patterns (*c*) and (*d*) must inevitably lead to severe deficiency as the nutrients in the plant are progressively diluted by increasing amounts of carbon and this cannot continue indefinitely. To survive, the plant must either extract previously unavailable quantities of the nutrient or exploit new volumes of soil which is not, of course, possible, in pot experiments. It is also questionable with respect to patterns (*a*) and (*b*), how the plants continue to take up certain nutrients in excess of or in proportion to the carbon assimilated, from a declining resource as in a pot experiment. Possibly the capital is not fixed and the plants extract progressively more nutrients from the soil, even though it is of low fertility. The dilution of nutrients through growth in (*c*) and (*d*) leads to apparent increase in growth per unit of nutrient (i.e. nutrient use efficiency), which appears to increase in response to an increase in ambient CO_2 concentration for nutrients such as nitrogen. This apparent increase in efficiency is, however, no more than an indication of the limited ability of the plants to extract nutrients from a limited source.

F. Extraction of Nutrients

Mechanisms by which plants extract nutrients from a limited source were investigated in the same series of papers. In a number of experiments a greater proliferation of fine roots was observed in response to an increase of CO_2 concentration and this was, in some cases, associated with the increase in the amount of nutrients taken up (Norby *et al.*, 1986a; Luxmore *et al.*, 1987; O'Neill *et al.*, 1987a).

In *Q. alba* and *P. echinata*, this increase in fine roots was associated also with increase in the rate of establishment and density of mycorrhizal symbiosis (O'Neill *et al.*, 1987a; Norby *et al.*, 1987). Increased root exudation of carbohydrates may explain this, as there was a temporary increase in exudation of carbon-containing compounds from roots of *P. echinata* (Norby *et al.*, 1987). The observed increase in carbon allocation to roots and concomitant root exudation may have stimulated rhizosphere microbial and symbiotic activity and, hence, nutrient availability, although conclusive proof of this point was not provided by Norby *et al.* (1987) and no response of the microbial populations in the rhizosphere of *L. tulipifera* was found by O'Neill *et al.* (1987a). Luxmore *et al.* (1986) suggested that increased acidification of the rhizosphere of *P. virginiana* as a result of increase in the atmospheric CO_2 concentration might have been responsible for increase in the uptake of certain nutrients such as Zn, as a result of proton exchange.

None of these studies demonstrated in a convincing way how nutrients are taken up from a limited source of low availability in response to an increase in atmospheric CO_2 concentration, so that the internal plant concentration of nutrient does not change. A possibility, not so far considered, is that root surface enzymes, such as phosphatases, may be effective in solubilising bound nutrients, enabling them to be extracted more effectively by the expanded network of fine roots and mycorrhizas.

An enhanced role of soil microbial activity in the release of nutrients through mineralisation and decomposition is a possibility, since there is some evidence that the composition of leaves abscissed from growing trees is affected by CO_2 fertilization. Leaves abscissed from seedlings of *Q. alba* grown in increased CO_2 concentration were found to contain larger amounts of soluble sugars and tannin and a smaller amount of lignin. However, the differences were small and unlikely to increase significantly the rates of litter decomposition (Norby *et al.*, 1986b).

G. Nitrogen Fixation

Nitrogen-fixing trees are important for providing a stable long-term source of nitrogen in many silvicultural systems. Norby (1987) investigated the influence of elevated CO_2 on nodulation and nitrogenase activity of three nitrogen-fixing trees growing in an infertile forest soil and found that total nitrogenase activity per plant increased because of a larger root system, whereas specific activity was not influenced. Symbiotic nitrogen fixation is closely linked to photosynthetic capacity, and Norby concluded that the observed increase in nitrogenase activity per unit leaf area resulted from the increase in photosynthesis stimulated by CO_2 enrichment. Since the contribution of nitrogen fixation to the total nitrogen budget of the tree increases with age (Akkermans and van Dijk, 1975), the small increase in nodulation observed in young trees may have significantly larger effects in mature trees.

Two nitrogen-fixing species, *Robinia pseudoacacia* and *Alnus glutinosa*, also showed significant increases in growth, nodule weight and in nitrogenase activity in reponse to a doubling of atmospheric CO_2 concentration (Norby, 1987), perhaps because nitrogen fixation was carbohydrate-limited for other reasons.

H. Seedling Regeneration

Compensation for light stress, water stress and nutrient stress by seedlings, as observed in some of the experiments, is likely to enhance the ability of seedlings to establish and grow in a competitive environment and, hence, to promote seedling regeneration. It is evident from the foregoing, that seedlings of different species respond differently to an increase in atmospheric CO_2 concentration and it is possible, therefore, that the balance amongst regenerating species may change. This was investigated in an experiment in which three species from an upland habitat and three from a lowland habitat were grown together in competition on a fixed nutrient capital with increase in CO_2 concentration in low and high quantum flux densities (Williams *et al.*, 1986). Increase in CO_2 concentration did not increase the overall biomass of either population, but the relative weight of each species changed in a complex way depending on the atmospheric CO_2 concentration, photon flux density and the community. In this experiment, too, N and P became progressively diluted within the plants as a result of growth, so that nutrient acquisition and physiological response to nutrients may well have been responsible for the development of dominance by one or other species. In the upland community, *Quercus rubra* became dominant over *Carya ovata* and *L. tulipifera*, whereas in the lowland community *Fraxinus lanceolata* dominated *Platanus occidentalis* and *Acer saccharinum*.

Differences amongst provenances with respect to increase in CO_2 concentration are also likely. Surano *et al.* (1986), in their 2½-year experiment showed significant differences in the growth response of provenances of *P. ponderosa* from the Sierra Nevada and the Rocky Mountains to CO_2, but this experiment was complicated by leaf chlorosis and abscission, perhaps as a result of excessive temperatures in the open-topped chambers.

Changes in old-field secondary succession have been predicted from observations of *L. styraciflua* and *P. taeda* grown under elevated CO_2 with imposed drought (Tolley and Strain, 1984a). First-year survival of seedlings of *L. styraciflua* in drier sites, an increased tolerance to drought of *L. styraciflua* compared to *P. taeda* and the overall greater, CO_2-enhanced seedling growth of *L. styraciflua* may favour the establishment of *L. styraciflua* in drier habitats presently dominated by *P. taeda* (Tolley and Strain, 1985).

I. Growth at Timescales of Years and Decades

Only two experiments have been reported lasting for more than two years (Sionit *et al.*, 1985, and Fetcher *et al.*, 1988; Surano *et al.*, 1986) and it is difficult, if not dangerous, to extrapolate from these experiments to longer time scales. For the two species, *P. taeda* and *L. styraciflua*, differences in growth in response to increase in CO_2 concentration were established early on and persisted but did not increase with time. It seems very likely that this was a result of the constrained growing conditions, even though nutrients were added, and would, perhaps, not occur with unrestricted rooting.

It is readily shown that the quite small differences in relative growth rate observed in many of the experiments discussed, should lead to very large differences in plant size after quite short periods, if they persist. For example a 20% increase in relative growth rate from 0.8 to 1.0% per day should, after ten years, lead to trees that are over $\times$ 1000 larger. There is, however, at present no evidence as to whether the increases in seedling growth rates will persist and large differences in tree size will eventuate.

There is some indication in the experiments on seedlings that crown structure may change as a result of increase in CO_2 concentration. For example, shoot length increased in *P. abies* (Mortensen and Sandvik, 1987) and plants of *L. styraciflua* developed increased branching (Sionit *et al.*, 1985). However, the extent to which these differences might persist is unknown. A change in shoot structure and degree of branching is likely to lead to a change in the distribution of leaf area density within the tree crown and it has been shown with models that this can have a major influence on the total amount of crown photosynthesis in stands of *P. sitchensis* (Russell *et al.*, 1988; Wang, 1988).

In some cases, stem growth was increased and this could, conceivably, lead to enhanced harvest index later in life. In *P. radiata* seedlings Conroy *et al.* (1986a) found an increase in stemwood density but in the longer experiment with *P. taeda* (Sionit *et al.*, 1985) average stemwood density did not change, although there was an increase in latewood density (Telewski and Strain, 1987). Whether increase in atmospheric CO_2 concentration will lead to changes in harvest index and wood quality is, at the present time, entirely speculative.

V. THE PLANTATION AND WOODLAND SCALE

For practical and technical reasons, direct measurement of the effects of increase in CO_2 concentration on processes at this scale are extremely difficult, if not impossible, and at present there are no directly determined data and little likelihood of any in the immediate future. Purely on account of the large volume occupied by a tree, the scale of the problem in exposing large enough areas of forest to increased atmospheric CO_2 concentration is very much greater than with agricultural crops and it seems unlikely that exposure without enclosure will be possible.

Whilst the maintenance of increased CO_2 concentration around individual trees is technically feasible for extended periods, given adequate resources, the results would not bear directly on the functioning of the plantation or woodland system as a whole. Thus, at the present time, we have the problem of estimating the likely effects of an increase in CO_2 concentration on acclimating stands from measurements made on unacclimated stands at the present day CO_2 concentrations or from measurements made on seedlings and young trees partially acclimated to increased CO_2 concentrations in artificial surroundings. The use of models provides the only way forward in this situation.

A. At the Timescales of Hours and Days

Measurements of the exchanges of energy, momentum, water vapour, carbon dioxide and various pollutants by a range of plantation and woodland canopies in current atmospheric conditions have been made (Jarvis *et al.*, 1976; Jarvis, 1986; Verma *et al.*, 1986; Baldocchi *et al.*, 1987) and these data can be used to test models of the influence of environmental, structural and physiological variables on stand carbon dioxide balance and water use efficiency. A suitable model for this purpose exists (Grace *et al.*, 1987; Wang, 1988). This model, called MAESTRO, is straightforward to run for an existing canopy structure, stand structure and environment and can, in principle,

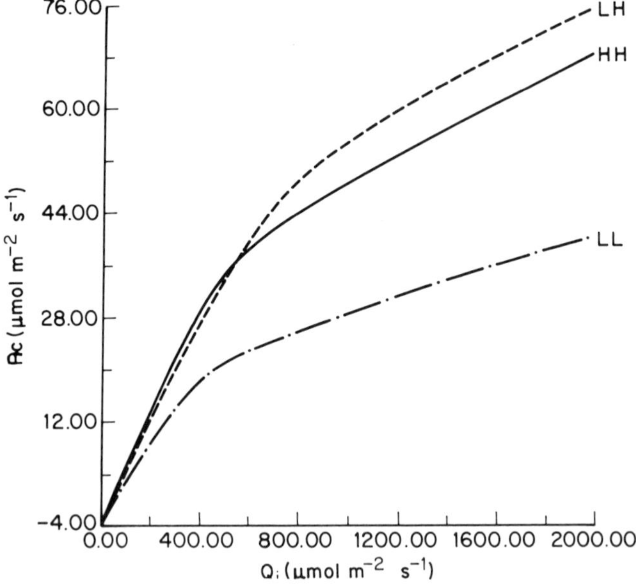

Fig. 7a. Response of canopy CO_2 assimilation (A_c) to the incident photon flux density (Q_i) calculated using MAESTRO for a pole stage stand of *Picea sitchensis* in central Scotland with appropriate parameters for the current ambient CO_2 concentration (345 μmol mol^{-1}) (LL) and double the current ambient CO_2 concentration (HH). The line LH shows what is obtained if current parameters are used in combination with a doubling in the atmospheric CO_2 concentration. The beam fraction of the incident quantum flux density was assumed to be 0·5, the leaf area index 9·0 and the zenith angle of the sun 45°. Other assumptions and a list of parameters are given by Wang (1988) (From Jarvis, 1989).

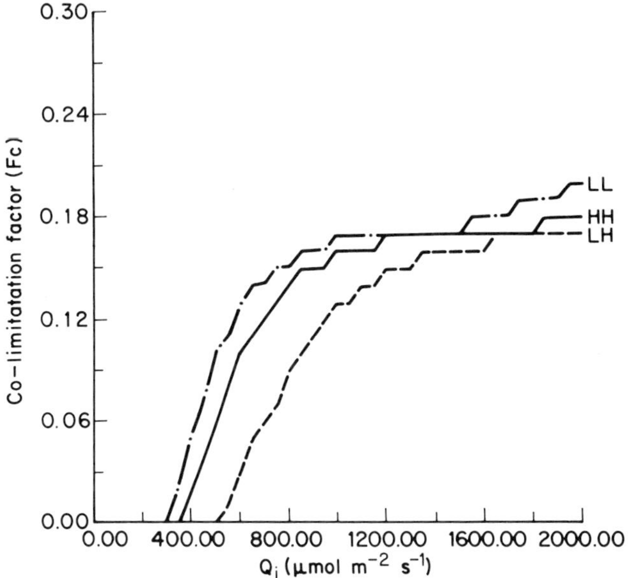

Fig. 7b. The calculated increase in co-limitation by RUPB-carboxylase in relation to photon flux density for the conditions in 7a. Limitation by electron transport is absolute ($A_c = 0$) up until an incident photon flux density of *ca* 400 µmol m^{-2} s^{-1} at which limitation by the carboxylase begins. A change in ambient CO_2 concentration has little or no effect on co-limitation.

provide answers to the question: what would be the consequence for assimilation, transpiration and water use efficiency of a stand of a doubling of the ambient CO_2 concentration? It is difficult, however, to know what aspects of acclimation to the increased CO_2 concentration should be taken into account. It is clear that acclimation of photosynthetic and stomatal parameters at the leaf scale should be considered, but it is far from clear as to what other aspects of tree or stand structure, canopy leaf area density distribution, soil properties, decomposition processes etc. need to be included. Figure 7 shows calculated canopy CO_2 assimilation for a stand of *Picea sitchensis* at the current ambient CO_2 concentration, using appropriate leaf parameters in an assimilation sub-model based on that of von Caemmerer and Farquhar (1981), in comparison with estimates in an atmosphere of double CO_2 concentration, again using appropriate parameters. The parameters were determined on seedlings grown in controlled environment rooms at current and twice the current ambient CO_2 concentrations. Canopy assimilation is increased substantially by a doubling of the atmospheric CO_2 concentration, but not as much as would be expected if parameters for unacclimated plants were used in the predictions at the doubled CO_2

concentration. Such predictions cannot be verified directly but depend for their acceptance on adequate verification of the model against measurements of canopy processes in the current CO_2 environment. MAESTRO, for example, has been shown to predict accurately the radiation environment beneath the canopy of stands of *Pinus radiata* and *Picea sitchensis* (Wang and Jarvis, 1989), and the predicted response of CO_2 assimilation to photon flux density with current parameters in the current CO_2 environment agrees reasonably well with measurements of CO_2 influx by a similar stand of *P. sitchensis* at similar, small water vapour saturation deficits (Jarvis and Sandford, 1986).

B. At the Timescales of Weeks and Months

In 1977 Monteith introduced a simple model of the growth of agricultural and horticultural tree crops as a function of intercepted radiation and the efficiency of utilization of that intercepted radiation. This model has been extended to plantations and woodlands (Jarvis and Leverenz, 1983) and has been shown to be applicable to stands of conifers and broadleaves (Linder, 1985, 1987), to biomass plantations of willow and poplar (Cannell *et al.*, 1987, 1988) and to commercial pole-stage plantations of *P. sitchensis* in Scotland (Wang, 1988). The model predicts that the rate of increase of dry matter is linearly proportional to the amount of photosynthetically active radiation (PAR) absorbed by the crop over the same period (Russell *et al.*, 1988).

For *P. sitchensis* in Tummel Forest in Central Scotland the average value of the dry matter: radiation absorption coefficient (efficiency of utilization of radiation) is 0.41 g of dry matter above ground per mole of photosynthetically active quanta and the relationship is linear over the range of radiation experienced. Thus, we may suppose that if an increase in CO_2 concentration leads to an increase in the efficiency with which PAR is utilized without compensating increases in respiration or turnover of fine roots, leaves and other plant parts, there will be an equally large increase in growth rate, i.e. a 10% increase in the net efficiency of PAR utilisation as a result of the increase in CO_2 concentration, will lead to a 10% increase in growth rate. If, in addition, the effect of an increase in CO_2 concentration is to increase the leaf area that the canopy can sustain (because an increase in CO_2 concentration may compensate for shading of the lower leaves in the canopy and reduce leaf senescence), so that more PAR is absorbed per hectare, then there will be a further increase in growth. This is exemplified in Figure 8 where Point A defines the present growth rate in relation to absorbed radiation and Point Z indicates the likely growth rate as a result of increase in both the efficiency of utilization of radiation (Point Y) and the amount of radiation absorbed (Point X). That the growth of forest may increase in relation to the increase

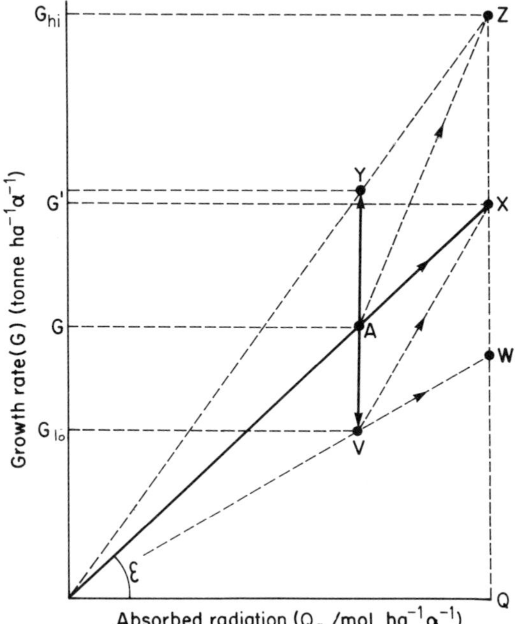

Fig. 8. Hypothetical relationship between stand growth rate (G) and absorbed photon flux density (Q_a). For a thinning and fertilizer experiment in a pole-stage stand of *Picea sitchensis* in central Scotland, $\varepsilon = 0{\cdot}3$ to $0{\cdot}5$ with an overall mean of 0·41 g of above-ground dry matter per mole of absorbed PAR. An increase in ambient CO_2 concentration, leading to an increase in either ε or the quanta absorbed could lead to an increase in growth rate from G to G′ (A→Y or A→X): an increase in both would lead to a further rise in G to G_{hi} (A→Y). A decline in ε, without a compensating increase in the quanta absorbed, could lead to a reduction in stand growth rate to G_{lo} (A→V). The likely magnitude of any such effects is unknown.

in CO_2 concentration, as a result of increases in both these processes, seems likely but is speculative.

To take this approach further, the two major processes discussed here require to be dissected into constituent, partial processes and the effects of an increase in CO_2 concentration on those processes investigated in detail, so that the likely changes in both efficiency of utilization and light absorption can be made quantitative. At the stand scale, we must necessarily be concerned about the consequences of an increase in CO_2 concentration on mineralization processes in the soil and on nutrient cycling through the system, since these processes may ultimately limit the extent to which photosynthetic efficiency can be increased or leaf populations maintained. Similarly, an increase in stand leaf area will lead to increases in both interception loss and transpiration loss of water (Jarvis and McNaughton, 1986), so that the availability of water may act to limit the projected increase in growth resulting from the increase in CO_2 concentration. This point is further complicated, since increases in water use efficiency and total leaf area per plant may increase, with opposing effects upon total plant water use. More complex models of stand processes than are available at present are required to take these possible feedbacks into account. Models for this purpose are under development at several places at the present time.

C. At the Timescales of Years and Decades

Accelerated growth during early life of a stand may lead to earlier canopy closure and ultimately to a reduced rotation length. At the present time, for example, this can readily be achieved by more intensive fertilizer applications (Axelsson, 1985; Linder, 1987) but this is costly and generally not in accordance with the economic strategy underlying current forest management practice. It is likely that small gains in this regard will result from the increase in CO_2 concentration but these gains are likely to be very much smaller and less immediate than the gains that could be obtained from more intensive stand management. Models capable of predicting the likely reduction in rotation length as a result of an increase in CO_2 concentration, with an appropriate mechanistic content, are not available, but there are, of course, adequate models to predict the economic benefits resulting from a reduction of rotation lengths of, say, 10% or five years.

It seems unlikely that the increase in CO_2 concentration will change the size-class distribution of stemwood within the stand, but this is of some economic importance and should be approached through the development of a suitable model.

It is probable that in natural woodlands an increase in the atmospheric CO_2 concentration will lead to a change in the species composition of both overstorey and understorey, but a natural woodland is a very complex ecological system and models adequate to test this are presently unavailable.

VI. THE FOREST AND ECOSYSTEM SCALE

A. At the Timescales of Years and Decades

Substantial changes are presently occurring in both the European and North American countryside, largely as a result of policy instruments of national governments and the European Community. Changes in forest management are likely to occur rapidly in the near future, as a result of the expansion of forest onto better quality, ex-agricultural soils, the introduction of new genotypes, a wider range of species and improved fertilizer regimes. In the British Isles, for example, the area of forest is expanding rapidly and is likely to continue to do so, and with the advent of progressively more second-rotation forest, the diversity of species and of stand and compartment composition is likely to increase substantially. Future growth enhancement resulting from increase in CO_2 concentration is likely to be subsumed into growth enhancement resulting from these and other changes, so that there are unlikely to be any directly consequential changes in forest management practice: the effects of an increase in CO_2 concentration will be superimposed upon, and difficult to distinguish from, the major changes in land management presently in progress. The scale of the increase in productivity over the next 25 years that may result from the increase in atmospheric CO_2 concentration is likely to be of the same magnitude as the increases in productivity that have occurred over the last 25 years, as the result of changes in fertilizer application regimes, for example. A change of this order of magnitude certainly could be expected to improve the economics of afforestation and may enhance the prospects for industrial utilization of wood in particular areas, but is unlikely to have a socio-economic impact of the same magnitude as the present expansion in area of the temperate forest estate.

There are other well-known consequences of afforestation, for stream-flow and the yield of water, for amenity, wildlife and recreation. It seems likely that the effects of increase in CO_2 on these aspects of forests will also not be noticeable in relation to the large changes resulting from fiscal policy and political instruments.

None the less, there is a strong probability that CO_2 fertilisation, accompanied by nitrogen fertilisation resulting from anthropogenic emissions, has been responsible in part for the substantial increases in temperate forest productivity over the past 100 years that have been observed in many places (e.g. the French Vosges region). Conversely, there is some evidence that the temperate and boreal forests are now affecting the global atmospheric CO_2 concentration (see discussion by Jarvis, 1989).

B. Timescales of a Century or More

Over much longer timescales we may expect the increase in CO_2 concentration to have a noticeable impact on the species composition and character of natural woodlands as a result of changes in the competitive balance among species, leading to ecological selection and, possibly, ultimately to natural selection, and evolution (Shugart *et al.*, 1986). What will happen in the future can, at present, only be guessed at and it is difficult to see how simplistic experiments in which three or more species are grown together in controlled environments can lead to useful predictions as to what may happen in natural woodland systems in the long term. Our understanding of natural ecosystems is still so elementary and likely to remain so for the foreseeable future, that sensible predictions about the likely effects of increase in CO_2 concentration on ecosystems are out of the question (Jarvis, 1987).

VII. SUMMARY

A. Assessing the Impact of the Rise in CO_2

A doubling of the present atmospheric CO_2 concentration is likely to occur by the end of the next century and will have many effects on trees and forests. There may be major changes in climate, especially temperature and precipitation, that will affect the growth and ecology of trees and forests: there will be substantial effects on the growth and ecology of trees and forests as a result of the direct effects of CO_2 on the physiology of trees. To a considerable extent the consequences of an increase in atmospheric CO_2 concentration on climate are speculative, whereas many of the direct effects of an increase in CO_2 concentration on the physiology and growth of plants have been established through experimentation and are quite well known and understood.

There are several reasons for showing active concern about the consequences of the rise in CO_2 concentration for trees and forests: to enhance knowledge, at a fundamental level, about the functioning of woodlands and forests; to assess the likely impact on their ecology, productivity and value; and to anticipate any significant downstream, socio-economic consequences.

The consequences of the rise in CO_2 concentration may be assessed at a range of temporal and spatial scales: from seconds to centuries, from cell to region. Biological information about the effects of CO_2 on plants is generally available at short time scales and small spatial scales, whereas ecological and socio-economic concern is largely expressed with respect to the much larger scales of the stand and the forest over tens or hundreds of years.

There are major problems with the data currently available. The environmental conditions pertaining in the majority of the experiments (e.g. soil type, temperature) differ significantly from forest conditions throughout the boreal and temperate regions. Secondly, the use of juvenile trees in growth cabinets and rooms makes extrapolation to the field situation difficult. Finally, the use of pot-grown plants with a limited soil volume, and all the problems of nutrient depletion that this entails, imposes severe constraints on the interpretations of changes in plant nutrient contents, assimilation rates and growth.

B. Photosynthesis and Stomatal Action

The primary role of CO_2 as a substrate for photosynthesis is well understood and there are no reasons to suppose that effects of CO_2 concentration on the primary carboxylation are any different in trees from other plants with the C_3 pathway. However, the role of CO_2 as an effective activator and regulator of rubisco is only currently being unravelled.

In general, photosynthesis increases substantially, but less than proportionately (up to 75%), in response to a doubling in ambient CO_2 concentration, but a lack of response, or even a reduction in rate of photosynthesis, has been observed. There is increasing evidence that maintenance of active sinks for assimilate is necessary for the effective stimulation of photosynthesis by an increase in CO_2 concentration. Constrained rooting of pot-grown trees and the lack of alternative sinks as a result of determinate growth, may be the reason why a strong, positive response has not always been seen and why the assimilation rate declines with time in many of the experiments. Closure of stomata in response to stress may also negate the increase from CO_2 fertilization.

When plants are grown for extended periods in elevated CO_2, so that they may become acclimated, the increase in rate of photosynthesis is less than in short-term experiments. Analysis of A/C_i response functions indicates that this may result from a reduction in the activity of rubisco, although it is not clear whether the amount or the activation state of the enzyme is reduced, or from Pi limitation.

Stomatal conductance generally declines with an increase in CO_2 concentration to between 0 to 70% of the control. There is considerable variation in response and in some trees, particularly conifers, stomatal conductance is rather unresponsive to changes in ambient CO_2 concentration. The lack of a mechanistic understanding of the role of CO_2 in stomatal action prevents explanation at present.

There is some evidence that a reduction in the number of stomata per unit area or per unit epidermal cell may also contribute to the reduction in stomatal conductance of plants grown at high CO_2 concentrations, but both increases and decreases have been observed in trees. It is not clear whether these changes are the result of acclimation of developmental processes, ecological selection or genetic adaptation.

There may be other presently unknown effects of CO_2 concentration on growth processes at the cellular level. In particular, the effects of CO_2 concentration on leaf growth may not be explicable solely in terms of substrate supply and there may be effects of CO_2 on leaf initiation, development and expansion in other ways.

C. Tree Growth

Growth of seedlings and young trees of both broadleaves and conifers is increased by a doubling of the ambient CO_2 concentration. The observed magnitude of the response ranges from 20 to 120% with a median of *ca* 40%. Both root and shoot growth are increased. Young trees grow faster in high CO_2 but normal ontogenetic processes occur.

In general, the experiments suffer from a number of inadequacies and this may explain the variability of response. In particular nearly all the experiments have been short-term (less than 12 months) on very young trees, that are often pot-bound and with growth restricted by lack of sinks and in a nutrient-deficient condition. Very low relative growth rates indicate far from good growth conditions.

Increases in growth in response to increase in CO_2 concentration occur in stress conditions and can ameliorate to a large extent the influence of stresses (water, nutrient, low light) on growth. Nutrient deficiency enhances biomass partitioning to the roots, whereas with free nutrient supply, an increase in CO_2 concentration leads to a relative enhancement of leaf growth.

Acclimation to nutrient depletion, evident as increased nutrient use efficiency, may occur but its extent and long-term significance remains uncertain. Enhanced fine-root mass and increased mycorrhizal infection may be significant in this respect, in addition to changes in root uptake processes.

The combination of enhanced assimilation and reduced stomatal conductance at elevated CO_2 leads to an increase in water use efficiency at the leaf and seedling scales in controlled environments, but it is not known whether this occurs at the stand scale in the field.

Because trees are perennial, a small increase in relative growth rate of 0.1% per day, at the prevailing low growth rates in many of the experiments, should lead to a large difference in individual tree size after a number of years. Whether CO_2 fertilization will have this effect is, however, uncertain.

A major gap in our knowledge is related to the influence of elevated CO_2 levels on yield quality (wood density, knottiness, etc.).

D. Stand Processes and Production

The experimental approach is not practical with older trees, stands and forests. The effects of CO_2 on assimilation, water use and other processes at the stand or larger scales, can only be assessed through models. Models of stand processes in relation to a doubling of CO_2 concentration, taking into account acclimation to CO_2 at the leaf scale, indicate a substantial increase in stand CO_2-assimilation and in water use efficiency. However, the information available to parameterize such models may not be appropriate because it has been derived from seedlings that were only partially acclimated.

Experimentation on mature trees is technically feasible but logistically and financially difficult. A more feasible approach available may be to use portions of mature trees, such as branches, in large cuvettes to obtain data required for the parameterization of stand-scale models that are used to make predictions of the likely consequences of the increase in the atmospheric CO_2 concentration for stand functioning. Micro-meteorological changes within the cuvettes and changes in feedback loops between the enclosed portion and the atmosphere, must, of course, be taken into account as well as the physiological consequences of treating only part of the plant. Such models require to be carefully tested at the present CO_2 concentrations and in relation to other environmental variables, since they cannot be tested in a doubled CO_2 environment in forests.

Changes in tree canopy structure may also influence canopy functioning as well as understorey species composition and growth: changes in the pattern of branching and in internode length, as a result of enhanced CO_2 concentrations, have been observed, for example. However, models that include feedbacks such as interactions between CO_2 and crown structure, growth and respiration of trees, or leaf composition and rates of litter decomposition, are not presently available. Whether changes in harvest index and wood quality will ensue at the stand scale cannot be predicted at present.

E. Forest and Region

At larger scales, a very complex network of processes must be taken into consideration and there is quite insufficient information about the effects of CO_2 on these processes to permit reasonable predictions. Future changes in forest management practice, the selection of new genotypes, extensive fertilizer treatments and an increase in the quality of land available for afforestation, are likely to lead to significant changes in forest production. The changes attributable to elevated CO_2 concentrations will be superimposed upon those caused by these other factors. It does, however, seem that the effects of the rise in CO_2 may be relatively small in relation to impending changes in land use and management practices.

VIII. RECOMMENDATIONS FOR FUTURE RESEARCH

In 1981 Kramer wrote: "We cannot make reliable predictions concerning the global effects of increasing CO_2 concentration, until we have information based on long-term measurements of plant growth from experiments in which high CO_2 concentration is combined with water and nitrogen stress on a wide range of species."

Whilst some progress has been made in this direction, as evidenced by the work referred to here, this conclusion of Kramer's is still generally valid, and it is particularly valid with respect to trees and forests.

Arising from this review, we have identified particular needs for research on trees and forests with respect to the increase in the global atmospheric CO_2 concentration in the following areas:

(*a*) The basic action of CO_2 on rubisco activity. Exploration of direct effects of CO_2 on leaf initiation and growth.

(*b*) Characterisation of the acclimation of photosynthetic processes to elevated CO_2 with associated development of leaf scale models. Investigation of the role of CO_2 in stomatal action. Comparison of the effects of CO_2 on seedling and mature foliage using bagged branches on trees and potted grafts of mature branches.

(*c*) The *long-term* effects of exposure to CO_2 on the growth and growth processes of forest species in appropriately controlled conditions; controlled environment rooms or well-ventilated open-topped chambers, with experimental control of nutrient supply rate according to Ingestad principles, control of water supply rate, and unconstrained rooting. Long-term acclimation of photosynthesis, stomatal conductance, water use efficiency, biomass partitioning, leaf and root dynamics and crown development in relation to high CO_2. Development of models of growth and partitioning to predict likely effects of high CO_2.

(*d*) Development and parameterization of models of stand processes incorporating multiple feedbacks. Rigorous testing of such models at the stand scale using eddy correlation measurements of water and CO_2 fluxes in present conditions to give confidence to predictions. Further development of more empirical stand-scale light-interception and growth models and parameterization with respect to predictions regarding elevated CO_2 effects.

(*e*) Tests of genotype/CO_2 interactions and selection of genotypes for growth in high CO_2 environment.

(*f*) Finally, we see the development of models as a matter of particularly urgent priority where forests are concerned and recommend the development of a new modelling framework to encompass a wide range of ecosystem processes for the purpose of assessing the consequences of the increase in CO_2 on forest stands at the present time.

ACKNOWLEDGEMENTS

We are grateful to Professor B. R. Strain for useful comments on the manuscript and to Dr R. A. Houghton and many other colleagues for helpful discussions and assistance with the literature. The substance of this review was first produced for the Department of the Environment (UK) in connection with the 1988 Toronto Conference. We gratefully acknowledge their financial sponsorship. The Department of the Environment is not in any way responsible for the opinions expressed here.

REFERENCES

Ågren, G. I. and Ingestad, T. (1987). Root:shoot ratio as a balance between nitrogen productivity and photosynthesis. *Plant, Cell and Environment*, **10**, 579–586.

Akkermans, A. D. L. and van Dijk, C. (1975). The formation and nitrogen-fixing ability of *Alnus glutinosa* under field conditions. In: P.S. Nutman (ed.), Symbiotic Nitrogen Fixation in Plants, pp. 511–520. Cambridge University Press: Cambridge.

Alden, T. (1971). Influence of CO_2, moisture and nutrients on the formation of lammas growth and prolepsis in seedlings of *Pinus sylvestris* L. *Studia Forestalia Suecica*, **93**, p. 21.

Allen, L. M., Beladi, S. E. and Shinn, J. H. (1985). Modelling the feasibility of free-air carbon dioxide releases for vegetation response research. *17th Conference on Agriculture and Forest Meteorology and 7th Conference on Biometeorology and Aerobiology, Scottsdale, Arizona.* A&F 9.4, 161–164. American Meteorological Society, Boston.

Arovaara, H., Hari, P. and Kuusela, K. (1984). Possible effect of changes in atmospheric composition and acid rain on tree growth. *Communicationes Instituti Forestalis Fennicae*, **122**, 15.

Ausubel, J. H. and Nordhaus, W. D. (1983). A Review of Estimates of Future CO_2 Emissions. In: Changing Climate. Report of the CO_2 assessment committee, National Academy Press: Washington.

Axelsson, B. (1985). Increasing Forest Productivity and Value. In: R. Ballard, P. Farnum, G. A. Ritchie and J. K. Winjum (eds), *Forest Potentials, Productivity and Value*. Weyerhaeuser Science Symposium 4, 5–37. Weyerhaeuser Co: Centralia.

Axelsson, E. and Axelsson, B. (1986). Changes in carbon allocation patterns in spruce and pine trees following irrigation and fertilization. *Tree Physiology*, **2**, 189–204.

Baldocchi, D. D., Verma, S. B. and Anderson, D. E. (1987). Canopy photosynthesis and water use efficiency in a deciduous forest. *Journal of Applied Ecology*, **24**, 251–260.

Beadle, C. L., Jarvis, P. G. and Neilson, R. E. (1979). Leaf conductance as related to xylem water potential and carbon dioxide concentration in Sitka spruce. *Physiologia Plantarum*, **45**, 158–166.

Beadle, C. L., Neilson, R. E., Jarvis, P. G. and Talbot, H. (1981). Photosynthesis as related to xylem water potential and CO_2 concentration in Sitka spruce. *Physiologia Plantarum*, **52**, 391–400.

Berry, J. A. and Downton, W. J. S. (1982). Environmental Regulation of Photosynthesis. In: Govindjee (ed.) pp. 263–344, *Photosynthesis II. Development, Carbon Metabolism, and Plant Productivity*, Academic Press: New York.

Bolin, B. (1977). Changes of land biota and their importance to the carbon cycle. *Science*, **196**, 613–615.

Bolin, B. (1986). How much CO_2 will remain in the atmosphere? In: B. Bolin, B. R. O. Doos, J. Jager and R. A. Warrick (eds), *The Greenhouse Effect, Climatic Change and Ecosystems*. Scope 29, pp. 93–156. Wiley: Chichester.

Brix, H. (1968). The influence of light intensity at different temperatures on rate of respiration of Douglas-fir seedlings. *Plant Physiology*, **43**, 389–393.

Brown, K. and Higginbotham, K. O. (1986). Effects of carbon dioxide enrichment and nitrogen supply on growth of boreal tree seedlings. *Tree Physiology*, **2**, 223–232.

Caemmerer, S. von, and Farquhar, G. (1981). Some relationships between the biochemistry of photosynthesis and the gas exchange of leaves. *Planta*, **153**, 376–387.

Caemmerer, S. von, and Farquhar, G. (1984). Effects of partial defoliation, changes in irradiance during growth, short term water stress and growth at enhanced p(CO_2) on photosynthetic capacity of leaves of *Phaseolus vulgaris*. *Planta*, **160**, 320–329.

Canham, A. E. and McCavish, W. J. (1981). Some effects of CO_2, day length and nutrition on the growth of young forest tree plants. I. In the seedling stage. *Forestry*, **54**, 169–182.

Cannell, M. G. R., Milne, R., Sheppard, L. J. and Unsworth, M. H. (1987). Radiation interception and productivity of willow. *Journal of Applied Ecology*, **24**, 261–278.

Cannell, M. G. R., Sheppard, L. J. and Milne, R. (1988). Light use efficiency and woody biomass production of poplar and willow. *Forestry*, **61**, 125–136.

Chang, C. W. (1975). Carbon dioxide and senescence in cotton plants. *Plant Physiology*, **55**, 515–519.

Conroy, J., Barlow, E. W. R. and Bevege, D. I. (1986a). Response of *Pinus radiata* seedlings to carbon dioxide enrichment at different levels of water and phosphorus: growth, morphology and anatomy. *Annals of Botany*, **57**, 165–177.

Conroy, J. P., Smillie, R. M., Kuppers, M., Bevege, D. I. and Barlow, E. S. (1986b). Chlorophyll *a* fluorescence and photosynthetic growth responses of *Pinus radiata* to P deficiencies, drought stress and high CO_2. *Plant Physiology*, **81**, 423–429.

Conway, T. J., Tans, P., Waterman, L. S., Thoning, K. W., Masarie, K. A. and Gammon, R. M. (1988). Atmospheric carbon dioxide measurements in the remote global troposphere, 1981–1984. *Tellus*, **40B**, 81–115.

Crane, A. J. (1985). Possible effects of rising CO_2 on climate. *Plant, Cell & Environment*, **8**, 371–379.

Cure, J. D. and Acock, B. (1986). Crop responses to CO_2 doubling: a literature survey. *Agricultural and Forest Meteorology*, **38**, 127–145.

Davis, J. E., Arkebauer, T. J., Norman, J. M. and Brandle, J. R. (1987). Rapid field measurement of the assimilation rate versus internal CO_2 concentration relationship in green ash (*Fraxinus pennsylvanica* Marsh): the influence of light intensity. *Tree Physiology*, **3**, 387–392.

De Lucia, E. H. (1987). The effect of freezing nights on photosynthesis, stomatal conductance and internal CO_2 concentration in seedlings of Engelmann spruce (*Picea Engelmannii* Parry). *Plant, Cell and Environment*, **10**, 333–338.

DeLucia, E. H., Sasek, T. W. and Strain, B. R. (1985). Photosynthetic inhibition after long term exposure to elevated levels of CO_2. *Photosynthesis Research*, **7**, 175–184.

Downton, W. J. S., Grant, W. J. R. and Loveys, B. R. (1987). Carbon dioxide enrichment increases yield of Valencia orange. *Australian Journal of Plant Physiology*, **14**, 493–501.

Eamus, D. (1986a). Further evidence in support of an interactive model in stomatal control. *Journal of Experimental Botany*, **37**, 657–665.

Eamus, D. (1986b). The response of leaf water potential and leaf diffusive resistance to abscisic acid, water stress and low temperature in *Hibiscus esculentus*. The effect of water stress and ABA pre-treatments. *Journal of Experimental Botany*, **37**, 1854–1862.

Edwards, G. and Walker, D. (1983). *C_3, C_4: Mechanisms, and Cellular and Environmental Regulation, of Photosynthesis.* Blackwell Scientific Publications: Oxford.

Ehret, I. L. and Jollife, P. A. (1985). Leaf injury to bean plants grown in carbon dioxide enriched atmospheres. *Canadian Journal of Botany*, **63**, 2015–2020.

Evans, J. R. (1988). Photosynthesis and nitrogen relations in leaves of C_3 plants. *Oecologia*, **78**, 9–19.

FAO (1982). *World Forest Products Demand and Supply 1990 and 2000*. FAO: Rome.

Fetcher, N., Jaeger, C. H., Strain, B. R. and Sionit, N. (1988). Long-term elevation of atmospheric CO_2 concentration and the carbon exchange rates of saplings of *Pinus taeda* L. and *Liquidambar styraciflua* L. *Tree Physiology*, **4**, 255–262.

Fifield, R. (1988). Frozen assets of the ice cores. *New Scientist*, **1608**, 28–29.

Funsch, R. W., Mattson, R. H. and Mowry, G. R. (1970). CO_2-supplemented atmosphere increases growth of *Pinus strobus* seedlings. *Forest Science*, **16**, 459–460.

Gates, D. M. (1983). An overview. In: E. R. Lemon (ed), pp. 7–20. *CO_2 and Plants: The Response of Plants to Rising Levels of Atmosphere Carbon Dioxide* AAAS Selected Symposia, **84**, Westsview Press: Boulder.

Gaudillère, J.-P. and Mousseau, M. (1989). Short-term effect of CO_2 enrichment in leaf development and gas exchange of young poplars (*Populus euramericana*) cv I 214). *Acta Oecologia/Oecologia Plantarum*, **10**, 95–105.

Gifford, R. M. (1982). Global Photosynthesis in Relation to our Food and Energy Needs. In: *Photosynthesis, Vol. 2, Development, Carbon metabolism and Plant Production.* Govindjee (ed). Academic Press: London, 459–495.

Grace, J. C., Jarvis, P. G. and Norman, J. M. (1987). Modelling the interception of solar radiant energy in intensively managed stands. *New Zealand Journal of Forestry Science*, **17**, 193–209.

Hamburg, S. P. and Cogbill, C. U. (1988). Historical decline of red spruce populations and climatic warming. *Nature*, **331**, 428–430.

Hårdh, J. E. (1967). Trials with carbon dioxide, light and growth substances on forest tree plants. *Acta Forestalia Fennica*, **82**, 1–10.

Hari, P. and Arovaara, H. (1988). Detecting CO_2-induced enhancement in the radial increment of trees. Evidence from the Northern timber line. *Scandinavian Journal of Forest Research*, **3**, 67–74.

Higginbotham, K. O., Mayo, J. M., L'Hirondelle, S. and Krystofiak, D. K. (1985). Physiological ecology of lodgepole pine (*Pinus contorta*) in an enriched CO_2 environment. *Canadian Journal of Forest Research*, **15**, 417–421.

Hollinger, D. Y. (1987). Gas exchange and dry matter allocation responses to elevation of atmospheric CO_2 concentration in seedlings of three tree species. *Tree Physiology*, **3**, 193–202.

Houghton, R. A. (1987). Terrestrial metabolism and CO_2 concentrations. *BioScience*, **37**, 672–678.

Houghton, R. A., Hobbie, J. E., Melillo, J. M., Moore, B., Peterson, B. J., Shaver, G. R. and Woodwell, G. M. (1983). Changes in the carbon content of terrestrial biota and soils between 1860 and 1980: a net release of CO_2 to the atmosphere. *Ecological Monographs*, **53**, 235–262.

Houghton, R. A., Boone, R. D., Fruci, J. R. *et al.* (1987). The flux of carbon from terrestrial ecosystems to the atmosphere in 1980 due to changes in land use: geographic distribution of the global flux. *Tellus*, **39B**, 122–139.

Ingestad, T. (1982). Relative addition rate and external concentration: Driving variable used in plant nutrition research. *Plant, Cell and Environment*, **5**, 443–453.

Ingestad, T. (1987). New concepts in soil fertility and plant nutrition as illustrated by research on forest trees and stands. *Geoderma*, **40**, 237–252.

Ingestad, T. and Ågren, G. I. (1988). Nutrient uptake and allocation at steady-state nutrition. *Physiologia Plantarum*, **72**, 450–459.

Ingestad, T. and Kahr, M. (1985). Nutrition and growth of coniferous seedlings at varied relative nitrogen addition rates. *Physiologia Plantarum*, **65**, 109–116.

Ingestad, T. and Lund, A-B. (1986). Theory and techniques for steady state mineral nutrition and growth of plants. *Scandinavian Journal of Forest Research*, **1**, 439–453.

Jarvis, P. G. (1985a). Increasing productivity and value of temperate coniferous forest by manipulating site Water Balance. In: R. Ballard, P. Farnum, G. A. Ritchie and J. K. Winjum (eds), *Forest Potentials, Productivity and Value*. Weyhaeuser Science Symposium, 4, 39–74. Weyerhaeuser Co: Centralia.

Jarvis, P. G. (1985b). Transpiration and assimilation of tree and agricultural crops: the 'omega factor'. In *Attributes of Trees as Crop Plants*, eds. M. G. R. Cannell and J. E. Jackson, pp. 460–480. Institute of Terrestrial Ecology: Abbots Ripton.

Jarvis, P. G. (1986). Coupling of carbon and water interactions in forest stands. *Tree Physiology*, **2**, 347–368.

Jarvis, P. G. (1987). Water and carbon fluxes in ecosystems. *Ecological Studies*, **61**, 50–67.

Jarvis, P. G. (1989). Atmospheric carbon dioxide and forests. In: P. G. Jarvis, J. L. Monteith, M. H. Unsworth and J. Shuttleworth (eds), Forests, Weather and Climate. *Philosophical Transactions of the Royal Society, Series B* (in the press).

Jarvis, P. G., James G. B. and Landsberg, J. T. (1976). Coniferous Forest. In: J. L. Monteith (ed), *Vegetation and the Atmosphere, Vol. 2 Case Studies* 171–240. Academic Press: London.

Jarvis, P. G. and Jarvis, M. S. (1964). Growth rates of woody plants. *Physiologia Plantarum*, **17**, 654–666.

Jarvis, P. G. and Leverenz, J. W. (1983). Productivity of Temperate, Deciduous and Evergreen Forest. In: O. L. Lange, P. S. Nobel, C. B. Osmond and H. Ziegler (eds) *Encyclopedia of Plant Physiology New Series Vol 12, Physiological Plant Ecology IV*, 233–280. Springer–Verlag: Berlin.

Jarvis, P. G. and McNaughton, K. G. (1986). Stomatal control of transpiration. *Advances in Ecological Research*, **15**, 1–29.

Jarvis, P. G. and Sandford, A. P. (1986). Temperate forests. In: N. R. Baker and S. P. Long (eds), *Photosynthesis in Contrasting Environments*, pp. 199–236. Elsevier: Amsterdam.

Jia, H-J. and Ingestad, T. (1984). Nutrient requirements and stress response of *Populus simonii* and *Paulownia tomentosa*. *Physiologia Plantarum*, **62**, 117–124.

Johnson, J. D. and Ferrell, W. K. (1983). Stomatal response to vapour pressure deficit and the effect of plant water stress. *Plant, Cell and Environment*, **6**, 451–456.

Jones, H. G. and Fanjul, L. (1983). Effects of water stress on CO_2 in apple. In: R. Marcelle, H. Clijsters and M. van Poucke (eds), *Effects of Stress on Photosynthesis*. pp. 75–93. Martinus Nijhoff/Dr W. Junk: The Hague.

Jones, P. D., Wigley, T. M. L. and Wright, P. B. (1986). Global temperature variations between 1861 and 1984. *Nature*, **322**, 430–434.

Jones, P. D., Wigley, T. M. L., Folland, C. K. *et al.* (1988). Evidence for global warming in the past decade. *Nature*, **332**, 790.

Keepin, W., Mintzer, I. and Kristoferson, L. (1986). Emission of CO_2 into the atmosphere. The rate of release of CO_2 as a function of future energy developments. In: B. Bolin, B. R. Doos, J. Jager *et al.* (eds), *The Greenhouse Effect, Climatic Change and Ecosystems*. Scope 29, pp. 35–92. Wiley: Chichester.

Kienast, F. and Luxmore, R. J. (1988). Tree-ring analysis and conifer growth responses to increased atmospheric CO_2-levels. *Oecologia*, **76**, 487–495.

Koch, K. E., Jones, P. H., Avigne, W. T. and Allen, L. H. (1986). Growth, dry matter partitioning, and diurnal activities of RuBP carboxylase in citrus seedlings maintained at two levels of CO_2. *Physiologia Plantarum*, **67**, 477–484.

Kramer, P. J. (1981). Carbon dioxide concentration, photosynthesis, and dry matter production. *BioScience*, **31**, 29–33.

Kramer, P. J. (1983). *Water Relations of Plants*. Academic Press: New York.

Kramer, P. J. and Sionit, N. (1987). Effects of increasing CO_2 concentration on the physiology and growth of forest trees. In: W. E. Shands and J. S. Hoffman (eds), *The Greenhouse Effect, Climate change and U.S. Forests*. The Conservation Foundation, Washington DC.

Krizek, D. T., Zimmerman, R. H., Klueter, H. H. and Bailey W. A. (1971). Growth of crab apple seedlings in controlled environments. Effect of CO_2 levels, and time and duration of CO_2 treatment. *Journal of American Society for Horticultural Science*, **96**, 285–288.

Ku, S. B., Edwards, G. E. and Tanner, C. B. (1977). Effects of light, carbon dioxide and temperature on photosynthesis, oxygen inhibition of photosynthesis and transpiration in *Solanum tuberosum*. *Plant Physiology*, **59**, 868–872.

Laiche, A. J. (1978). Effects of refrigeration, CO_2 and photoperiod on the initial and subsequent growth of rooted cuttings of *Ilex cornuta* Lindl. et Paxt. cv. Burfordii. *Plant Propagation*, **24**, 8–10.

Lamarche, V. C., Graybill, D. A., Fritts, H. C. and Rose, M. R. (1984). Increasing atmospheric carbon dioxide: tree ring evidence for growth enhancement in natural vegetation. *Science*, **225**, 1019–1021.

Lin, W. C. and Molnar, J. M. (1982). Supplementary lighting and CO_2 enrichment for accelerated growth of selected woody ornamental seedlings and rooted cuttings. *Canadian Journal of Plant Science*, **62**, 703–707.

Linder, S. (1985). Potential and Actual Production in Australian Forest Stands. In: J. J. Landsberg and W. Parsons (eds), *Research for Forest Management*. pp. 22–35. CSIRO: Melbourne.

Linder, S. (1987). Responses to water and nutrients in coniferous ecosystems. *Ecological Studies*, **61**, 180–202.

Linder, S., McMurtrie, R. E. and Landsberg, J. J. (1986). Growth of eucalyptus: a mathematical model applied to *Eucalyptus globulus*. In: P. M. A. Tigerstedt, P. Puttonen and V. Koshi, (eds), *Crop Physiology of Forest Trees*, pp. 107–127. Helsinki University Press: Helsinki.

Ludlow, M. M. and Jarvis, P. G. (1971). Photosynthesis in Sitka spruce (*Picea sitchensis* (Bong.) Carr.) I. General characteristics. *Journal of Applied Ecology*, **8**, 925–953.

Luxmore, R. J. (1981). CO_2 and phytomass. *BioScience*, **31**, 626.

Luxmore, R. J., O'Neil, E. G., Ellis, J. M. and Rogers, H. H. (1986). Nutrient uptake and growth responses of Virginia pine to elevated atmospheric carbon dioxide. *Journal of Environmental Quality*, **15**, 244–251.

Luxmore, R. J., Tharp, M. L. and West, D. C. (1989). Simulating the physiological basis of tree ring responses to environmental changes. In: *Forest Growth: Process Modelling of Responses to Environmental Stress*. Timber Press: Alabama.

Makela, A. (1986). Partitioning coefficients in plant models with turn-over. *Annals of Botany*, **57**, 291–297.

Marland, G. (1988). *The prospect of solving the CO_2 problem through global reforest-*

ation. US Department of Energy TRO39, pp. 66. US Department of Commerce: Springfield, Virginia.

Mauney, J. R., Guinn, G., Fry, K. E. and Hesketh, J. D. (1979). Correlation of photosynthetic carbon dioxide uptake and carbohydrate accumulation in cotton, soyabean, sunflower and sorghum. *Photosynthetica*, **13**, 260–266.

McMurtrie, R. E. and Wolf, L. (1983). Above- and below-ground growth of forest stands: a carbon budget model. *Annals of Botany*, **52**, 437–448.

Monteith, J. L. (1977). Climate and efficiency of crop production in Britain. *Philosophical Transactions of the Royal Society of London, Series B*, **281**, 277–294.

Morison, J. I. L. (1985). Sensitivity of stomata and water use efficiency to high CO_2. *Plant, Cell and Environment*, **8**, 467–474.

Morison, J. I. L. and Gifford, R. M. (1983). Stomatal sensitivity to CO_2 and humidity. *Plant Physiology*, **71**, 789–796.

Mortensen, L. M. (1983). Growth response of some greenhouse plants to environment. VIII. Effect of CO_2 on photosynthesis and growth of Norway spruce. *Meldinger fra Norges Landbrukshogskole*, **62 (10)**, 1–13.

Mortensen, L. M. and Sandvik, M. (1987). Effects of CO_2 enrichment at varying photon flux density on the growth of *Picea abies* (L.) Karst. seedlings. *Scandinavian Journal of Forest Research*, **2**, 325–334.

Mott, K. A. (1988). Do stomata respond to CO_2 concentrations other than intercellular? *Plant Physiology*, **86**, 200–203.

Mousseau, M. and Enoch, H. Z. (1989). Effect of doubling atmospheric CO_2 concentration on growth, dry matter distribution and CO_2 exchange of two-year-old sweet chestnut seedlings (*Castanea sativa* Mill.). *Plant, Cell and Environment* (in the press).

Neftel, A., Moor, E., Oeschger, H. and Stauffer, B. (1985). Evidence from polar ice cores for the increase in atmospheric CO_2 in the past 2 centuries. *Nature*, **315**, 45–47.

Norby, R. J. (1987). Nodulation and nitrogenase activity in nitrogen-fixing woody plants stimulated by CO_2 enrichment of the atmosphere. *Physiologia Plantarum*, **71**, 77–82.

Norby, R. J., O'Neill, E. G. and Luxmore, R. J. (1986a). Effects of atmospheric CO_2 enrichment on the growth and mineral nutrition of *Quercus alba* seedlings in nutrient poor soil. *Plant Physiology*, **82**, 83–89.

Norby, R. J., Pastor, J. and Melillo, J. (1986b). Carbon-nitrogen interactions in CO_2-enriched white oak: physiological and long-term perspectives. *Tree Physiology*, **2**, 233–241.

Norby, R. J., O'Neill, E. G., Hood, W. G., and Luxmore, R. J. (1987). Carbon allocation, root exudation and mycorrhizal colonization of *Pinus echinata* seedlings grown under CO_2 enrichment. *Tree Physiology*, **3**, 203–210.

Oberbauer, S. F., Sionit, N., Hastings, S. J. and Oechel, W. R. (1986). Effects of CO_2 enrichment and nutrition on growth, photosynthesis and nutrient concentrations of Alaskan tundra species. *Canadian Journal of Botany*, **64**, 2993–2998.

Oberbauer, S. F., Strain, B. R. and Fetcher, N. (1985). Effect of CO_2 enrichment on seedling physiology and growth of two tropical species. *Physiologia Plantarum*, **65**, 352–356.

O'Neill, E. G., Luxmore, R. J. and Norby, R. J. (1987a). Elevated atmospheric CO_2 effects on seedling growth, nutrient uptake, and rhizosphere bacterial populations of *Liriodendron tulipifera* L. *Plant and Soil*, **104**, 3–11.

O'Neill, E. G., Luxmore, R. J. and Norby, R. J. (1987b). Increases in mycorrhizal colonization and seedling growth in *Pinus echinata* and *Quercus alba* in an enriched CO_2 atmosphere. *Canadian Journal of Forest Research*, **17**, 878–883.

Pearcy, R. W. and Björkman, O. (1983). Physiological Effects. In: E. R. Lemon (ed), *CO_2 and Plants: The Response of Plants to Rising Levels of Atmospheric Carbon Dioxide* AAAS Selected Symposia Vol. 84, pp. 65–105. Westview Press: Boulder.

Pearman, G. I., Etheridge, D., de Silva, F. and Fraser, P. J. (1986). Evidence of changing concentrations of CO_2, N_2O and CH_4 from air bubbles in Antarctic ice. *Nature*, **320**, 248–250.

Peet, M. M., Huber, S. C. and Patterson, I. T. (1985). Acclimation to high CO_2 in monoecious cucumber. II. Alterations in gas exchange rates, enzyme activities and starch and nutrient concentrations. *Plant Physiology*, **80**, 63–67.

Porter, M. A. and Grodzinsky, B. (1984). Acclimation to high CO_2 in bean. *Plant Physiology*, **74**, 413–416.

Raper, P. C. and Peedin, G. F. (1978). Photosynthetic rate during steady state growth as influenced by CO_2 concentration. *Botanical Gazette*, **139**, 147–149.

Regehr, D. C., Bazzaz, F. A. and Boggess, W. R. (1975). Photosynthesis, transpiration and leaf conductance of *Populus deltoides* in relation to flooding and drought. *Photosynthetica*, **9**, 52–61.

Reynolds, J. F. and Thornley, J. H. M. (1982). A shoot:root partitioning model. *Annals of Botany*, **49**, 585–597.

Roberts, D. R. and Dumbroff, E. F. (1986). Relationships among drought resistance, transpiration rates and abscisic acid levels in 3 northern conifers. *Tree Physiology*, **1**, 161–167.

Roberts, J. (1983). Forest transpiration: a conservative process? *Journal of Hydrology*, **66**, 133–141.

Rogers, H. H., Thomas, J. F. and Bingham, G. E. (1983). Response of agronomic and forest species to elevated atmospheric carbon dioxide. *Science*, **220**, 428–429.

Russell, G., Jarvis, P. G. and Monteith, J. L. (1988). Absorption of Radiation by Canopies and Stand Growth. In *Plant Canopies: their growth form and function* (eds. G. Russell, B. Marshall and P. G. Jarvis), pp. 21–39. Cambridge University Press: Cambridge.

Rutter, A. J. (1957). Studies in the growth of young plants of *Pinus sylvestris*. I. The annual cycle of assimilation and growth. *Annals of Botany*, **21**, 399–426.

Sage, R. F. and Sharkey, T. D. (1987). The effect of temperature on the occurrence of O_2 and CO_2-insensitive photosynthesis in field grown plants. *Plant Physiology*, **84**, 658–664.

Sage, R. F., Sharkey, T. D. and Seemann, J. R. (1988). The in-vivo response of the ribulose–1,5-bisphosphate carboxylase activation state and the pool sizes of photosynthetic metabolites to elevated CO_2 in *Phaseolus vulgaris* L. *Planta*, **174**, 407–416.

Sage, R. F., Sharkey, T. D. and Seeman, J. R. (1989). Acclimation of photosynthesis to elevated CO_2 in five C_3 species. *Plant Physiology* **89**: 590–596.

Sasek, T. W., DeLucia, E. H. and Strain, B. R. (1985). Reversibility of photosynthetic inhibition in cotton after long-term exposure to elevated CO_2 concentrations. *Plant Physiology*, **78**, 619–622.

Schulte, P. J. and Hinckley, T. M. (1987). Abscisic acid relations and the response of *Populus trichocarpa* stomata to leaf water potential. *Tree Physiology*, **3**, 103–113.

Schulte, P. J., Hinckley, T. M. and Stettler, R. F. (1987). Stomatal responses of *Populus* to leaf water potential. *Canadian Journal of Botany*, **65**, 255–260.

Shugart, H. H. Antonovsky, M. Ja., Jarvis, P. G. and Sandford, A. P. (1986). CO_2, climatic change and forest ecosystems. In: B. Bolin, B. R. Doos, J. Jager and R. A. Warrick (eds), *The Greenhouse Effect, Climatic Change and Ecosystems*. Scope 29, pp. 475–521 Wiley: Chichester.

Simon, J.-P., Potvin, G. and Strain, B. R. (1984). Effects of temperature and CO_2 enrichment on kinetic properties of phospho-enol-pyruvate carboxylase in two ecotypes of *Echinochloa crus-galli* (L). Beaur. a C4 weed grass species. *Oecologia*, **63**, 145–152.

Sionit, N. and Kramer, P. J. (1986). Woody plant reaction to CO_2 enrichment. In: H. Z. Enoch and B. A. Timball (eds), *CO_2 Enrichment and Greenhouse Crops*, Vol. II pp. 69–85. CRC Press: Boca Raton.

Sionit, N., Strain, B. R. and Hellmers, H. (1981). Effects of different concentrations of atmospheric CO_2 on growth and yield components of wheat. *Journal of Agricultural Science*, **79**, 335–339.

Sionit, N., Strain, B. R., Hellmers, H., Riechers, G. H. and Jaeger, C. H. (1985). Long-term atmospheric CO_2 enrichment affects the growth and development of *Liquidambar styraciflua* and *Pinus taeda* seedlings. *Canadian Journal of Forest Research* 15, 468–471.

Siren, G. & Alden, T. (1972). *CO_2 supply and its effect on the growth of conifer seedlings grown in plastic greenhouses.* Research Note 37, pp. 15. Department of Reforestation, Royal College of Forestry, Sweden.

Smagorinsky, J. (1983). Effects on Climate. In: *Changing Climate*, Report of the CO_2 Assessment Committee. National Research Council, USA, National Academy Press, USA.

Strain, B. R. (1985). Physiological and ecological control on carbon sequestering in ecosystems. *Biogeochemistry*, **1**, 219-232.

Strain, B. R. and Cure, J. D. eds. (1986). Direct Effects of atmospheric CO_2 on plants and ecosystems: a bibliography with abstracts. ORNL/CDIC-13. NTIS, US Department of Commerce: Springfield.

Strain, B. R. and Bazzaz, F. (1983). Terrestrial plant communities. In: E. R. Lomon (ed), *CO_2 and Plants: The response of Plants to Rising Levels of Atmospheric Carbon Dioxide*, AAAS Selected Symposium Vol. 34, pp. 177–222. Westview Press: Boulder.

Strand, M. & Öquist, G. (1985). Inhibition of photosynthesis by freezing temperatures and high light levels in cold acclimated seedlings of Scots pine (*Pinus sylvestris*). *Physiologia Plantarum*, **64**, 425–430.

Surano, K. A., Daley, P. F., Houpis, J. L. J. *et al.* (1986). Growth and physiological responses of *Pinus ponderosa* Dougl. ex P. Laws. to long-term elevated CO_2 concentrations. *Tree Physiology*, **2**, 243–259.

Telewski, F. W. and Strain, B. R. (1987). Densiometric and ring width analysis of 3-year-old *Pinus taeda* L. and *Liquidambar styraciflua* L. grown under three levels of CO_2 and two water regimes. In: G. C. Jacoby, and J. W. Hornbeck, (eds). *Proceedings of the International Symposium on Ecological Aspects of Tree Ring Analysis.* DoE CONF-8608144. NTIS: Springfield, Virginia.

Tenhunen, J. D., Lange, O. L., Gebel, J., Beyschlag, W. and Weber, J. A. (1984). Changes in photosynthetic capacity, carboxylation efficiency, and CO_2 compensation point associated with midday stomatal closure and midday depression of net CO_2 exchange of leaves of *Quercus suber*. *Planta*, **162**, 193–203.

Teskey, R. O., Fites, J. A., Samuelson, L. J. and Bongarten, B. C. (1986). Stomatal and non-stomatal limitations to net photosynthesis in *Pinus taeda* under different environmental conditions. *Tree Physiology*, **2**, 131–142.

Thomas, J. F. and Harvey, C. N. (1983). Leaf anatomy of four species grown under continuous long-term CO_2 enrichment. *Botanical Gazette*, **144**, 303-309.

Tinus, R. W (1972). CO_2-enriched atmosphere speeds growth of ponderosa pine and blue spruce seedlings. *Tree Planters Notes*, **23**, 12–15.

Tolbert, N. E. and Zeltich, I. (1983). Carbon metabolism. In: E. R. Lemon (ed.) *CO_2 and Plants: The Response of Plants to Rising Levels of Atmospheric Carbon*

Dioxide. AAAS Selected Symposia Vol. 84, pp. 21–64. Westview Press: Boulder.

Tolley, L. C. and Strain, B. R. (1984a). Effects of CO_2 enrichment on growth of *Liquidambar styraciflua* and *Pinus taeda* seedlings under different irradiance levels. *Canadian Journal of Forest Research*, **14**, 343–350.

Tolley, L. C. and Strain, B. R. (1984b). Effects of CO_2 enrichment and water stress on growth of *Liquidambar styraciflua* and *Pinus taeda* seedlings. *Canadian Journal of Botany*, **62**, 2135–2139.

Tolley, L. C. and Strain, B. R. (1985). Effects of CO_2 enrichment and water stress on gas exchange of *Liquidambar styraciflua* and *Pinus taeda* seedlings grown under different irradiance levels. *Oecologia*, **65**, 166–172.

Verma, S. B., Baldocchi, D. D., Anderson, D. E., Matt, O. R. and Clement, R. J. (1986). Eddy fluxes of CO_2, water vapour and sensible heat over a deciduous forest. *Boundary-Layer Meteorology*, **367**, 71–91.

Wang, Y.-P. (1988). *Crown structure, radiation absorption, photosynthesis and transpiration*. PhD Thesis, University of Edinburgh.

Wang, Y.-P. and Jarvis, P. G. (1989). Description and analysis of an array model—MAESTRO. *Agricultural and Forest Meteorology* (submitted).

Warrick, R. A., Gifford, R. M. and Parry, M. L. (1986). Assessing the response of food crops to the direct effects of increased CO_2 and climatic change. In: B. Bolin, B. R. Doos, J. Jaeger & R. A. Warrick (eds), *The Greenhouse Effect, Climatic Change and Ecosystems*. Scope 29, pp. 393–473. Wiley: Chichester.

Warrit, B., Landsberg, J. J. and Thorpe, M. R. (1980). Responses of apple leaf stomata to environmental factors. *Plant, Cell and Environment*, **3**, 13–22.

Watson, R. L., Landsberg, J. J., Thorpe, M. R. (1978). Photosynthetic characteristics of the leaves of golden delicious apple trees. *Plant, Cell and Environment*. **1**, 51–58.

Watts, W. R. and Neilson, R. E. (1978). Photosynthesis in Sitka spruce (*Picea sitchensis* (Bong.) Carr.) VIII. Measurements of stomatal conductance and $^{14}CO_2$ uptake in controlled environments. *Journal of Applied Ecology*, **15**, 245–255.

Wigley, T. M. L., Jones, P. O. and Kelly, P. M. (1986). Empirical climate studies. In: B. Bolin, B. R. Doos, J. Jager and R. A. Warrick (eds), *The Greenhouse Effect, Climatic Change and Ecosystems*. Scope 29, pp. 271-322. Wiley: Chichester.

Williams, W. E., Garbutt, K., Bazzaz, F. A. and Vitousek, P. M. (1986). The response of plants to elevated CO_2 IV. Two deciduous forest communities. *Oecologia*, **69**, 454–459.

Wong, S. C., Cowan, I. R. and Farquhar, G. D. (1978). Leaf conductance in relation to assimilation in *Eucalyptus pauciflora* Sieb. ex Spreng. Influence of irradiance and partial pressure of carbon dioxide. *Plant Physiology*, **62**, 670–674.

Wong, S. C. and Dunin, F. X. (1987). Photosynthesis and transpiration of trees in a eucalypt forest stand: CO_2, light and humidity responses. *Australian Journal of Plant Physiology*, **14**, 619–632.

Woodward, F. I. (1987). Stomatal numbers are sensitive to increases in CO_2 from pre-industrial levels. *Nature*, **327**, 617–618.

Woodward, F. I. and Bazzaz, F. A. (1988). The responses of stomatal density to CO_2 partial pressure. *Journal of Experimental Botany*, **39**, 1771–1781.

Woodwell, G. M., Whittaker, R. M., Reiners, W. A., Likens, G. E., Delwiche, C. C. and Botkin, D. B. (1978). The biota and the world carbon budget. *Science*, **199**, 141–146.

Wulff, R. D. and Strain, B. R. (1981). Effects of CO_2 enrichment on growth and photosynthesis in *Desmodium paniculatum*. *Canadian Journal of Botany*, **60**, 1084–1091.

Yeatman, C. W. (1970). CO_2-enriched air increased growth of conifer seedlings. *Forestry Chronicle*, **46**, 229–230.

On the Evolutionary Pathways Resulting in C_4 Photosynthesis and Crassulacean Acid Metabolism (CAM)

RUSSELL K. MONSON

ADVANCES IN ECOLOGICAL RESEARCH Vol. 19
ISBN 0-12-013919-7

I. SUMMARY

This review covers the literature concerning C_3–C_4 intermediate photosynthesis and the various modes of Crassulacean acid metabolism, in relation to the evolution of fully-expressed C_4 photosynthesis and CAM. Several hypotheses are proposed to explain the paths that have been taken during the evolution of these two pathways. In the case of C_4 photosynthesis it is proposed that the initial evolutionary stages were driven by selection for reduced photorespiration rates and the influence of such reduction on the net CO_2 assimilation rate. Reductions in the photorespiration rate occurred through the evolution of C_4-like anatomy, with increased concentrations of photorespiratory organelles and the photorespiratory enzyme glycine decarboxylase, in bundle-sheath cells. Such compartmentation facilitated the recycling of photorespired CO_2, without imposing costs on CO_2 assimilation by the mesophyll. The presence of C_4-like biochemistry is only known from two genera containing C_3–C_4 intermediates. Initially it evolved in a fairly inefficient form, exhibiting evidence of futile energy-driven cycles of carboxylation and decarboxylation. These inefficient cycles presumably represented the precursor to the CO_2-concentrating mechanism in fully-expressed C_4 plants. The presence of an inefficient C_4 cycle during the evolution of C_4 photosynthesis, appears to have resulted in an increased quantum requirement for CO_2 assimilation, which might have acted as a barrier to the evolution of C_4 photosynthesis in shaded habitats. Presumably, those shade-adapted C_4 species that have been identified evolved from sun adapted, fully-expressed C_4 species. Upon evolution of the CO_2-concentrating mechanism, improvements in water-use efficiency and nitrogen-use efficiency took over as primary "driving forces" in the evolution of C_4 photosynthesis.

The assimilation of respired CO_2 during the night, with assimilation of atmospheric CO_2 during the day (known as CAM-cycling) is proposed as a precursor to the evolution of fully-expressed CAM in some species. Thus, C_4 photosynthesis and CAM may have both originated from mechanisms that improved the plant's carbon balance by assimilating internally generated CO_2. The adaptive significance of CAM-cycling is thought to be water conservation, a role that is consistent with the adaptive significance of fully expressed CAM. CAM-cycling may have evolved from the organic-acid metabolism associated with ion uptake in plants (particularly in the guard cells of stomata), or from the organic-acid metabolism characteristic of aquatic plants. Following the evolution of CAM-cycling, stomatal opening patterns must have evolved an inverted pattern—opening during the night and closing during the day. The evolution of such an inverted pattern would have resulted in large improvements in water-use efficiency, and presumably accelerated the rate of evolution to the full-CAM stage.

II. INTRODUCTION

Long before the reader has arrived at this part of my work, a crowd of difficulties will have occurred to him. Some of them are so serious that to this day I can hardly reflect on them without being in some degree staggered: but to the best of my judgment, the greater number are only apparent, and those that are real are not, I think, fatal to the theory.

These difficulties and objections may be classed under the following heads:—First, why, if species have descended from other species by fine gradations, do we not everywhere see innumerable transitional forms? Why is not all nature in confusion, instead of the species being, as we see them, well defined?

Charles Darwin (1859) *On the Origin of Species*

Evolutionary biology has always been plagued by a lack of clear paths to follow in delineating evolutionary patterns. This has led to a significant degree of subjectivity in our interpretations of evolutionary patterns. The apparent lack of transitional types is especially perplexing in those cases where the same evolutionary outcome has been repeated several times independently, and the trait appears to have evolved in recent geological time. Until recently, such a case was found in the evolution of C_4 photosynthesis and Crassulacean acid metabolism (CAM) in plants. Since the days when the metabolic details of these pathways were first being identified, there has been agreement that they evolved following the advent of the angiosperms, and originated from more primitive C_3 photosynthetic-type plants (Hatch and Slack, 1970; Evans, 1971), although there is recent evidence that CAM has also evolved in more primitive plant groups (Hew and Wong, 1974; Wong and Hew, 1976; Keeley, 1981; Keeley *et al.*, 1984; Raven *et al.*, 1988). There has also been consensus on the suggestion that these more advanced pathways evolved independently within the angiosperms on several different occasions (Hatch and Slack, 1970; Evans, 1971; Kluge and Ting, 1978; Smith *et al.*, 1979; Teeri, 1982a).

With such an evolutionary story it seemed as though transitional forms should be evident (Björkman, 1976). Indeed forms intermediate to C_4 photosynthesis began to be discovered in the mid- to late 1970s (Kennedy and Laetsch, 1974; Brown and Brown, 1975). Although these intermediate forms exhibited anatomical, and in some cases biochemical, traits of both C_3 and C_4 plants there has not been consensus on whether they represent evolutionarily transitional forms or the products of ancient hybridization events between C_3 and C_4 ancestors. Although caution about readily accepting the evolutionary intermediacy of these plants is required (Monson *et al.*, 1984), evidence from a variety of sources—taxonomical, ecological, physiological, and biochemical—strongly suggests that these intermediates have indeed acquired their C_4 traits through the evolutionary processes of mutation and natural selection (Monson and Moore, 1989).

Recently, there have also been a number of studies published that describe CAM-like function in some plants at a level of integration below that of fully expressed CAM (e.g. Ting and Rayder, 1982; Martin *et al.*, 1988). Although the evolutionary significance of such carbon-assimilation modes has been largely unstated (but see Teeri, 1982a; Teeri, 1982b), they may represent transitional stages in the evolution of fully expressed CAM.

(A note of clarification should be added at this point. By describing these plants as intermediate forms, it is not implied that they will necessarily evolve into fully expressed C_4 or CAM species. But rather that they have acquired their unique carbon-assimilation traits through the evolutionary process and thus represent 'models' with which we can examine the evolutionary paths that have produced C_4-like and CAM-like traits. Additionally, it cannot be known for certain that fully-expressed C_4 and CAM plants passed through the same evolutionary stages as these intermediate forms. Thus, one could argue that these intermediate forms have limited potential to provide insight into the pathways taken during the evolution of C_4 photosynthesis and CAM. This is a stand that I cannot refute. It represents a position that one could take on virtually any evolutionary issue, and is only refuted by the same subjective interpretations that were alluded to above. Having explained the possible positions that I could take in interpreting the existence of these photosynthetic intermediates, I will venture out on a long limb and interpret the intermediate forms as true evolutionary transitional forms that can be used as valid models to identify the evolutionary pathways that have been taken during the evolution of fully expressed C_4 photosynthesis and CAM.)

The purpose of this review is to assemble the literature on C_3–C_4 intermediate photosynthesis and CAM-like function into an evolutionary framework that can be used to evaluate some possible paths that may have been taken during the evolution of fully-expressed C_4 photosynthesis and CAM. The review is structured such that the known limitations to C_3-cycle assimilation are discussed first, followed by a discussion of how the evolution of C_4-like traits and CAM-like traits alleviate these limitations. In the second section the characteristics of C_3–C_4 intermediate photosynthesis will be described along with some of the apparent adaptive advantages conferred by such photosynthetic traits. Also in that section the evidence will be presented to establish the argument that these intermediates represent true evolutionary intermediates. In the third section the case of CAM-like traits will be discussed, including the evidence that the CAM-like transitional forms also represent evolutionary intermediates. In the fourth, and final section, some discussion is devoted to the costs associated with the evolution of C_4 photosynthesis and CAM, and how those costs may have constrained evolutionary patterns.

III. COMPARATIVE ASPECTS OF CO_2 ASSIMILATION BY C_3, C_4, AND CAM PLANTS

A. Limitations to Carbon Assimilation Through the C_3 Cycle

Two recent reviews have provided thorough synopses of the factors thought to limit CO_2 assimilation in the C_3 cycle (Sharkey, 1985; Woodrow and Berry, 1988). This section therefore merely summarizes the limitation of the C_3 cycle in the light of what must be accomplished through the evolutionary process to result in the advantages of C_4 photosynthesis and CAM.

Limitations to C_3-cycle assimilation are caused by five prominent features of the cycle (Figure 1).

(1) The availability of O_2 relative to CO_2. Because of the oxygenase activity of ribulose 1, 5-bisphosphate carboxylase/oxygenase (rubisco), the presence of atmospheric concentrations of O_2 results in competitive inhibition of CO_2 assimilation (Ogren, 1984).

(2) The availability of CO_2. The C_3 cycle is thought to have evolved in an atmosphere characterized by much higher CO_2 concentrations than exist today (Cloud, 1968; Broda, 1975). As atmospheric CO_2 concentrations decreased during the past several million years, evolutionary changes in rubisco's active site resulted in only partial compensation through increased substrate affinity and turnover capacity (Jordan and Ogren, 1981). The lack of full compensation means that the carboxylase activity of rubisco operates at approximately half of its maximum velocity—representing a significant substrate limitation.

(3) The activity of rubisco. It has recently been suggested that the largest limitation to CO_2 assimilation is the concentration and activation state of rubisco (Woodrow and Berry, 1988). The activation state of rubisco appears to be modified *in vivo* to match rubisco activity with the other limitations to CO_2 assimilation. This maintains a balance among all limiting factors and leads to an efficient steady-state condition (Woodrow, 1986).

(4) Regeneration rate of ribulose 1, 5-bisphosphate (RuBP). This process, which is driven by ATP and NADPH production through the light reactions of photosynthesis, has been shown to be a factor co-limiting the rate of rubisco activity at normal atmospheric CO_2 concentrations (von Caemmerer and Farquhar, 1981; Evans, 1986). Recently, however, several studies have demonstrated that the concentration of RuBP in the chloroplast stroma is not below the concentration of rubisco binding sites during steady-state photosynthesis (Badger *et al.*, 1984; Mott *et al.*, 1984; Dietz and Heber, 1984; Makino *et al.*, 1985; von Caemmerer and Edmondson, 1986). Only during certain transient changes in environmental conditions (such as photon flux density) do RuBP concentrations become limiting. This limitation is typically alleviated quickly through changes in the activation state of rubisco (Mott *et*

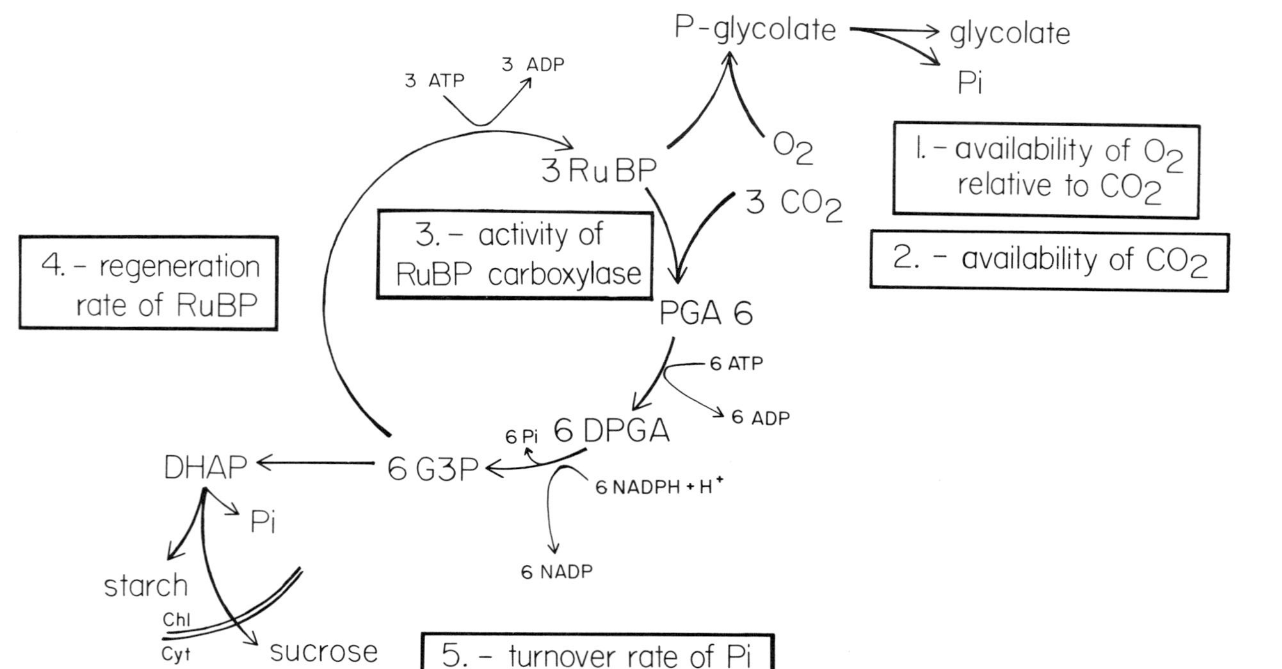

Fig. 1. A schematic diagram of the principal processes involved in photosynthetic CO_2 assimilation through the C_3 cycle (reductive pentose phosphate cycle). The five major potential limitations to the rate of CO_2 assimilation are identified in the boxes. See text for a discussion of the limitations. *Abbreviations*: P_i = inorganic phosphate, PGA = 3-phosphoglyceric acid, DPGA = 1,3-diphosphoglyceric acid, G3P = 3-glyceraldehyde phosphate, DHAP = dihydroxyacetone phosphate, RuBP = ribulose 1,5-bisphosphate, Chl = chloroplast, Cyt = cytoplasm.

al., 1984; von Caemmerer and Edmondson, 1986). Limitations caused by the rate of RuBP regeneration appear to be most important at CO_2 concentrations in excess of normal ambient values. However, there is a primary uncertainty concerning the potential chelation of Mg^{2+} by RuBP, which influences the free RuBP concentration (von Caemmerer and Edmondson, 1986). Thus, the status of this limitation is at present uncertain.

(5) The turnover rate of inorganic phosphate (P_i). Recent studies have demonstrated that a balance exists between the rate at which phosphate is freed from organic metabolite pools through sucrose and starch synthesis and the rate of P_i consumption during the photosynthetic production of sugar-phosphate molecules (Walker and Herold, 1977; Herold, 1980; Sharkey, 1985). Under near-optimal environmental conditions this balance probably does not have a role in limiting CO_2 assimilation. However, under conditions of suboptimal temperatures or elevated CO_2 concentrations an imbalance may occur in the rate at which P_i is consumed or freed, resulting in limitations to the rate of photosynthesis (Leegood and Furbank, 1986; Stitt, 1986; Sage and Sharkey, 1987).

As the primitive atmosphere of the earth evolved from a condition of high CO_2 and low O_2 to one of low CO_2 and high O_2, these limitations must have had primary roles in constraining the evolution of the photosynthetic systems. For example, the decreased availability of CO_2 and increased competition between O_2 and CO_2 for active sites of rubisco presumably underlay the evolution of increased specificity of the active site of rubisco for CO_2 (Jordan and Ogren, 1983).

Compensatory changes in the kinetic properties and activity of rubisco, however, appear to only have been partly effective in attaining maximum photosynthesis rates in the modern atmosphere. This is evidenced by the fact that rubisco only carboxylates RuBP at approximately half its maximum velocity at the current normal atmospheric CO_2 concentration (350 μmol mol^{-1}).

The most effective evolutionary changes in the photosynthetic system, with respect to allowing rubisco to approach more closely its maximum carboxylation velocity, have been the evolution of mechanisms that place rubisco in an anatomical micro-environment of high CO_2 concentration, relative to O_2 concentration, an atmospheric environment that more closely resembles that of the primitive earth. This is exemplified by the evolution of C_4 photosynthesis, CAM, and the CO_2-concentrating function of many aquatic autotrophs (e.g. Raven *et al.*, 1985).

It is intriguing that the evolutionary option most often observed in plants, with respect to substantially improving the catalytic capacity of rubisco, involves complex changes in the anatomical architecture and biochemistry of individual leaves or stems in the case of C_4 photosynthesis and CAM, and individual cells in the case of aquatic organisms, rather than changes in the active site characteristics of rubisco. At the present time, it is not known why evolution at the active site of rubisco has not resulted in eliminating the oxygenase activity. Lorimer and Andrews (1981) proposed that the oxygenation of RuBP might be an unavoidable consequence of carboxylating RuBP. According to this hypothesis, the chemical mechanisms involved in catalysis produce an intermediate that can either react with CO_2 or O_2. The hypothesis requires an assumption of ordered substrate binding to the active site, with RuBP binding first. This assumption has received some experimental support (Badger *et al.*, 1980; Pierce *et al.*, 1986; Pierce, 1986). It is not known which part of the enzyme, if any, interacts with CO_2 during carboxylation of the enolized RuBP. There is some evidence that it interacts with the divalent metal ion (see Pierce *et al.*, 1986). If there is little interaction between CO_2, O_2, and a binding site on the enzyme, then there is little room for evolutionary solutions to the oxygenation reaction. It is not clear, however, how this hypothesis is reconciled with the observations that evolution has at least to some extent, resulted in less specificity for O_2 in more advanced taxa (Jordan and Ogren, 1981).

B. The Evolution of CO_2-Concentrating Mechanisms in C_4 and CAM Plants

The presence of CO_2-concentrating mechanisms in C_4 and CAM plants relieves, to some extent, the substrate limitations on rubisco activity. The details of these CO_2-concentrating mechanisms have been described in recent reviews (Edwards and Huber, 1981; Osmond and Holtum, 1981), and will not be repeated here. However, it will be useful to discuss some of the evolutionary patterns that must have occurred during the appearance of these mechanisms.

The complex anatomical and biochemical integration of these CO_2-concentrating mechanisms suggests that they evolved in stages, rather than in single events. However, there has been some suggestion that several of the enzymes involved in the CO_2-concentrating mechanisms of both C_4 and CAM plants are encoded by linked genes that regulate stomatal guard-cell metabolism (Cockburn, 1981; Teeri, 1982a; Cockburn, 1983). Similar arguments can be developed for those aspects of plant metabolism that involve the regulation of pH and ion balance in *all* plant cells, both root and shoot. If this is the case, then some aspects of the CO_2-concentrating mechanisms may have evolved by mutations of the regulatory gene(s) that control these blocks of functional genes.

There are certain patterns of C_4-cycle evolution that must have been followed during the appearance of both C_4 and CAM plants. For example, the CO_2 concentration at the active sites of rubisco must have increased so as to elevate the rate of photosynthesis and lead to the improved water- and nitrogen-use efficiencies of CO_2 assimilation typically associated with these plants. In modern C_4 and CAM plants this is accomplished through two features: (1) the velocity of the C_4 cycle is greater than that of the C_3 cycle until a steady-state, elevated CO_2 concentration is reached, at which point the velocities approach each other and come to equilibrium with the rate of CO_2 leakage from the C_3 cycle, and (2) a diffusion barrier exists that reduces the rate of CO_2 leakage from the C_3 cycle and uncouples the CO_2 pool at the site of rubisco from that in the outside atmosphere. In C_4 plants the diffusion barrier is represented by the thickened, occasionally suberized, cell walls of the bundle-sheath cells. In CAM plants the diffusion barrier is formed by closed stomata during the daylight hours. In either case, it seems intuitive that the diffusion barriers evolved after the CO_2-concentrating mechanism since there would be little advantage to the former without the latter. In fact, without the presence of the CO_2-concentrating mechanism the diffusion barriers would have represented a hindrance to supplying CO_2 to the C_3 cycle. Similarly, there would have been little benefit to the CO_2-concentrating mechanism unless there was some barrier to diffusion of CO_2 away from the C_3 cycle. This may have occurred in the early stages of C_4-cycle evolution through intracellular compartmentation. Decarboxylation of C_4-acids may have occurred in the chloroplast (or mitochondria located next to chloroplasts), and thus faced a short diffusion distance to rubisco and a long diffusion distance (with membrane barriers) to the atmosphere of the intercellular air spaces of the leaf. Scenarios such as these are only speculative and depend upon several assumptions, including the status of enzymes such as carbonic anhydrase in the cytoplasm (Burnell and Hatch, 1988) and whether the components of the C_4 cycle had adaptive functions other than pumping CO_2 to rubisco (see Davies, 1979; Latzko and Kelly, 1983 for a discussion of the many functions of PEP carboxylase). These speculations are only discussed here to point out that although one can construct intuitive scenarios to describe evolutionary patterns in photosynthetic pathways, the requirement for multiple assumptions to support the proposed patterns render them tenable. A more fruitful approach appears to be the examination of plants that are in transitional stages of photosynthetic evolution.

C. Ecological Traits Associated with the CO_2-Concentrating Mechanism in C_4 and CAM Plants

In C_4 plants, the CO_2-concentrating mechanism results in photosynthetic advantages in terms of improved water-use efficiencies, higher temperature optima, improved nitrogen-use efficiencies, and improved quantum efficiencies at high photon flux densities (Edwards *et al.*, 1985). Although CAM is typically discussed in terms of improved water-use efficiency, the same list of advantages discussed for C_4 plants would generally apply to CAM plants (Winter, 1985). These gas-exchange traits have been compiled from studies on isolated photosynthetic organs. There has been little effort to examine the importance of these traits in relation to the growth and reproductive success of plants in their natural environments (although see Nobs *et al.*, 1972; Caldwell *et al.*, 1977b; and review by Fischer and Turner, 1978).

One of the more common approaches to examining the ecological significance of these gas-exchange advantages in C_4 and CAM plants has been to examine geographical distributions in relation to environmental features. Through such an approach one can determine whether C_4 and CAM plants are typically found in habitats in which the gas-exchange traits described above are truly advantageous. The results provide a complex picture with only a few clear-cut generalizations. Several studies have found significant correlations between the proportion of a flora that is composed of C_4 monocots and the presence of warm midsummer temperatures (Teeri and Stowe, 1976; Teeri *et al.*, 1980; Hattersley, 1983; Cavagnaro, 1988). In these and other studies, however, it is not always clear whether the relative distributions of C_3 and C_4 grasses are most closely associated with temperature gradients or the moisture gradients that usually accompany temperature gradients (see also Tieszen *et al.*, 1979; Rundel, 1980). In at least one study it was found that C_3 and C_4 grass distribution patterns were more strongly associated with water availability than prevailing temperature (Chazdon, 1978). In North American C_4 dicots precipitation patterns, rather than prevailing temperature, seem to be more important in influencing species abundance patterns (Stowe and Teeri, 1978), although warm midsummer temperatures have secondary importance. Most C_4 species are native to unshaded environments, although several exceptions exist in the form of shade-adapted C_4 species (Robichaux and Pearcy, 1980; Winter *et al.*, 1982a; Ward and Woolhouse, 1986; Smith and Martin, 1987).

From these studies, it is clear that although temperature cannot always be identified as the principal determinant of C_3 and C_4 distribution patterns, C_4 plants are usually associated with warm growth-season habitats. Long (1983) reviewed the literature describing the distributions of C_3 and C_4 plants, and concluded that C_4 species are completely excluded from areas with mean summer minima below 8°C. It is probable that the positive correlation between C_4 abundance and warm growth-season temperatures is due to the fact that most C_4 species evolved from ancestors native to the primitive floras of tropical or sub-tropical regions (Long, 1983). There appear to be no inherent weaknesses in the design of the C_4 system with respect to functioning at low temperatures (Caldwell *et al.*, 1977a). It is also apparent that although C_3 plants tend to exhibit greater inhibition of photosynthesis by high, but sub-lethal, temperatures, there are many C_3 species native to warm habitats. Thus, the C_3 design, with its inherent photorespiration, does not preclude successful existence in warm climates. Given the lack of definitive disadvantages of either C_4 or C_3 photosynthesis in most environments, it is unlikely that C_4 mutations that appear in large populations would have advantages of such magnitude that they would over-ride the traditional mode of C_3-cycle assimilation and become rapidly fixed. It is more likely that C_4 traits become most rapidly established in small, isolated populations. The lack of large numbers of C_4 species in floras native to cool environments, however, also makes clear the influence of warm environments on plant growth following the mutations that initiate the evolution of C_4 characteristics. It is likely that C_4 species native to cool or shaded habitats originated from C_4 ancestors native to warmer sites.

The distribution of CAM plants seems to be better defined along environmental gradients, being most abundant in areas of relatively low water availability (Teeri *et al.*, 1978). Thus, in CAM plants there is a clear correlation between geographic distribution patterns and the observed gas-exchange advantages of improved water-use efficiency. The habitats of low water availability that are occupied by CAM plants are characterized by low precipitation amounts or by thin substrates that hold little moisture. The latter case is best exemplified by epiphytes from tropical and sub-tropical areas (Smith *et al.*, 1986a; Winter *et al.*, 1983). Although on a broad level the occurrence of CAM is strikingly correlated with moisture deficit, there are some specialized ecological situations in which CAM has evolved. These include (1) submerged and partially submerged aquatic isoetid plants, in which CAM is an adaptation to low CO_2 availability (Keeley, 1983; Keeley *et al.*, 1984; Raven *et al.*, 1988) and (2) the roots of leafless orchids, in which the advantages of CAM are unknown (Goh *et al.*, 1983; Cockburn *et al.*, 1985).

Studies of competitive interactions between C_3 and C_4 plants have provided some additional insight into how the gas-exchange advantages described above translate into ecological advantages that may ultimately affect fitness. Pearcy *et al.* (1981) conducted de Wit replacement-series studies with the C_3 weed *Chenopodium album* and the C_4 weed *Amaranthus retroflexus*. The C_4 species had significant growth advantages over the C_3 species when grown in mixed culture at relatively high temperatures (34/28°C day/night), but significant disadvantages when grown at cooler temperatures (17/14°C). There were no apparent differences in growth between the species when grown over a range of water availabilities. These studies support a dominant role for temperature in affecting the ecological advantages of C_4 photosynthesis.

Christie and Detling (1982) conducted similar de Wit type studies with the C_3 grass *Agropyron smithii* and the C_4 grass *Bouteloua gracilis*. They found that, once again, temperature had a dominant role in regulating competitive abilities of the C_3 versus C_4 species. Temperature had a much greater influence on growth under competitive conditions than did soil nitrogen supply. In fact, the C_3 species had a higher relative crowding coefficient (an index of competitive ability) than the C_4 species at low soil nitrogen when grown at cool temperatures (20/12°C). In recent studies conducted with *Amaranthus retroflexus* and *Chenopodium album* it was shown that even in warm midsummer temperature regimes the C_3 species exhibited greater growth at low nitrogen availability (Sage and Pearcy, 1987). Similar results were observed in comparisons of the C_3 species, *Atriplex hortensis*, with *Amaranthus retroflexus* (Gebauer *et al.*, 1987). Brown (1978) suggested that the higher photosynthetic nitrogen-use efficiencies observed in individual leaves of C_4 plants would provide whole-plant growth advantages in nitrogen-limited soils. The studies cited above would weaken such arguments. However, in defense of Brown's theory, the studies on *A. retroflexus* and *C. album* may not be readily applicable to non-weedy species. Weeds typically occupy nitrogen-rich soils and presumably have not evolved the range of adaptive traits required to support maximum growth rates in nitrogen-poor soils (Chapin, 1980). Brown's theory should be tested with C_3 and C_4 species native to nitrogen-poor soils. Thus, the influence of photosynthetic pathway type would be assessed without limitations caused by other growth-related factors (such as nitrogen and carbon partitioning patterns). Aside from the studies by Christie and Detling (1982), there have been no definitive C_3/C_4 comparisons conducted on plants native to nitrogen-poor soils.

The studies by Sage and Pearcy (1987) raised an interesting new perspective on the adaptive significance of the CO_2-concentrating mechanism and improved photosynthetic nitrogen-use efficiencies in C_4 plants. These workers observed that the C_4 species exhibited consistently better growth at high nitrogen availability, compared to the C_3 species. In weedy C_4 species, the improved photosynthetic nitrogen-use efficiency may relieve the nitrogen requirement for photosynthesis, allowing more nitrogen to be allocated to new growth and leaf expansion, compared to C_3 plants (Sage and Pearcy, 1987). This could give C_4 species a growth advantage in nitrogen-rich sites, allowing them to more rapidly establish canopy and root space and compete more effectively for light, moisture, and nutrient resources (Sage and Pearcy, 1987). Such traits would be most advantageous in the open, disturbed sites typically colonized by weedy species. Thus, CO_2-concentrating mechanisms do not only confer advantages to plants at times of resource limitations.

IV. C_3–C_4 INTERMEDIATE PLANTS AS EVOLUTIONARY INTERMEDIATES

A. C_3–C_4 Intermediate Mechanisms

1. Recycling of Photorespired CO_2

C_3–C_4 intermediates have now been reported for 23 species from seven genera (Table 1). Most of the species are weedy and most common in disturbed habitats. All C_3–C_4 intermediates exhibit reduced levels of photorespiration. Early studies with *Panicum milioides*, a C_3–C_4 intermediate, revealed that it exhibited CO_2 compensation points below normal C_3 values (Brown and Brown, 1975), but no evidence of reduced competition between O_2 and CO_2 at the active site of rubisco (Brown, 1980). This suggested that CO_2 was being produced through photorespiration, but that the photorespired CO_2 was not being released from the leaf (see Brown, 1980). Similar conclusions were derived for the C_3–C_4 intermediate *Moricandia arvensis* (Holaday *et al.*, 1981; Winter *et al.*, 1982b). In neither of these species was a CO_2-concentrating mechanism detected, nor was there any evidence of significant CO_2 assimilation through the C_4 cycle (Winter *et al.*, 1982; Edwards *et al.*, 1982; Holaday and Chollet, 1983).

The mechanisms involved in recycling photorespired CO_2 and producing reduced CO_2 compensation points became more clear as a result of careful observations of leaf anatomy and ultrastructure in these species. It was initially observed that both *P. milioides* and *M. arvensis* contain well defined bundle-sheath cells with large and numerous mitochondria, chloroplasts, and peroxisomes (Kanai and Kashiwagi, 1975; Holaday *et al.*, 1981; Winter *et al.*, 1982b; Brown *et al.*, 1983a). (Similar anatomical features had been reported for the C_3–C_4 intermediate *Mollugo verticillata* by Laetsch (1971) although at the time the intermediate nature of this species was not fully realized; see Kennedy and Laetsch, 1974.) Brown *et al.* (1983a) cautiously proposed that these anatomical traits could underlie the recycling of photorespired CO_2, if there were a high concentration of peroxisomes and mitochondria in the bundle-sheath cells, relative to the mesophyll cells. Thus, by simply concentrating the photorespiratory "machinery" in the innermost tissue layer of the leaf, photorespired CO_2 would have a greater chance of being reassimilated by chloroplasts before it escaped from the leaf. These workers presented convincing data that in three C_3–C_4 intermediate *Panicum* species approximately 40% of the leaf's mitochondria and 32% of the leaf's peroxisomes but only 17% of the leaf's chloroplasts were concentrated in the bundle-sheath layer (Brown *et al.*, 1983a). In C_3 congeners only 14% of the leaf's mitochondria, 6% of the leaf's peroxisomes, and 9% of the leaf's chloroplasts were located in the bundle-sheath cells. Additionally, there were very

Table 1
C_3–C_4 intermediate species and typical habitats in which they are found

Species	Habitat
Aizoaceae	
Mollugo verticillata L. *Mollugo nudicaulis* Lam.	Native to open, sandy pastures and other disturbed areas of tropical America; introduced weed in open disturbed sites of the United States and Europe.
Amaranthaceae	
Alternanthera ficoides L.R.Br.R. *Alternanthera tenella* Colla.	Widely distributed in dry disturbed sites in tropical America; West Indies, to Mexico and south to Argentina.
Asteraceae	
Flaveria angustifolia (Cav.) Pers. *Flaveria anomala* B.L. Robinson *Flaveria brownii* A.M. Powell *Flaveria chloraefolia* A. Gray *Flaveria floridana* J.R. Johnson *Flaveria linearis* Lag. *Flaveria oppositifolia* (D.C.) Rydb. *Flaveria pubescens* Rydb. *Flaveria ramosissima* Klatt *Flaveria sonorensis* *Flaveria vaginata* B.L. Robinson & Greenman	Native to open, disturbed sites in northern Mexico, Caribbean Islands, and southern United States. Most often found in moist, alkaline or gypseous soils.
Parthenium hysterophorus L.	Widespread along roadsides and in disturbed fields in tropical America.
Cruciferae	
Moricandia arvensis (L.) DC. *Moricandia spinosa* Pomel *Moricandia sinaica* Boiss.	Native to open fields and banks in Mediterranean region.
Poaceae	
Neurachne minor S.T. Blake	Native to dry hillside and rocky outcrops in western Australia.
Panicum milioides Nees. ex. Trin. *Panicum decipiens* Nees. ex. Trin. *Panicum schenkii* Hack.	Native to moist, lowlying, open grassland sites in northern South America.

Habitats were identified through surveys of regional Floras.
From: Monson and Moore, 1989 with permission.

close associations observed between mitochondria (the CO_2-evolving organelle in photorespiration) and chloroplasts (Brown *et al.*, 1983a; Brown *et al.*, 1983b), in the bundle-sheath cells of the intermediate species, suggesting that reassimilation of photorespired CO_2 could occur very efficiently in some intermediate plants. (However, these workers also noted that there were some observations of close associations between mitochondria and chloroplasts in C_3 *Panicum* species.) Later studies by Holbrook and Chollet (1986), revealed that the reduced photorespiration rates in the *Moricandia* intermediates were dependent upon intact leaf anatomy. They demonstrated that reduced photorespiration rates were not evident in isolated protoplasts from these species. Studies with photorespiration and photosynthesis inhibitors were also consistent with the hypothesis that the reduced photorespiration rates in *M. arvensis* and *P. milioides* are due to a recycling of photorespired CO_2 through the C_3 cycle (Holbrook *et al.*, 1985). Bauwe *et al.* (1987) have examined the capacity for C_3–C_4 *Flaveria* species to recycle photorespired CO_2 using a procedure in which the photosynthetic and photorespiratory metabolite pools in intact leaves are labelled with ^{14}C and then flushed with ^{12}C. During the flush treatment, the rate of $^{14}CO_2$ evolution was measured directly with an ionization chamber. Using this procedure, these workers confirmed that the C_3–C_4 intermediate species released less CO_2 from their leaves, compared to C_3 species.

Following the anatomical observations by Brown and co-workers, Monson *et al.* (1984) reviewed the literature on C_3–C_4 intermediate photosynthesis and suggested that in addition to an anatomical component to the CO_2-recycling mechanism, a biochemical component might exist in the form of differential compartmentation of glycine decarboxylase between mesophyll and bundle-sheath cells. Thus, higher activities of glycine decarboxylase in the bundle-sheath cells, relative to the mesophyll cells, could result in a greater amount of CO_2 being released in the innermost photosynthetic tissue of the leaf and increase the chances of refixation through the C_3 cycle before it diffuses from the leaf. Conclusive support for the hypothesis that glycine decarboxylase was differentially compartmentalized between mesophyll and bundle-sheath cells, came from immunogold labelling studies with leaves of *M. arvensis* (Rawsthorne *et al.*, 1988). These workers found virtually 100% of the leaf's glycine decarboxylase activity to be localized in the bundle-sheath cells! In studies by Moore *et al.* (1988), techniques were employed to separate mesophyll and bundle-sheath protoplasts from the C_3–C_4 intermediate, *Flaveria ramosissima*. Analysis of photosynthetic and photorespiratory enzymes in these protoplast fractions revealed that glycine decarboxylase activity was approximately three times higher in the bundle-sheath fraction. The exact contribution of glycine decarboxylase to reduced photorespiration in this latter species, however, is not known since it also has a limited, but functional, C_4 cycle (Rumpho *et al.*, 1984; Monson *et al.*, 1986, also see next section). The status of glycine decarboxylase localization in other C_3–C_4 species is currently not known. In those species described above, however, efficient recycling of photorespired CO_2 appears to be due at least in part (and perhaps completely in some species) to unique anatomical and biochemical differentiation of the photorespiratory "machinery". It is important to note that the reduced photorespiration rates observed for these

C_3–C_4 intermediates are due to fundamentally different mechanisms than those in C_4 species.

2. The Biochemistry of C_3–C_4 Intermediate Photosynthesis

The first report of a functional C_4 cycle in an otherwise C_3–C_4 intermediate plant was for *Mollugo verticillata* (Sayre and Kennedy, 1977). However, subsequent studies revealed the lack of significant C_4-acid decarboxylase activity in this species (Sayre *et al.*, 1979). Without such activity it is dificult to imagine how a C_4 cycle might function, and so the status of this species as having a true C_4 cycle is equivocal. One genus has now been conclusively shown to have the features of a limited, but functional, C_4 cycle. This is *Flaveria* (Asteraceae), a genus native to Mexico and the Caribbean (see Powell, 1978, for a systematic monograph). Significant activities of C_4-cycle enzymes have also been reported in *Neurachne* (Hattersley and Stone, 1986), although conclusive pulse-chase data establishing a functional C_4 cycle have not yet been reported. *Neurachne* (Poaceae), has one known C_3–C_4 species (*N. minor*) that is endemic to Australia (see Blake, 1972 for a systematic description).

The most extensive studies of the nature of the C_4 cycle in intermediates have been conducted with the genus *Flaveria*. This genus contains 21 species, with at least 11 exhibiting characteristics of biochemical C_3–C_4 intermediacy (see Monson and Moore, 1989 for a list). It is difficult to categorize some species accurately as intermediates or fully expressed C_4 species, since they assimilate more than 90%, but less than 95%, of their atmospheric CO_2 through a C_4 cycle (see Moore *et al.*, 1987b). It might be most appropriate to refer to these species as C_4-like C_3–C_4 intermediates. In that case, after studying 18 of the 21 species, only two can be comfortably classified as C_3 species and two as C_4 species (B. Moore and M. Ku, Washington State University, personal communication). Thus, this appears to be a group of plants that have undergone widespread diversification with respect to photosynthetic metabolism.

All of the C_3–C_4 species in the genus *Flaveria* that have been examined exhibit CO_2 assimilation into the C_4 acids malate and aspartate (Rumpho *et al.*, 1984; Bassüner *et al.*, 1984; Monson *et al.*, 1986; Moore *et al.*, 1987b). Within these studies, however, variability has been observed with respect to the level of C_4 cycle expression, the degree of integrated function between the C_3 and C_4 cycles, and the effectiveness of the C_4 cycle in concentrating CO_2. Some of the C_3–C_4 *Flaveria* species also exhibit the ability to reassimilate photorespired CO_2 (Bauwe *et al.*, 1987), presumably through the differential compartmentation of glycine dicarboxylase between mesophyll and bundle-sheath cells (Moore *et al.*, 1988). At present, it is not known to what degree photorespiration rates in the *Flaveria* intermediates are reduced by a limited capacity to concentrate CO_2 or by the recycling of photorespired CO_2.

To some extent, there is a natural gradation in the overall expression of C_4

photosynthesis in *Flaveria* and its effectiveness in concentrating CO_2 and reducing the competitive component of O_2 inhibition of photosynthesis. There is one species, *F. linearis*, that exhibits the least advanced C_4 characteristics of the intermediates, in terms of both anatomy (Holaday *et al.*, 1984) and biochemistry (Monson *et al.*, 1986). (One other species, *F. cronquistii*, may be even less advanced than *F. linearis*, but there is still debate about whether it should be classified as an intermediate or a C_3 species (see Monson *et al.*, 1986 versus Holaday *et al.*, 1988).)

Several other C_3–C_4 species exhibit considerable C_4 cycle assimilation of atmospheric CO_2 (*F. floridana, F. anomala*, and *F. pubescens*), but based on quantum-yield measurements in 2 and 21% O_2, there is no significant reduction of the competition between O_2 and CO_2 for active sites of rubisco (Monson *et al.*, 1986). The C_3 and C_4 cycles do not exhibit the intercellular compartmentation typical of C_4 plants, since enzymes of both cycles are found in mesophyll and bundle-sheath cells (Bauwe, 1984; Reed and Chollet, 1985). These species also appear to be capable of concentrating CO_2 in the light (Moore *et al.*, 1987b), although the chemical form and site of increased CO_2 concentration is not known. At first consideration, it seems contradictory that these species can concentrate CO_2, yet we have not yet observed evidence of reduced competition between CO_2 and O_2 for the active site of rubisco. This apparent enigma might be reconciled if the CO_2-concentrating capacity is not accompanied by reductions in the activity of carbonic anhydrase in the photosynthetic cells containing the C_3 cycle, reducing the rate of conversion of the concentrated CO_2 from the free form (the substrate of rubisco) to the bicarbonate form (see Burnell and Hatch, 1988). To date, it is not known if this is the case for these species.

Another species, *F. ramosissima*, exhibits considerable C_4 cycle activity (Rumpho *et al.*, 1984; Monson *et al.*, 1986; Moore *et al.*, 1987b), the capacity to concentrate CO_2 (Moore *et al.*, 1987b), and based on quantum-yield measurements, reduced competition between CO_2 and O_2 (Monson *et al.*, 1986). However, the fact that the carbon isotope ratio exhibits $\delta^{13}C$ values in the range -22 to -28% (values typical of C_3 plants), suggests that the C_4 cycle functions in a highly inefficient manner (Monson *et al.*, 1988). This species also exhibits activities of C_3- and C_4-cycle enzymes in both mesophyll and bundle-sheath cells, suggesting an inefficient co-ordination between the two cycles (Moore *et al.*, 1988).

The species *F. brownii* typically assimilates greater than 75% of its carbon through the C_4 cycle. It exhibits low levels of O_2 inhibition of photosynthesis (2–13%, Monson *et al.*, 1987; Holaday *et al.*, 1988), and $\delta^{13}C$ values of -16 to -17% (values close to those typical of C_4 plants, Smith and Powell, 1984; Monson *et al.*,1988). Additionally, this species possesses isozymes of PEP carboxylase (an enzyme with a primary role in C_4 photosynthesis as well various metabolic functions in C_3 plants) that are intermediate to the C_3 and C_4 forms in kinetic characteristics (Bauwe and Chollet, 1986). This enzyme, however, is not exclusively localized in mesophyll cells, as in C_4 plants (Reed and Chollet, 1985; Holaday *et al.*, 1988). Thus although this species is a C_3–C_4 intermediate, it has evolved to a level very close to that of fully expressed C_4 photosynthesis.

Finally, there are a group of species that assimilate over 90%, but never greater than 95%, of their carbon through the C_4 pathway in $^{14}CO_2$ pulse-feeding experiments designed to evaluate the proportions of C_3 versus C_4 assimilation at infinitely small time intervals (B. Moore, Washington State University, personal communication). These include *F. palmeri* and *F. australasica*. There is currently confusion over whether these species should be classified as advanced C_3–C_4 intermediates or C_4 species with some inefficiency in C_4 cycle function.

B. Systematic, Ecological, and Biochemical Evidence for Evolutionary Intermediacy

Although it is not currently possible to develop unequivocal statements supporting the evolutionary intermediacy of C_3–C_4 species, there are lines of evidence that can be used to argue strongly in favour of accepting such a position (see Monson and Moore, 1988). In this chapter the term "evolutionary intermediacy" is used to describe the C_3–C_4 species as the products of evolution from ancestral C_3 taxa to the current state with photosynthetic and photorespiratory characteristics of both C_3 and C_4 plants. Conclusive statements of the validity of evolutionary intermediacy for these species will probably only come through studies of DNA homologies among the various related C_3, C_4, and C_3–C_4 species.

One approach to establishing the probable origins of C_3–C_4 species is to develop and evaluate alternatives to the proposed evolutionary pattern, and establish their relative strengths and weaknesses. Monson and Moore (1989) took such an approach and proposed three possible alternatives to assuming that the intermediates could be used as models to study the evolution of C_4 photosynthesis.

(1) The intermediates may represent the products of reverse evolution—from fully expressed C_4 plants towards fully expressed C_3 plants.

(2) The intermediates may represent the products of earlier hybridizations between C_3 and C_4 ancestors, that have become stabilized at an intermediate photosynthetic stage.

(3) Although these species acquired their C_3–C_4 traits through evolutionary processes, they may have reached an evolutionary "dead-end" with respect to further photosynthetic evolution. That is, they cannot progress to the point of fully expressed C_4 plants, even if given the proper selection regimes—they possess inherent barriers to further photosynthetic evolution.

This latter possibility does not rule out that these plants represent evolutionary products, but rather it suggests that they cannot be used to examine the paths of C_4 evolution.

There is considerable evidence that several of the C_3–C_4 species do not represent the products of reverse evolution. In the case of the *Moricandia* (Cruciferae) intermediates there are no known C_4 species in the Cruciferae. This fact not only argues strongly against the possibility of reverse evolution, but also the hybridization alternative for these species. A similar argument can be applied to the recently described C_3–C_4 species, *Parthenium hysterophorus* (Asteraceae), although in this case C_4 species exist in the family Asteraceae, but not in the genus *Parthenium* (Moore *et al.*, 1987a).

In the case of the *Flaveria* intermediates, there is evidence against reverse evolution of photosynthetic traits if one considers photosynthetic mode along with life-cycle and breeding-system traits. All but two of the *Flaveria* intermediates are perennials with relatively high levels of self-incompatibility (Powell, 1978). The two exceptions, *F. ramosissima* and *F. anomala*, are annual species, but they too exhibit high levels of self-incompatibility (Powell, 1978). The known C_3 species in this genus are perennials with high levels of self-incompatibility, whereas the known C_4 species are annuals with extremely low levels of self-incompatibility (Powell, 1978). If one assumes a pattern of reverse evolution in photosynthetic-pathway type, then it must have been accompanied by reverse evolution in life-cycle type and breeding-system type. Reversals in the latter two traits have not been reported and it has been argued that, if they do occur, they would be extremely rare (Stebbins, 1974; Jain, 1976).

A final weakness in the reverse-evolution hypothesis is that there are no apparent differences in the habitats typically occupied by C_4 species and those occupied by C_3–C_4 species (Monson and Moore, 1989). Both photosynthetic types occupy hot environments which are thought to be favourable for C_4 photosynthesis. Thus, it seems most logical that the intermediates would have advantages by possessing C_4 traits and tend not to experience selection against such traits.

With respect to the possibility that the intermediates represent the products of hybridization, we can gain some insight through recent studies in which plants of various photosynthetic types have been successfully crossed. Studies of hybridizations between C_3 and C_4 species in the genera *Atriplex* and *Flaveria* have revealed that some of the progeny exhibit reduced CO_2 compensation points, presumably reflecting reduced photorespiration rates compared to the C_3 parent, despite the absence of functional C_4 cycles (Björkman *et al.*, 1971; Holaday *et al.*, 1985). (The C_4 parent used in the *Flaveria* cross is now considered to be a C_4-like C_3–C_4 intermediate: see Monson *et al.*, 1987; Holaday *et al.*, 1988.) The basis for these reduced compensation points is not entirely clear. Although the hybrids exhibit some inheritance of C_4-like anatomy and ultrastructure, the high concentrations of organelles and centripetal organelle arrangement that are seen in the naturally occurring C_3–C_4 intermediates are weakly expressed or not present in the hybrids (Björkman *et al.*, 1971; Holaday *et al.*, 1985). Out of 28 hybrids from these two genera for which there is published data, only one (*Atriplex* F_2 #7740–5) exhibits a CO_2 compensation point that is as low as those observed for C_3–C_4 species (Björkman *et al.*, 1971; Holaday *et al.*, 1988). Thus, although C_4 traits can be transferred to C_3 plants at a limited level, it is rare that they result in the levels of reduced photorespiration and ultrastructural modification that are observed for naturally occurring C_3–C_4 species.

Many hybridizations between C_3 and C_4 plants result in progeny that are less vigorous than their parents. For example, the *Atriplex* hybrids exhibited lower photosynthetic rates than either parent (Björkman *et al.*, 1971). In contrast, the *Flaveria* hybrids exhibited photosynthetic rates that were comparable to their parents (Holaday *et al.*, 1988). However, as pointed out above, the C_4 parent was actually an advanced C_3–C_4 species. When crosses were attempted between a fully expressed C_4 *Flaveria* (*F. trinervia*) species and the C_3 species the progeny grew slowly and died before flowering (Holaday *et al.*, 1988). In the genus *Flaveria* there are cases where C_4 species are sympatric with C_3–C_4 species, but hybridization has never been observed (Powell, 1978). Also with respect to the *Flaveria* intermediates, the fact that over 70% of the species can be classified at a level between fully expressed C_3 or C_4 photosynthesis (see Edwards and Ku, 1987; Monson and Moore, 1988) suggests that if these intermediate levels were reached through hybridization events, floral and leaf morphological traits that are intermediate between existing C_3 and C_4 *Flaveria* species should exist—but they don't! Taken together, these studies provide no substantial evidence that the naturally occurring C_3–C_4 intermediates represent the products of hybridization.

One of the keys to establishing whether C_3–C_4 species are evolutionary intermediates or the products of hybridization may lie in the properties of PEP carboxylase. This enzyme is present in both C_3 and C_4 species, although it exhibits different kinetic and electrophoretic properties depending on its origin (Ting and Osmond, 1973; O'Leary, 1982). Analysis of the *Atriplex* hybrids revealed that both the C_3 and C_4 forms of the enzyme are inherited (Ting and Osmond, 1973). In those individuals that exhibited the highest activities of PEP carboxylase, the C_4 form was most prevalent (Ting and Osmond, 1973). Thus, we should be able to obtain some insight into the origins of C_4 traits in the C_3–C_4 species by analyzing the properties of PEP carboxylase in these species. In a comprehensive survey of the kinetic properities of PEP carboxylase in C_3, C_4, and C_3–C_4 species of *Flaveria*, Bauwe and Chollet (1986) observed that most of the C_3–C_4 species possessed the C_3 form of the enzyme. One exception was for the C_4-like intermediate, *F. brownii*, in which PEP carboxylase exhibited kinetic characteristics intermediate to the C_3 and C_4 forms. Adams *et al.* (1986) found that, on the basis of peptide-fragment mapping, PEP carboxylase in the C_3–C_4 species, *F. floridana*, exhibited greater similarity with the C_3 form than the C_4 form. Thus, even those C_3–C_4 species that exhibit considerable C_4 cycle activity, do not exhibit evidence of inheriting the C_4 photosynthetic apparatus from C_4 ancestors. Future efforts to sequence PEP carboxylase from C_3, C_4, and C_3–C_4 species should provide additional insight into the question of hybridization versus evolution.

Edwards and Ku (1987) discussed the results of preliminary electrophoretic studies of mobilities for 12 enzymes representing a wide range of metabolic functions. The studies were aimed at examining if such an experimental approach could be used to discern whether C_3–C_4 intermediate *Flaveria* species are the products of hybridization between extant C_3 and C_4 *Flaveria* species. The studies were apparently successful in ruling out hybrid origins for some of the *Flaveria* intermediates.

As discussed above, one cannot be sure that the naturally occurring intermediates are continuing to evolve towards fully expressed C_4 plants. It is entirely possible that the intermediates are at a stable point with respect to C_4 evolution and will not evolve into C_4 plants. Using them as models to study the evolution of C_4 photosynthesis does not, however, require that they eventually evolve into C_4 plants—only that they are capable of undergoing such evolution if exposed to the proper selection regime. To date, there is no evidence to suggest that the majority of C_3–C_4 species cannot evolve fully expressed C_4 photosynthesis. Only two of the *Flaveria* intermediates (*F. anomala* and *F. linearis*) that were examined with $^{14}CO_2$ pulse-chase techniques exhibited dysfunction in transferring assimilated ^{14}C from the C_4 cycle to the C_3 cycle (Monson *et al.*, 1986). Thus the C_4 cycle operates normally in most of the intermediates, it is just present at a reduced level, compared to C_4 plants.

The evidence supporting an evolutionary intermediacy for C_3–C_4 species is, therefore, equivocal and mostly based on negative evidence against the alternative hypotheses. However, it is also clear that there is as yet no compelling evidence to interpret the intermediates as anything other than evolutionary intermediates. It should be noted that there is also no reason to suspect that all of the intermediates have followed the same indentical paths during the evolution of C_4-like traits. There is variability in the co-occurence of C_4 traits among C_3–C_4 intermediates. For example, *Moricandia arvensis* has evolved a bundle-sheath anatomy characterized by centripetally arranged organelles—an anatomical feature of many fully expressed C_4 species (Holaday *et al.*, 1981; Winter *et al.*, 1982b). However, it does not exhibit evidence of the complete C_4 cycle (Holaday and Chollet, 1983; Winter *et al.*, 1982b). Thus, this species has evolved to a point near the extreme of C_4-like anatomy and ultrastructure, but without development of C_4-like biochemistry. In the C_3–C_4 species *Flaveria linearis* the bundle-sheath cells exhibit higher organelle concentrations than C_3 species, but lack a centripetal arrangement (Holaday *et al.*, 1984). This is in contrast to the rest of the C_3–C_4 *Flaveria* species that have been examined, in that they exhibit the centripetal arrangement. Despite the fact that bundle-sheath anatomy and ultrastructure in *F. linearis* has not evolved to the C_4 extreme, this species exhibits evidence of having evolved a partial C_4 cycle (Monson *et al.*, 1986). Thus in *F. linearis*, unlike *M. arvensis*, there is evidence that C_4-like biochemistry evolved along with C_4-like anatomy.

C. Water- and Nitrogen-Use Efficiencies of C_3–C_4 Intermediates

Two of the most commonly stated advantages of C_4 photosynthesis are the improved water- and nitrogen-use efficiencies that accompany the CO_2-concentrating mechanism. Thus, it is worth considering whether C_3–C_4 intermediate plants exhibit such improvements. Analyses of water- and nitrogen-use efficiencies have been conducted for C_3–C_4 species in the genera *Panicum* and *Flaveria* (Brown and Simmons, 1979; Bolton and Brown, 1980; Monson, 1989). In the *Panicum* intermediates there were no significant improvements in nitrogen-use efficiency or water-use efficiency, relative to C_3 species, when analyzed under optimal growth conditions and normal atmospheres (320–340 μmol mol^{-1} l^{-1} CO_2 and 21% O_2). However, improvements in water-use efficiency were noted for the intermediate *P. milioides* at low intercellular CO_2 concentrations (Brown and Simmons, 1979). For the most part, a similar pattern was observed in the *Flaveria* intermediates (Monson, 1989). In all of the three C_3–C_4 species that were analyzed, potential improvements in water-use efficiency were observed only when photosynthetic rates were below 12.5 μmol m^{-2} s^{-1} (approximately 50% of the maximum rate). This pattern was observed whether the photosynthetic rate was reduced by low availability of CO_2 (Fig. 2) or low availability of nitrogen (Fig. 3). At the maximum photosynthetic rates observed in normal atmospheres the C_3–C_4 species exhibited water-use efficiencies equal to, or less than, a C_3 species. Photosynthetic nitrogen-use efficiencies measured over a range of leaf nitrogen contents were not higher in two of the three intermediates, compared to the C_3 species (Fig. 4). In one of the intermediates, *F. ramosissima*, photosynthetic nitrogen-use efficiencies were significantly higher than the C_3 species, but lower than the C_4 species. As mentioned above, the latter species exhibits evidence of an effective partial CO_2-concentrating mechanism (see Monson *et al.*, 1986), which may explain the improved nitrogen-use efficiency relative to the other two intermediate species.

The improved water-use efficiencies observed at low photosynthetic rates

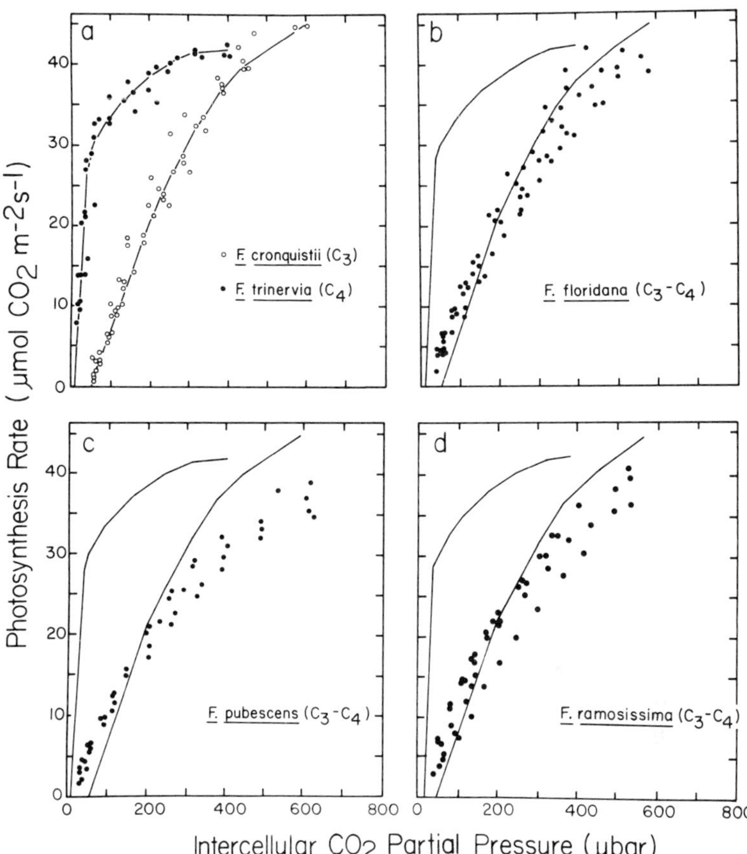

Fig. 2. The response of net photosynthetic rate to intercellular CO_2 partial pressure in a C_3, a C_4, and three C_3–C_4 species of *Flaveria.* In panel a, responses typical of C_3 and C_4 species are seen for *F. cronquistii* and *F. trinervia*, respectively. In panels *b, c*, and *d* the patterns for *F. cronquistii* and *F. trinervia* are repeated as the solid lines, and the data for each of the C_3–C_4 species are presented as individual data points. In each of the C_3–C_4 intermediates, photosynthetic rates were typically higher than the C_3 species at intercellular CO_2 concentrations below 200–250 µbar, but lower than the C_3 species at higher values. (From Monson (1989) with permission).

in the *Flaveria* intermediates are probably the result of recycling photorespired CO_2. Reductions in the rate of photorespiratory CO_2 loss should result in increases in the rate of net photosynthesis per unit stomatal conductance, resulting in reduced intercellular CO_2 concentrations and improved water-use efficiencies. The fact that improved water-use efficiencies are not observed at higher photosynthetic rates means that any potential increases in photosynthetic rates due to recycling photorespired CO_2 are

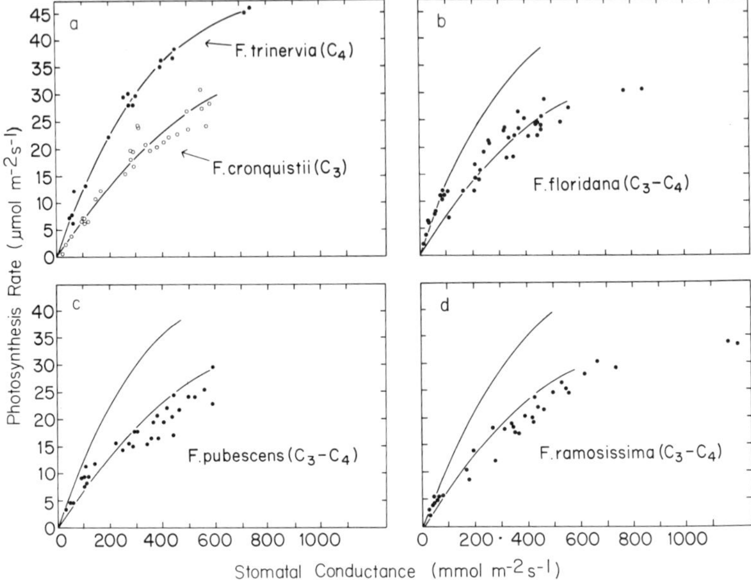

Fig. 3. The relationship between net photosynthetic rate and stomatal conductance in a C_3, a C_4, and three C_3–C_4 species of *Flaveria*. The higher photosynthetic rates at any given stomatal conductance for *F. trinervia* are indicative of higher water-use efficiencies for this C_4 species, compared to the C_3 species. In panels *b*, *c*, and *d* the pattern for *F. cronquistii* and *F. trinervia* are repeated as the solid lines, and the data for each of the C_3–C_4 species are presented as individual data points. In each of the C_3–C_4 intermediates, water-use efficiencies were higher than the C_3 species when photosynthetic rates were below 12.5 μmol m^{-2} s^{-1}, but equal to, or lower than, the C_3 species at higher photosynthetic rates. The range of values used to construct these patterns were obtained from leaves of different ages in plants grown with low or high nitrogen. Thus, the lowest photosynthetic rates and stomatal conductances are the result of nitrogen limitations. (From Monson (1989) with permission.)

offset by proportionately greater increases in stomatal conductance. This would maintain the intercellular CO_2 concentration (C_i) at an equal or higher value compared to C_3 plants. The reasons for such stomatal compensation are not clear at this time, although they may be due to a relatively high gain in the stomatal CO_2-feedback loop such that perturbations to the operational C_i value caused by the evolution of a CO_2 recycling mechanism are compensated for by increased stomatal opening. Thus, a set-point is maintained for the operational C_i value. Indeed, it is probably not coincidental that nearly all C_3 species that have been examined maintain operational C_i values within a fairly narrow range (Wong *et al.*, 1985; Yoshie, 1986). The operational C_i value is likely to be a trait with high adaptive significance. At low nitrogen availability the set-point C_i value apparently changes to lower

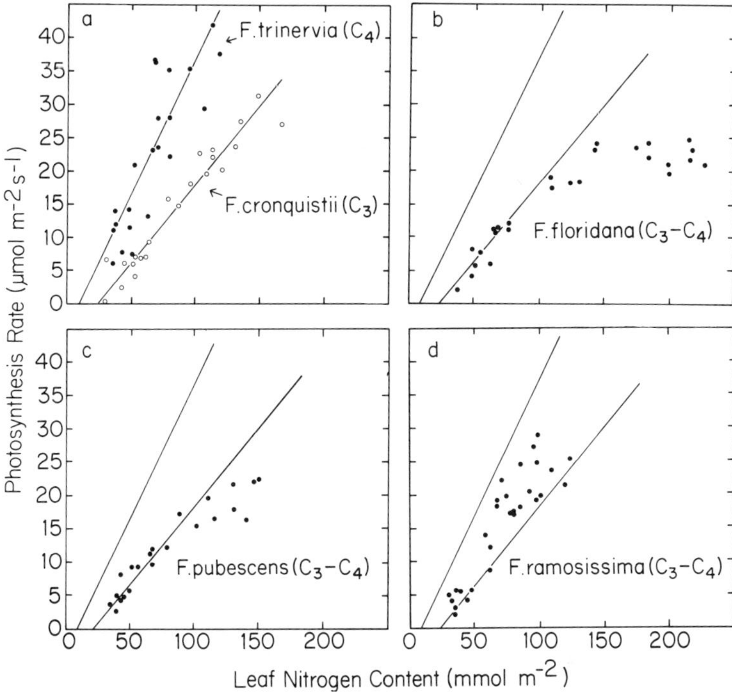

Fig. 4. The relationship between net photosynthetic rate and leaf nitrogen content in a C_3, a C_4, and three C_3–C_4 species of *Flaveria.* The higher photosynthetic rates at any given leaf nitrogen content for *F. trinervia* are indicative of higher nitrogen-use efficiencies for the C_4 species, compared to the C_3 species. In panels *b*, *c*, and *d* the patterns for *F. trinervia* and *F. cronquistii* are repeated as the solid lines, and the data for each of the C_3–C_4 species are presented as individual data points. The C_3–C_4 species *F. ramosissma* exhibited higher photosynthetic nitrogen-use efficiencies than the C_3 species, but lower than the C_4 species. The other two C_3–C_4 species exhibited photosynthetic nitrogen-use efficiencies that were equal to, or lower than, the C_3 species. (From Monson (1989) with permission.)

values as photosynthetic rates are reduced in these intermediate plants, resulting in the improved water-use efficiencies.

It is also evident that at low intercellular CO_2 concentrations, the intermediates exhibit advantages in terms of higher photosynthetic rates (Fig. 2). Once again this is probably due to the recycling of photorespired CO_2. As photosynthetic rates increase, however, this advantage also disappears, at least at the 30°C leaf temperatures used in these experiments. Apparently, the photosynthetic benefits that result from decreased photorespiration rates are offset by some disadvantage of the C_3–C_4 intermediate condition at higher C_i values. It should be noted that these results have only been obtained for the

Flaveria intermediates, so that it is difficult to generalize at this time. One possible explanation for this pattern is that, as C_i values increase, the C_3 cycle in these intermediates begins to compete for available ATP with the C_4 cycle. This would be a consequence of the pyruvate P_i, dikinase reaction in the C_4 cycle not being localized in a different cell type from the C_3 cycle (see above for discussion of incomplete intercellular compartmentation in the *Flaveria* intermediates). At high C_i values limiting ATP could cause the C_3 cycle to be rate-limited either through limitations to the rate of RuBP regeneration (von Caemmerer and Farquhar, 1981; Evans, 1986) or through decreased activation of rubisco (Woodrow and Berry, 1988). There is obviously a need for much greater research into the costs and benefits of incomplete intercellular compartmentation between the C_3 and C_4 cycles in the *Flaveria* intermediates.

At leaf temperatures of 35°C the C_3–C_4 *Flaveria* species exhibited higher photosynthetic rates at all C_i values up to 300 μbar, compared to a C_3 congener (Monson, 1989). Thus, at this higher temperature, the benefits of decreased photorespiration rates are not offset by the costs proposed above. This temperature dependence is probably due to the strong correlation between temperature and photorespiration rate observed for C_3 plants (Monson *et al.*, 1982b). Accordingly, benefits in terms of reduced photorespiration rates that influence the overall rate of photosynthesis would have their greatest effect at higher temperatures.

D. What Do C_3–C_4 Intermediates Reveal About the Evolution of C_4 Photosynthesis?

1. Reduced Photorespiratory CO_2 Loss as an Initial "Driving Force" for C_4 Evolution in Warm Environments

The studies described above reveal that two possible gas-exchange advantages can be identified in all C_3–C_4 species that have been examined—improved water-use efficiencies at low photosynthetic rates and reduced photorespiration rates. Monson (1989) argued that in the case of the *Flaveria* intermediates improved water-use efficiencies at low photosynthetic rates would be of limited adaptive value since these are weedy species that have probably experienced selection for rapid growth rates and high reproductive outputs. Thus, adaptive traits that are restricted to conditions of low photosynthetic rates probably do not influence plant fitness in this genus as much as traits that enhance photosynthesis and growth when the potential for rapid growth is high. This argument is derived from Grime's (1979) theory that ruderal species (such as most of the *Flaveria* species) typically lack adaptations for stress tolerance, being better adapted to rapid growth and high reproductive outputs under less stressful conditions. Many *Flaveria* species are native to arid areas of the southwestern United States and northern Mexico (Powell, 1978). Thus, one could argue that improved water-use efficiencies represent an adaptation for maintaining moisture in the rooting zone for extended periods between stochastic precipitation events. In response to this I note that although these species occur in regions with arid macroclimates, they are most often found in ditches and moist drainage areas along roadsides and recently disturbed fields.

After observing two *Flaveria* species in their natural habitats for several years, I am convinced that at least in Florida the plants are poor competitors, exhibiting instead a pattern of rapid establishment following a disturbance, accompanied by rapid and prolific reproduction, before being eliminated from the site by increased competitive pressures. In recent observations of populations of one C_3–C_4 species, *F. linearis* in southern Florida, it has been observed that populations can exhibit dramatic changes in plant numbers in just three years (R. K. Monson, unpublished observations). During 1984 a large population of approximately 1000 individuals of this species was examined along a recently disturbed roadcut near Key West, Florida. The plants were relatively large with numerous flowers and fruits. Three years later, in 1987, not a single plant of this species was found! The roadcut had been invaded by several other species of herbaceous perennials that formed a dense canopy. Similarly, in 1985, a small population of approximately 30 flowering individuals was observed along a roadcut outside Naples, Florida. In 1986 the population had increased to approximately 100 plants with many seedlings present, but by 1987 all individuals had disappeared. There was no evidence of further disturbance to the population; rather they appeared to have succumbed to the increased shade and competitive nature of the site as other herbaceous species became established.

If in fact reduced photorespiration rates represent the initial physiological change during the evolution of C_4-like traits in C_3–C_4 intermediates, it is of interest to ask, what are the benefits of reduced photorespiration to these plants? All of the C_3–C_4 species that have been identified are native to warm habitats (Table 1; Monson and Moore, 1989), ranging from wet, low-lying areas of the tropics (e.g. *Panicum milioides*), to dry, arid regions (e.g. *Neurachne minor*). All of the species tend to be native to unshaded habitats, and most are weedy in their life-history traits. Given the high daily temperatures that most of these species must experience in their native habitats, mechanisms that reduce photorespiratory CO_2 loss should improve the plant's carbon balance. If the recovered CO_2 is partitioned into growth it should enhance the plant's growth rate.

Ignoring herbivory, the carbon balance of a leaf will be determined by the rate of net CO_2 assimilation minus the carbon that is lost during night-time respiration and the carbon that is transported from the leaf to other parts of the plant. The rate of net CO_2 assimilation per unit leaf area can be increased by increasing the rate of gross photosynthesis, by decreasing the rate of photorespiratory CO_2 loss, or by decreasing the rate of CO_2 loss from the tricarboxylic acid cyle. Recycling photorespired CO_2 will result in an increased rate of net assimilation only if it does not occur at the expense of mesophyll-cell CO_2 assimilation. For each molecule of photorespired CO_2 assimilated by rubisco in the mesophyll cells, assimilation of a molecule of atmospheric CO_2 is precluded. In the C_3–C_4 intermediates, it appears as though much of the recycling of photorespired CO_2 occurs in the bundle-sheath cells following decarboxylation by glycine decarboxylase (although an unequivocal demonstration of the site of recapturing the photorespired CO_2 is lacking). If the capacity for recycling CO_2 in the bundle-sheath cells has not occurred at the expense of rubisco activity in the mesophyll cells, then the leaf's rate of net assimilation should be increased. Thus, the bundle-sheath tissue, which serves no significant photosynthetic function in C_3 plants, contributes to the rate of net CO_2 assimilation in C_3–C_4 plants.

One of the weakest parts of the hypothesis presented above, is lack of an explanation for why such a complex mechanism to recycle CO_2 has evolved in these species as a means of increasing net photosynthetic rates, as opposed to increases in the concentration or activity of mesophyll-cell rubisco. Two possible reasons are: (1) increases in the activity of rubisco in mesophyll cells would require that the stomatal conductance be increased to supply CO_2 to the enzyme, increasing the transpiration rate, and (2) there are as yet undiscovered advantages of increasing the activity of rubisco in bundle-sheath cells that can recycle photorespired CO_2, rather than in mesophyll cells that assimilate atmospheric CO_2. With respect to the first reason, although I argued above that improved water-use efficiency was not a principal advantage underlying the evolution of C_3–C_4 intermediacy in *Flaveria*, it would be quite premature to assume that these plants can withstand evolutionary changes that result in increased transpiration rates. With respect to the second reason, it is possible that CO_2 concentrations in the bundle-sheath cells of C_3–C_4 intermediates are increased, relative to those in the intercellular air spaces. This might result from the preferential transport of photorespiratory intermediates to bundle-sheath cells and subsequent decarboxylation by glycine decarboxylase at a greater rate than reassimilation of the photorespired CO_2 by rubisco. Such increases in bundle-sheath CO_2 concentration would result in improved photosynthetic rates in the intermediates, relative to C_3 plants, under those conditions that enhance photorespiration (such as low intercellular CO_2 concentrations and high leaf temperatures). Enhanced photosynthetic rates were indeed observed in the *Flaveria* intermediates under these conditions (Fig. 2; Monson, 1989).

2. *Improved Water- and Nitrogen-use Efficiencies as Late-Stage "Driving Forces" for C_4 Evolution*

All of the C_3–C_4 species that have been identified exhibit C_4-like anatomy, whereas only some species in *Flaveria* exhibit evidence of a limited, but functional C_4 cycle. Possible reasons underlying the evolution of the anatomical traits were discussed above. The reasons underlying the evolution of the biochemical traits associated with the C_4 cycle are more difficult to establish. Monson and Moore (1989) argued that the increased activity of PEP carboxylase which has been observed in many of the C_3–C_4 species, compared to C_3 congeners (Kestler *et al.*, 1975; Sayre *et al.*, 1979; Holaday and Black, 1981; Holaday *et al.*, 1981; Ku *et al.*, 1983; Nakamoto *et al.*, 1983; Bauwe and Chollet, 1986; Hattersley and Stone, 1986), may have a role in recycling photorespired CO_2. This suggestion is supported by recent observations that PEP carboxylase activity is inversely correlated with the CO_2 compensation point in hybrids obtained from crosses among *Flaveria* species (Brown *et al.*, 1986). Increased PEP carboxylase activities may have driven the evolution of increases in other C_4 cycle enzyme activities as the need to supply PEP substrate and process C_4-acid products increased.

Whatever the reasons underlying the evolution of the C_4 cycle in plants, when such a cycle is present with C_4-like anatomy that exhibits substantial amounts of bundle-sheath chloroplasts, mitochondria, and peroxisomes, as well as differentiated functions between mesophyll and bundle-sheath cells, the potential exists for further evolution to the fully expressed C_4 state. If the C_3–C_4 *Flaveria* species represent the typical pattern in the evolution of the C_4 cycle, the initial stages are characterized by little, or no, differential compartmentation between the C_3 and C_4 cycles. This results in substantial futile cycling of CO_2 between carboxylation and decarboxylation events (see Monson *et al.*, 1986; Monson *et al.*, 1988). Evolutionary increases in the expression of such futile cycles is one of the most difficult patterns to explain in C_4 evolution. It is also difficult to explain how futile C_4 cycles evolved into effective CO_2-concentrating mechanisms. The latter step would have required evolutionary changes in compartmentation patterns, the relative activities of C_4 cycle enzymes, as well as the activities of enzymes such as carbonic anhydrase.

Once the C_4 cycle became effective in concentrating CO_2 to the point where the carboxylase activity of rubisco increased in velocity, the water- and nitrogen-use efficiencies of CO_2 assimilation would have been increased substantially. These latter increases would have represented significant advantages to C_3–C_4 plants, and probably took over as the predominant driving force for the evolution of fully expressed C_4 photosynthesis.

In Figure 5 I have presented a possible evolutionary progression from fully expressed C_3 plants to fully expressed C_4 plants along with representative

Proposed Evolutionary Sequence From C_3 to C_4

	C_3			C_4
Evolutionary Modification	Increased concentration of organelles in BSCs and differential compartmentation of glycine and decarboxylase between MC and BSC	Increased PEP carboxylase activity in both MC and BSC	Increased expression of C_4 biochemical steps in both MC and BSC	Differential compartmentation of C_3 and C_4 biochemistry between BSC and MC
Advantage of Evolutionary Modification	Increased efficiency of refixing photorespired CO_2 without impairing mesophyll assimilation of atmospheric CO_2	Increased trapping of photorespired CO_2 before it escapes from the cell	Processing of C_4 products produced from PEP carboxylation and supply of PEP substrate to PEP carboxylase	CO_2-concentrating function and improved photosynthetic water- and nitrogen-use efficiencies
Representative Species at Their Respective Ranks	*Parthenium hysterophorus*	*Flaveria linearis*	*Flaveria pubescens*	
			Flaveria floridana	
			Flaveria anomala	
		Moricandia arvensis		*Flaveria ramosissima*
		Panicum milioides		*Flaveria brownii*
		Mollugo verticillata		*Flaveria palmeri*
	Alternanthera ficoides		*Neurachne minor*	
	Alternanthera tenella			

Fig. 5. An evolutionary scheme to explain the steps taken during the evolution of C_4 plants from C_3 plants. Representative C_3–C_4 intermediates are presented at their respective positions along the proposed evolutionary sequence. Not all of the proposed evolutionary traits have been described for the species listed under that trait. For example, the presence of differential compartmentation of glycine decarboxylase has only been reported for two species, *Flaveria ramosissima* and *Moricandia arvensis*.

C_3–C_4 intermediate species. In offering this scheme I do not intend to propose it as explaining evolutionary patterns for all cases of C_4 evolution. However, it does provide a useful framework with which to organize the numerous observations on distinct groups of C_3–C_4 intermediates, and creates some hypotheses to explain the driving-forces underlying C_4 evolution.

V. CAM-CYCLING AS AN INTERMEDIATE STAGE IN THE EVOLUTION OF CRASSULACEAN ACID METABOLISM

A. The Nature of CAM Variations

There has been general agreement on several criteria that distinguish CAM from other modes of CO_2 assimilation (see Kluge and Ting, 1978; Osmond, 1978; Ting, 1985). These include: (1) nocturnal accumulation of organic acids, (2) nocturnal depletion of glucan to produce substrate for the primary carboxylation reaction, (3) daytime decarboxylation of the stored organic acids and reassimilation of the freed CO_2 through the C_3 cycle, (4) buildup of the stored glucan pool during the day, and (5) nocturnal assimilation of atmospheric CO_2. Plants that exhibit these characteristics will be referred to as "full-CAM plants".

Recently, several variations on these criteria have been described. The first variation was originally described by Szarek and co-workers (Szarek *et al.*, 1973; Szarek and Ting, 1975) in the stem succulent *Opuntia basilaris*. During periods of extreme water stress this plant completely closes its stomata, stopping nocturnal CO_2 assimilation, but continuing to exhibit nocturnal accumulation of organic acids. The organic acid fluctuations are due to the night-time assimilation of respired CO_2 (Szarek *et al.*, 1973). This variation has been coined CAM-idling. (Although I will use the terms CAM-idling, CAM-cycling, and full-CAM in the interest of consistency with the past literature, it should be noted that many variations exist in organic acid metabolism among plants, and it is probably a more accurate reflection of reality to recognize the breadth of such variability, rather than to force it into the constraints of these categories.) The adaptive significance of CAM-idling has not been unequivocally demonstrated. There is some evidence that it may allow the photosynthetic systems to remain functional during rainless periods and poised to resume assimilation of atmospheric CO_2 rapidly upon rehydration (Rayder and Ting,. 1981). A principal aspect of remaining functional during the rainless periods is the protective function provided by CAM-idling from high-light damage to the photosystems (Osmond, 1982; Adams and Osmond, 1988). However, recent studies of *O. basilaris* in Death Valley and *Hoya australis* in eastern Australia have demonstrated that photoinhibition can occur despite the presence of CAM-idling (Adams *et al.*, 1987; Adams *et al.*, 1988). However, at the present time there exists uncertainty as to whether the experimental analyses of photoinhibition in these succulents detected actual damage to the photosystems or reductions in quantum efficiency brought about by increases in the dissipation of absorbed energy by carotenoids (Demmig *et al.*, 1987; Adams *et al.*, 1989). More work needs to be done to establish the adaptive significance of CAM-idling.

A second variation on the CAM theme is known as CAM-cycling. It is defined as a diurnal fluctuation of organic acids in plants that only exhibit daytime uptake of atmospheric CO_2 (Ting, 1985). As in CAM-idling, the nocturnal increase in organic acid concentration is a result of C_4 assimilation of respired CO_2 (Winter *et al.*, 1986; Patel and Ting, 1987). Most species that exhibit CAM-cycling occur in families that also contain genera with full-CAM (Martin *et al.*, 1988). However, there are exceptions, such as four genera in the Chenopodiaceae (Zalenskii and Glagoleva, 1981), and *Welwitschia mirabilis*, a gymnosperm native to arid regions of South Africa (Ting and Burk, 1983). There is evidence that CAM-cycling has been the precursor to the evolution of full-CAM in some taxa (see next section).

These variations of CAM-like metabolism can occur in a variety of combinations with each other (Ting and Rayder, 1982; Ting, 1985). For example, there are cases of entire shifts from fully expressed C_3 photosynthesis to full CAM within the same individual (e.g., *Mesembryanthemum crystallinum*, see Winter *et al.*, 1978). Other species have been shown to shift from CAM-cycling to full-CAM to CAM-idling, depending on leaf age and level of water stress (e.g., *Peperomia camptotricha*, see Sipes and Ting, 1985; and *Sedum telephium*, see Lee and Griffiths, 1987). These interchanging modes of carbon fixation within the same individual are an aspect of CAM that is not seen in C_4 plants. It reflects a greater environment × genotype interaction than is present in C_4 plants, and has probably facilitated more rapid evolutionary transitions among the various CAM modes, relative to C_4 photosynthesis.

Several other variations of CAM metabolism have been described recently that involve primitive taxa with unique ecological niches (e.g., *Isoetes* spp. and *Stylites* spp., see Keeley, 1983; Keeley *et al.*, 1984; Raven *et al.* 1988). These examples have been described in detail in previous reviews (Winter, 1985; Ting, 1985). Their importance to the current review is that they provide evidence that CAM has evolved for a variety of reasons in the plant kingdom. Although CAM is typically associated with high water-use efficiencies and adaptive advantages in arid environments, the discovery of CAM in these other ecological situations has expanded our interpretation of its adaptive value (Ting, 1985; Raven *et al.*, 1988).

B. The Evidence That CAM-Cycling is an Evolutionary Precursor to Full-CAM

The most complete studies of evolutionary patterns in taxa that exhibit CAM have been conducted by Teeri and colleagues with various genera in the family Crassulaceae (for reviews see Teeri, 1982a; Teeri, 1982b). These workers relied upon carbon-isotope data to examine the relative importance of daytime versus night-time CO_2 assimilation in several genera. It has been well established that night-time CO_2 assimilation through the C_4 cycle will result in $\delta^{13}C$ values between −10 and −18‰ (Osmond, 1978; O'Leary, 1981; Ting, 1985; O'Leary, 1988). Assimilation of CO_2 during the daytime typically occurs through the C_3 cycle and will result in $\delta^{13}C$ values of approximately −27‰. Combinations of daytime and night-time CO_2 assimilation will result in intermediate values, the exact value depending on the relative proportions of the two pathways. This interpretation is open to some errors due to leakage of CO_2 from the internal tissue to the atmosphere during the day, and assimilation of some respired CO_2 during the night. Teeri and co-workers have interpreted the degree to which $\delta^{13}C$ values resemble the C_3 versus the C_4 extremes as evidence of the level of evolutionary advancement—with the least negative values reflecting the greatest reliance on night-time CO_2 assimilation, and thus the greatest evolutionary advancement (Teeri, 1982a). This reasoning has support in the fact that independent analyses of evolutionary patterns in the Crassulaceae using morphological and cytogenetic traits complemented the patterns established from $\delta^{13}C$ values (Fig. 6). Thus, within the Crassulaceae there appears to be a pattern of less utilization of daytime atmospheric CO_2 assimilation and greater utilization of night-time atmospheric CO_2 assimilation as evolutionary divergence and speciation progressed. Some caution must be used in interpreting the carbon-isotope ratios presented in Figure 6 as indicative of the relative amounts of daytime and night-time atmospheric CO_2 assimilation. This is because the values reported in this figure were from field-collected tissue, and there is the possibility that water stress increased the stomatal limitations of photosynthesis and thus increased the contribution of diffusive fractionation to the carbon-isotope ratios (see Farquhar *et al.*, 1982). This could have resulted in less negative $\delta^{13}C$ values, even though all of the carbon was assimilated during the day through the C_3 cycle. However, in support of the $\delta^{13}C$ values in Figure 6 being good representations of the relative contributions of daytime and night-time atmospheric CO_2 assimilation, non-stressed greenhouse-grown plants of many of these species exhibit $\delta^{13}C$ values in the same range as the field-grown plants (Teeri, 1982a).

A prerequisite to establishing that the relative amounts of daytime and night-time atmospheric CO_2 assimilation are subject to selection is that they have a genetic basis. Evidence for this latter criterion is seen in: (1) differences

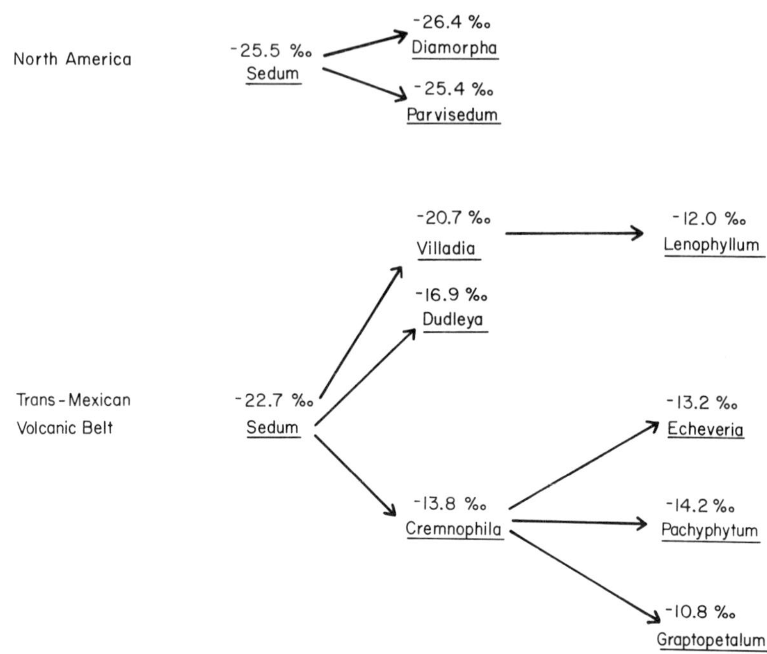

Fig. 6. Evolutionary patterns in two lines of Crassulaceae, one from North America and one from Mexico. The evolutionary patterns were constructed from morphological and cytogenetic evidence. Typical $\delta^{13}C$ values are presented for each genus. The results reveal an evolutionary pattern of less negative $\delta^{13}C$ values in the most advanced taxa in the Mexican group. (From Teeri (1982a) with permission.)

in carbon-isotope values among field grown plants of different genera persist during growth in a common greenhouse environment (Teeri *et al.*, 1981), and (2) intermediate $\delta^{13}C$ values in a hybrid between parents with C_3-like and C_4-like $\delta^{13}C$ values (Teeri and Gurevitch, 1984). However, recent studies on allopatric populations of *Sedum wrightii* have revealed substantial plasticity in the expression of daytime and night-time CO_2 assimilation (Kalisz and Teeri, 1986). The precise degree of genetic versus environmental control over the expression of daytime and night-time CO_2 assimilation in the Crassulaceae as a whole is unknown. Definitive heritability estimates for this trait have not yet been obtained.

All of the genera that were examined in Teeri's studies of the Crassulaceae exhibited night-time accumulation of malic acid, even those that exhibited exclusive daytime uptake of atmospheric CO_2 (Teeri, 1982b). Thus, even the most primitive taxa of this group exhibit evidence of CAM-cycling. Based on the cumulative observations of carbon-isotope ratios and patterns of nocturnal acid accumulation, Teeri (1982b) proposed a scheme for the evolution of complete utilization of night-time assimilation of CO_2 from complete utilization of daytime assimilation of CO_2, with the intermediate stages composed of various levels of CAM-cycling (Fig. 7).

The applicability of the evolutionary pattern presented in Figure 7 to other cases of CAM evolution is not known. There is evidence that similar patterns may have occurred in the Cactaceae. One of the most primitive genera, *Pereskia*, exhibits atmospheric CO_2 assimilation during the day through the C_3 cycle, yet accumulates organic acids at night (Rayder and Ting, 1981; Nobel and Hartstock, 1986). However, in other cases it is clear that a great deal of plasticity exists within a species with respect to the expression of CAM-cycling and night-time assimilation of CO_2 (e.g., Kalisz and Teeri, 1986; Martin *et al.*, 1988). *Peperomia camptotricha* exhibits daytime CO_2 uptake with CAM-cycling at night in young leaves, but shifts to increasing night-time uptake of CO_2 in older leaves (Sipes and Ting, 1985). *Peperomia scandens*, a tropical epiphyte, has been observed to undergo shifts from C_3 to CAM-cycling to full-CAM as leaves age (Holthe *et al.*, 1987). A recent report by Nishio and Ting (1987) describes a unique compartmentation of C_3 and C_4 activities in *P. camptotricha*. They found most of the night-time CO_2 assimilation to occur in the spongy mesophyll, whereas most of the daytime CO_2 assimilation occurred in the pallisade mesophyll. However, even the pallisade mesophyll exhibited some evidence of CAM-cycling. This unique compartmentation of the CAM processes may result in reduced photorespiration rates in the pallisade tissue (Nishio and Ting, 1987). The complexity of CAM expression in plants such as *P. camptotricha* suggests that there are no unique evolutionary patterns leading to CAM. Compared

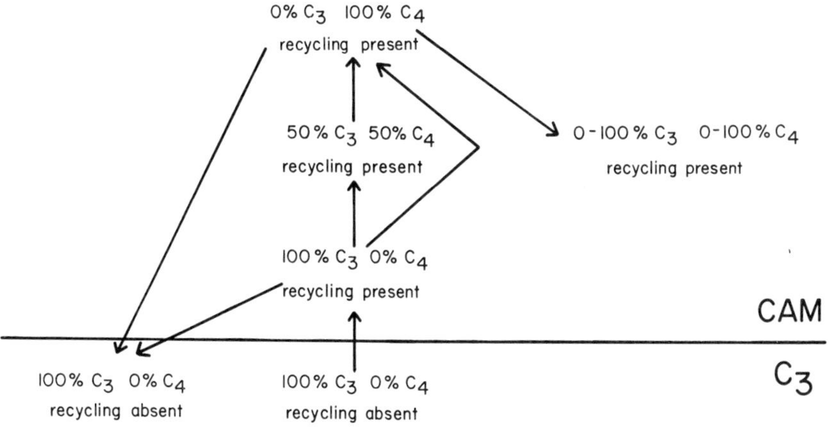

Fig. 7. A proposed evolutionary scheme for various types of organic-acid metabolism in the Crassulaceae. "Recycling" refers to night-time respiratory CO_2 assimilation through the C_4 pathway. (From Teeri (1982b) with permission.)

to C_4 photosynthesis (i.e. spatial compartmentation of the C_4 and C_3 cycles between bundle-sheath and mesophyll cells), CAM has been expressed with many more variations each of which may have unique adaptive advantages.

There is evidence that CAM-cycling is a primitive mode in some taxa, and probably acted as a precursor to the evolution of full-CAM, but to suggest that CAM-cycling has represented the primitive condition in all cases of CAM evolution would ignore the wide range of variations of organic-acid metabolism that occur in the plant kingdom, many of which appear capable of evolving to full-CAM. Certainly, there is no evidence that CAM-cycling, as it is known in terrestrial plants, was the precursor to the evolution of CAM in submerged aquatic plants. Most rooted, submerged aquatic plants, whether CAM or not, obtain significant amounts of inorganic carbon from the sediments in which they are rooted (e.g. Raven *et al.*, 1988). In these plants, CAM probably evolved as a means of extending the time for acquiring sediment-derived carbon into the night (Raven *et al.*, 1988).

C. The Adaptive Significance of CAM-Cycling

If indeed CAM-cycling represented the precursor to full-CAM in some species, then it would be useful to identify the adaptive advantages of CAM-cycling and gain some insight into the "driving forces" of the evolution of CAM. Martin *et al.* (1988) discussed the merits of three suggested advantages of CAM-cycling: (1) CAM-cycling is simply a precursor to CAM-idling (see Rayder and Ting, 1981; Sipes and Ting, 1985), (2) CAM-cycling conserves respired CO_2 that would otherwise be lost to the atmosphere (see Martin and Zee, 1983; Martin and Jackson, 1986), and (3) CAM-cycling serves to conserve water by taking advantage of an internal CO_2 source for carbon assimilation and avoiding the loss of water through the stomata (Cockburn, 1985). Martin *et al.* (1988) concluded that the water-conservation hypothesis is the most tenable, and they calculated that in the CAM-cycling species *Talinum calycinum* as much as 44% of the daily water budget could be conserved through CAM-cycling. The latter calculation used the assumption that all of the decarboxylated CO_2 obtained from the stored organic acids was converted into sugar phosphates during the daylight hours. The validity of this assumption may be questionable, given that stomatal opening during the day would allow for leakage of at least part of the decarboxylated CO_2. Even with such leakage, however, the fact that some of the recycled CO_2 is used in daytime photosynthesis should result in improved water-use efficiencies, relative to the case of no night-time recycling.

An alternative hypothesis has recently been proposed to explain the evolution of CAM (Gil, 1986). This hypothesis rests on the assumption that photorespiration has advantages to plants in that it serves as a means of dissipating excessive light energy in an orderly fashion and thus protects against photoinhibition (see review by Powles, 1984). According to Gil (1986), CAM may have evolved as a mechanism to reduce photorespiration while maintaining the protective role towards photoinhibition. The high internal-CO_2 concentrations generated during the daylight hours are known to provide some protection against photoinhibition in high light (Adams and Osmond, 1988). If one accepts CAM-cycling, in the form we currently understand, as a probable evolutionary precursor to full-CAM, then little support for the above hypothesis is currently available. This is because it has yet to be demonstrated that CAM-cycling provides enough internally generated CO_2 during the daylight hours to provide protection against photoinhibition. Nor has it been demonstrated that the daytime internal CO_2 concentrations generated during water stress in the CAM-cycling mode are sufficient to suppress photorespiration. Both of these criteria need to be established before the hypothesis proposed by Gil is consistent with the hypothesis of CAM-cycling as an evolutionary precursor to full-CAM. In defense of Gil's hypothesis, however, it has been demonstrated that in full-CAM plants undergoing CAM-idling, internal CO_2 concentrations can reach 2% during the day, a value that is surely high enough to reduce photorespiration to negligible levels, as well as provide some protection against photoinhibition (Cockburn *et al.*, 1979). In fact if, as suggested by Osmond *et al.* (1982), a two- to three-fold increase in intercellular CO_2 concentration is all that is needed to suppress photorespiration, and decarboxylation of 100–200 mol m^{-3} of malic acid produces an internal CO_2 concentration of 10 mbar (Cockburn *et al.*, 1979), then decarboxylation of only 20 mol m^{-3} of malic acid in a plant with closed stomata could be sufficient to suppress photorespiration for a limited time. In order to sustain the inhibition of photorespiration throughout the middle of the day, however, decarboxylation of more malic acid would be required. If internal CO_2 concentrations in the range 1–2 mbar are achieved in plants with CAM-cycling when stomata are closed during water stress (a condition that is very similar to CAM-idling in full-CAM plants), then arguments that CAM evolved as a means of reducing photorespiration while preventing photoinhibition are tenable.

If, in fact, the principal advantage of CAM-cycling in terrestrial habitats is improved water-use efficiency, and if CAM-cycling represents a transitional stage in the evolution of full-CAM, then a consistent "driving force" (i.e. improved water-use efficiency) can be identified from the initial stages through the final product in terrestrial CAM evolution. This is in contrast to the pattern proposed above for C_4 photosynthesis in which the initial advantages may have been reduced photorespiration rates, with improved water-use efficiencies evolving at a later stage. It is also interesting that the CO_2 recycling aspect of the proposed precursor to CAM (i.e. CAM-cycling) retains a function in full-CAM, being expressed as CAM-idling during periods of severe water stress. In full-CAM, however, the adaptive advantage of recycling CO_2 may have changed from that of improved water-use efficiency, to maintenance of the photosynthetic apparatus in an active state. As discussed above, in aquatic habitats, improved water-use efficiency is obviously not the principal driving force underlying the evolution of CAM.

D. Hypotheses Concerning the Evolutionary Origins of CAM Biochemistry

An hypothesis was proposed independently by Teeri (1982a) and Cockburn (1981; 1983) that the biochemical steps typical of CAM evolved from magnification of genes coding for stomatal guard cell biochemistry. This hypothesis originated from the observation that the guard cells of all plants and the mesophyll cells of CAM plants share many of the same biochemical sequences. For example, guard cells typically carboxylate phosphoenolpyruvate (PEP) to form malic acid, which is used as a counterion to balance the uptake of cations that drive stomatal opening. The PEP is produced from stored glucan, just as in CAM cells. There is also a decarboxylase in guard cells that can metabolize malic acid during stomatal closure (see Raschke, 1975, for a review of guard cell metabolism).

Organic acid metabolism is not restricted to guard cells in C_3 plants. Many root and shoot cells are capable of producing malate as a counterion to balance the uptake of cations (Smith and Raven, 1979). Thus, the genes that regulate the expression of many CAM-like biochemical traits are ubiquitous among plant cells. It not not surprising that evolutionary modifications in the patterns of expression of such genes may have occurred several times independent of one another. If the genes that code for organic acid metabolism, in guard cells or other cells, are linked and regulated by the same regulatory gene(s), then relatively few mutation events might be capable of altering the expression of these metabolic sequences. This seems to be a reasonable hypothesis for the origins of CAM biochemistry.

An alternative hypothesis explaining the evolution of CAM biochemistry has been suggested by Osmond (1984) and expanded by Cockburn (1985). According to this hypothesis, organic acid metabolism as occurs in CAM originated in primitive aquatic plants, presumably as a means of extending inorganic carbon uptake into the night. The genes for such metabolism then became expressed intermittently during the evolution of terrestrial plants. Organic-acid metabolism that resembles CAM has been reported in a number of submerged aquatic angiosperms, including *Hydrilla verticillata* (Holaday and Bowes, 1980) and *Scirpus subterminalis* (Beer and Wetzel, 1981), as well as more primitive taxa such as the Isoetoids discussed above (Keeley, 1981; Keeley *et al.*, 1984; Raven *et al.*, 1988). However, it is not known how such species relate to the evolution of terrestrial CAM plants. In fact many of the submerged aquatic vascular plants are thought to have acquired the aquatic habit secondarily, evolving from terrestrial ancestors (Raven, 1984). More information is needed on the diversity and type of organic acid metabolism in aquatic plants and their relationships to terrestrial plants before evolutionary patterns can be elucidated and this hypothesis properly evaluated.

While it is attractive to search for the origins of CAM among extant non-CAM metabolic systems, it is possible that the metabolic sequences of CAM evolved from independent metabolic sources. However, such evolution would require increased expression of several independent genes. To account for the polyphyletic origins of CAM (see Teeri, 1982a), such independent evolution would have to have occurred several times. This latter possibility seems less likely than the first two mentioned above.

E. The Evolution of an Inverted Diurnal Stomatal Pattern as a Means of Further Improvements in Water-Use Efficiency

Upon the evolution of CAM-like organic-acid metabolism, further improvements in water-use efficiency would have occurred as a result of the inverted diurnal stomatal opening found in full-CAM plants. Stomata of CAM plants are known to be sensitive to intercellular CO_2 concentration (Cockburn *et al.*, 1979). It has been proposed that the inverted diurnal stomatal pattern observed in CAM plants is a result of stomata responding to diurnal fluctuations in internal CO_2 concentrations (Cockburn and McAulay, 1977; Cockburn *et al.*, 1979). However, holding on to the assumption that CAM-cycling was an evolutionary precursor to full-CAM in some species, certain problems arise in interpreting the evolution of inverted stomatal patterns according to this hypothesis. First, it is not clear why reductions in internal CO_2 concentrations in plants that exhibit only CAM-cycling do not elicit night-time stomatal opening. Indeed, the definition of CAM-cycling includes the fact that night-time stomatal opening does not occur (Ting, 1985). Second, it is not clear why the decarboxylation of organic acids during the day in plants exhibiting CAM-cycling does not cause stomatal closure. One possibility is that CAM-cycling does not produce daytime internal CO_2 concentrations of sufficient magnitude to induce stomatal closure. As stated above we do not have accurate measurements of the influence of daytime decarboxylation on the internal CO_2 concentration in CAM-cycling plants. Alternatively, assuming that CAM-cycling plants evolved from fully expressed C_3 plants, they may not have evolved a sufficiently high gain in the stomatal response to CO_2 to respond to the subtle changes in internal CO_2 concentrations that might occur. It has been demonstrated that some C_3 plants exhibit reduced gains in the stomatal feedback response to CO_2, compared to C_4 plants (Sharkey and Raschke, 1981; Ramos and Hall, 1983; although also see Morison and Gifford, 1983). It is not unreasonable to propose that the sensitivity to internal CO_2 required to allow the stomata of CAM plants to exhibit an inverted stomatal pattern evolved after the metabolic patterns of CAM-cycling.

There is still too much unknown, however, concerning the function of stomata in full-CAM plants, as well as those exhibiting variations of CAM, to derive any conclusive perspectives about the evolution of inverted stomatal patterns.

VI. THE COSTS ASSOCIATED WITH THE EVOLUTION OF C_4 PHOTOSYNTHESIS AND CAM

Ehleringer (1978) proposed that one of the principal costs to conducting C_4 photosynthesis is the higher ATP requirement for each CO_2 molecule assimilated. This cost would be most prohibitive to carbon assimilation, and presumably growth, in low-light habitats where photogenerated-ATP limits the rate of CO_2 assimilation. Ehleringer's argument was based on differences in the quantum yield among C_3 and C_4 dicot species. The C_4 species exhibited higher quantum yields at leaf temperatures above 30°C, but lower quantum yields below 30°C. He used these differences to construct a model of carbon assimilation in C_3 and C_4 grasses growing along latitudinal temperature gradients. The results of the model predicted relative competitive abilities for C_3 and C_4 species that accurately reflected the different distribution patterns of C_3 and C_4 species along the temperature gradient (according to Teeri *et al.*, 1976). The crossover temperature of 30°C was important because C_3/C_4-distribution patterns were such that at those latitudes where C_3 species dominated local grass floras mean temperatures during the growing season were below 30°C. As mean temperatures increased above 30°C, C_4 species became dominant. Thus, there was good correlation between the crossover temperature observed with individual leaves and the apparent crossover for relative importance of C_3 versus C_4 species at the community level. In a later study, Ehleringer and Pearcy (1983) reported that the quantum yields of many C_4 grass species are higher than their C_3 counterparts when measured at 30°C. When considering quantum yield measurements only for grasses, the crossover temperature occurs at values lower than 30°C. The exact crossover point is variable depending on the decarboxylation-type of the C_4 grass, but often occurs at temperatures between 16 and 25°C (also see Monson *et al.*, 1982a). These results modify the conclusions of the study by Ehleringer (1978), and cause one to question the validity of concluding that the increased quantum requirement of C_4 photosynthesis represents the principal factor restricting the distribution of C_4 plants.

The quantum yields in C_3–C_4 intermediate plants that do not exhibit functional C_4 cycles are not significantly different from C_3 plants in those species that have been examined (Ehleringer and Pearcy, 1983). However, in those *Flaveria* intermediates that have been examined, there is evidence that the presence of inefficient C_4-cycle biochemistry results in reduced quantum yields for CO_2 uptake in some species (Fig. 8; Monson *et al.*, 1986). The reduced quantum yields in some of the intermediates presumably reflect inefficiency in the transfer of CO_2 from the C_4 cycle to the C_3 cycle. This results in energy-driven futile cycles of carboxylation and decarboxylation (see Monson *et al.*, 1986). Such inefficiency may represent a significant barrier to competitive vigor in low-light environments during the evolution of C_4 photosynthesis. It may explain the fact that shade-adapted C_4 species are rare, and suggests that such shade-adapted plants evolved from sun-adapted C_4 ancestors that possessed fully-integrated C_4 photosynthesis.

In addition to the increased ATP costs of assimilating CO_2 through the C_4 pathway, the evolution of C_4 photosynthesis would incur the energetic and biomass costs of constructing the anatomical and ultrastructural traits of C_4 leaves. These latter costs would include (1) any alterations to the bundle-sheath cell walls which might be required to reduce the leakage of CO_2 and HCO_3^-, and (2) provisioning the bundle-sheath cells with the organelles and enzymes required to establish the differential cellular compartmentation of C_4 photosynthesis. Brown *et al.* (1983a) have observed that the bundle-

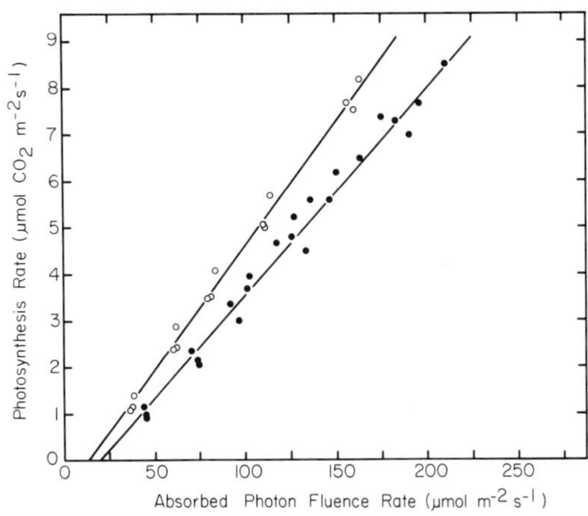

Fig. 8. The response of net photosynthetic rate to low absorbed photon fluence rates in a C_3 species (*Flaveria cronquistii*; ○) and a C_3–C_4 intermediate (*Flaveria pubescens*; ●). The slope of the response is the quantum yield for CO_2 uptake. It is evident that the C_3–C_4 species exhibits lower quantum yields (0.044 mol CO_2/mol quanta) relative to the C_3 species (0.053 mol CO_2/mol quanta). These data were determined in an atmosphere of 340 µbar CO_2 21% O_2. (From Monson *et al.* (1986) with permission.)

sheath cells of several C_3–C_4 intermediate *Panicum* species contain approximately 2 times as many mitochondria, 1·5 times as many chloroplasts, and 3 times as many peroxisomes, as in C_3 *Panicum* species. The high numbers of mitochrondria and chloroplasts in bundle-sheath cells has also been observed in C_3–C_4 intermediates of the genera *Neurachne* (Hattersley *et al.*, 1986), *Flaveria* (Holaday *et al.*, 1984), *Parthenium* (Moore *et al.*, 1987a), and *Moricandia* (Holaday *et al.*, 1981; Winter *et al.*, 1982b). As discussed earlier, these high concentrations of organelles probably have a role in recycling photorespired CO_2 in these species. The presence of high concentrations of organelles in the bundle-sheath cells in these C_3–C_4 species means that during the initial stages of C_4 evolution the cost of provisioning the bundle-sheath cells would have to be weighed against the benefits of recycling photorespired CO_2. The carbon-gain benefits would have to exceed the costs of producing the extra organelles, such that the increased increment of assimilated carbon creates an increase in overall plant growth rate and presumably fitness. As C_3–C_4 species evolve further towards the C_4 extreme, the costs of provisioning the bundle-sheath cells are presumably offset by decreased investments in the number and size of organelles and C_3-cycle enzymes in mesophyll cells, as well as the often expressed increase in the overall rate of CO_2 assimilation due to more effective compartmentation patterns of the C_3 and C_4 cycles. The carbon cost/benefit ratio, in terms of the evolutionary construction of C_4-like leaves, probably reaches a maximum during the intermediate stages of the evolutionary progression towards fully expressed C_4 species.

A similar scenario might exist for the nitrogen cost/benefit ratio. This is because during the evolution of C_4 photosynthesis, nitrogen investments to C_3- and C_4-cycle enzymes probably reach a maximum at the C_3–C_4 intermediate stages that exhibit C_4-cycle activity (i.e. *Flaveria* and possibly *Neurachne*), since both cycles are present in mesophyll and bundle-sheath cells. As discussed earlier, in the intermediates these cycles function with reduced efficiency (and reduced benefit in terms of the rate of CO_2 assimilation), compared to fully-expressed C_4 species. Thus, at the C_3–C_4 intermediate stage, there is a maximum level of nitrogen investment, and a minimum level of return for that investment. As C_3–C_4 intermediates evolve further towards the C_4 extreme, the level of nitrogen investment may stay the same or be reduced, as the C_3 cycle becomes restricted to bundle-sheath cells, and the CO_2-assimilating enzymes of the C_4 cycle become restricted to mesophyll cells. The level of benefit will increase as the CO_2-concentrating mechanism becomes more efficient and effective.

The most commonly described costs to conducting CAM are the slow rates of CO_2 assimilation and growth (Ting, 1985). The slow capacity for CO_2 assimilation has been ascribed to the constraints imposed by the amount of carbon available for night-time generation of PEP substrate, and the limitations imposed by vacuolar storage of malic acid during dark-phase CO_2

assimilation (Winter, 1985). In reference to the limitations imposed by vacuolar storage, however, Smith *et al.* (1986b) have found that several Bromeliad species accumulate as much as 474 mol m^{-3} of malic acid in tissue of which a significant fraction is non-CAM water-storage tissue. This represents an extremely high malic acid storage capacity, and suggests that vacuolar storage capacity should be re-evaluated as a limitation to growth in CAM plants.

In general, CAM plants exhibit slow instantaneous growth rates, compared to C_3 and C_4 plants. Pineapple, one of the most productive CAM plants, exhibits daily growth rates comparable to the slowest growing C_3 crop species (Bartholomew and Kadzimin, 1977), but most CAM plants exhibit much lower growth rates (see Osmond, 1978). When growth rates are considered on an annual basis, CAM plants often exhibit rates that are comparable with, and sometimes greater than, those for C_3 crop species (see Winter, 1985).

From an ecological perspective, slow instantaneous growth rates would reduce the competitiveness of CAM plants in resource-rich environments and, in fact, CAM plants are typically found in macro- or micro-habitats with low vegetation densities and restricted resource availability. In deserts this is reflected in the greatest abundance of CAM plants on open flats or hillsides (Smith and Nobel, 1987). In tropical rainforests, it is reflected in the greatest abundance of CAM plants on the sparsely vegetated trunks of trees (Winter, 1985). From an evolutionary perspective, it is not clear whether the slow instantaneous growth rates of CAM plants are a result of the evolutionarily derived constraints imposed by the temporal separation of carboxylation events, or are due to limitations not related to CAM. In other words, did CAM plants originate from fast-growing C_3 species that became burdened with a temporal separation of carboxylation events and concomitant slow rate of CO_2 assimilation? Or did they derive from slow-growing C_3 species that evolved CAM-cycling and CAM because they were water-conserving metabolic pathways that were compatible with the already-present slow growth rate? The existence of many slow-growing leaf-succulent genera with C_3 photosynthesis, CAM-cycling, and CAM suggests that the latter possibility is tenable. However, in the Cactaceae, where there appears to have been an evolutionary shift from faster growing, leafy, primitive C_3 taxa to slower-growing, stem-succulent CAM taxa (Nobel and Hartstock, 1986), the former alternative may be tenable.

An additional cost to the evolution of CAM is the reduced cellular volume that can be occupied by chloroplasts as a result of the large vacuolar volumes required for malic acid storage. Thus, the total photosynthetic capacity per unit of cellular volume is reduced. This cost is probably incurred, to some extent, by CAM-cycling species, and is thus an early cost during the evolution of C_3 taxa towards CAM taxa.

VII. CONCLUSION

After conducting this review, I am convinced that several breakthroughs will occur in the next few years concerning the evolution of photosynthetic metabolism in plants. Many of the significant breakthroughs will no doubt come as a result of molecular techniques applied to the genes and proteins contained in the C_3–C_4 and CAM intermediates described above. However, there are also several new approaches in the field of plant systematics (such as the cladistic and phenetic approaches that are currently popular) that have great potential to elucidate evolutionary patterns in those groups that exhibit intermediate and fully-expressed photosynthetic modes, be it CAM or C_4.

As with any treatise on biological evolutionary patterns, this one has many subjective interpretations, often with inconclusive data on which to base those interpretations. It is hoped that the reader will realize that I recognize the speculative nature of several of the hypotheses presented in this review. One of the few benefits of writing a review on evolutionary patterns is to allow one's imagination to play havoc with the available data. I plead guilty to taking full advantage of this benefit. The proposals of this review are meant to stimulate more arguments than they settle. In this spirit I make my escape.

ACKNOWLEDGEMENTS

The ideas in this paper have resulted from discussions, debates, and downright arguments with a number of colleagues. I thank B. Moore, G. Edwards, M. Ku, D. Ort, C. Martin, S. Szarek, A. Hall, J. Nishio, J. Keeley, J. Teeri, M. Grant, J. Mitton, J. Ehleringer, J. Karron, J. Jaeger, W. Bowman, and J. Miernyk for the ideas that they either knowingly, or unknowingly, placed in my head during the preparation of this manuscript. An anonymous reviewer provided an extremely enlightening review, for which I am grateful. I am also grateful for National Science Foundation Grants BSR-8407488 and BSR-8604960 which supported many of my studies reported here.

REFERENCES

Adams, C. A., Leung, F. and Sun, S. S. M. (1986). Molecular properties of phosphoenolpyruvate carboxylase from C_3, C_3–C_4 intermediate, and C_4 *Flaveria* species. *Planta* (Berl.) **167**, 218–225.

Adams, W. W., III, Diaz, M. and Winter, K. (1989). Diurnal changes in photochemical efficiency, the reduction state of Q, radiationless energy dissipation, and non-photochemical fluorescence quenching from cacti exposed to natural sunlight in northern Venezuela. *Oecologia* (Berl.) (in press).

Adams, W. W., III and Osmond, C. B. (1988). Internal CO_2 supply during photosynthesis of sun and shade grown CAM plants in relation to photoinhibition. *Plant Physiol.* **86**, 117–123.

Adams, W. W., III, Smith, S. D. and Osmond, C. B. (1987). Photoinhibition of the CAM succulent *Opuntia basilaris* growing in Death Valley: evidence from 77K fluorescence and quantum yield. *Oecologia* (Berl.) **71**, 221–228.

Adams, W. W., III, Terashima, I., Brugnoli, E. and Demmig, B. (1988). Comparisons of photosynthesis and photoinhibition in the CAM vine *Hoya australis* and several C_3 vines growing on the coast of eastern Australia. *Plant, Cell Environ.* **11**, 173–181.

Badger, M. R., Andrews, T. J., Canvin, D. T. and Lorimer, G. H. (1980). Interactions of hydrogen peroxide with ribulose bisphosphate carboxylase oxygenase. *J. Biol. Chem.* **255**, 7870–7875.

Badger, M. R., Sharkey, T. D. and Caemmerer, S. von (1984). The relationship between steady-state gas exchange of bean leaves and the levels of carbon-reduction cycle intermediates. *Planta* (Berl.) **160**, 310–313.

Bartholomew, D. P. and Kadzimin, S. B. (1977). Pineapple. *In* "Ecophysiology of Tropical Crops" (Eds. P. T. Alvim and T. T. Kozlowski), pp. 113–156. Academic Press, New York.

Bassüner, B., Keerberg, O., Bauwe, H., Pyarnik, T. and Keerberg, H. (1984). Photosynthetic CO_2 metabolism in C_3–C_4 intermediate and C_4 species of *Flaveria* (Asteraceae). *Biochem. Physiol. Pflanzen* **179**, 631–634.

Bauwe, H. (1984). Photosynthetic enzyme activities and immunofluorescence studies on the localization of ribulose-1, 5-bisphosphate carboxylase/oxygenase in leaves of C_3, C_4 and C_3–C_4 intermediate species of *Flaveria* (Asteraceae). *Biochem. Physiol. Pflanzen* **179**, 253–268.

Bauwe, H. and Chollet, R. (1986). Kinetic properties of phosphoenol pyruvate carboxylase from C_3, C_4, and C_3–C_4 intermediate species of *Flaveria* (Asteraceae). *Plant Physiol.* **82**, 695–699.

Bauwe, H., Keerberg, O., Bassüner, R., Parnik, T. and Bassüner, B. (1987). Reassimilation of carbon dioxide by *Flaveria* (Asteraceae) species representing different types of photosynthesis. *Planta* **172**, 214–218.

Beer, S. and Wetzel, R. G. (1981). Photosynthetic carbon metabolism in the submerged aquatic angiosperm *Scirpus subterminalis*. *Plant Sci. Lett.* **21**, 199–207.

Björkman, O. (1976). Adaptive and genetic aspects of C_4 photosynthesis. *In* "CO_2 Metabolism and Plant Productivity" (Eds R. H. Burris and C. C. Black), pp. 287–309. University Park Press, Baltimore.

Björkman, O., Nobs, M., Pearcy, R., Boynton, J. and Berry, J. (1971). Characteristics of hybrids between C_3 and C_4 species of *Atriplex*. *In* "Photosynthesis and Photorespiration." (Eds M. D. Hatch, C. B. Osmond and R. O. Slatyer), Wiley-Interscience, Sydney.

Blake, S. T. (1972). *Neurachne* and its allies (Graminae). *Contr. Qd. Herb.* **13**, 1–53.

Bolton, J. K. and Brown, R. H. (1980). Photosynthesis in grass species differing in carbon dioxide fixation pathways. V. Response of *Panicum maximum, Panicum milioides* and tall fescue (*Festuca arundinacea*) to nitrogen nutrition. *Plant Physiol.* **66**, 97–100.

Broda, E. (1975). "The Evolution of the Bioenergetic Processes." Pergamon, Oxford.

Brown, R. H. (1978). A difference in N use efficiency in C_3 and C_4 plants and its implications in adaptation and evolution. *Crop Sci.* **18**, 93–98.

Brown, R. H. (1980). Photosynthesis of grass species differing in carbon dioxide fixation pathways. IV. Analysis of reduced oxygen response in *Panicum miliodes* and *Panicum schenkii*. *Plant Physiol.* **65**, 346–349.

Brown, R. H., Bassett, C. L., Cameron, R. G., Evans, P. T., Bouton, J. H., Black, C. C., Sternberg, L. O. and DeNiro, M. J. (1986). Photosynthesis of F_1 hybrids between C_4 and C_3–C_4 species of *Flaveria*. *Plant Physiol.* **82**, 211–217.

Brown, R. H., Bouton, J. H., Rigsby, L. L. and Rigler, M. (1983a). Photosynthesis of grass species differing in carbon dioxide fixation pathways. VIII. Ultrastructural characteristics of *Panicum* species in the *Laxa* group. *Plant Physiol.* **71**, 425–431.

Brown, R. H. and Brown, W. V. (1975). Photosynthetic characteristics of *Panicum miliodes*, a species with reduced photorespiration. *Crop Sci.* **15**, 681–685.

Brown, R. H., Rigsby, L. L. and Akin, D. E. (1983b). Enclosure of mitochondria by chloroplasts. *Plant Physiol.* **71**, 437–439.

Brown, R. H. and Simmons, R. E. (1979). Photosynthesis of grass species differing in CO_2 fixation pathways. I. Water-use efficiency. *Crop Sci.* **19**, 375–379.

Burnell, J. N. and Hatch, M. D. (1988). Low bundle sheath carbonic anhydrase is apparently essential for effective C_4 pathway operation. *Plant Physiol.* **86**, 1252–1256.

Caemmerer, S. von and Edmondson, D. E. (1986). Relationship between steady-state gas exchange, in vivo ribulose bisphosphate carboxylase activity and some carbon reduction cycle intermediates in *Raphanus sativus*. *Aust. J. Plant Physiol.* **13**, 669–688.

Caemmerer, S. von and Farquhar, G. D. (1981). Some relationships between the biochemistry of photosynthesis and the gas exchange of leaves. *Planta* (Berl.) **153**, 376–387.

Caldwell, M. M., Osmond, C. B. and Nott, D. L. (1977a). C_4 pathway photosynthesis at low temperature in cold-tolerant *Atriplex* species. *Plant Physiol.* **60**, 157–164.

Caldwell, M. M., White, R. S., Moore, R. T. and Camp, L. B. (1977b). Carbon balance, productivity, and water-use of cold-winter desert shrub communities dominated by C_3 and C_4 species. *Oecologia* (Berl.) **29**, 275–300.

Cavagnaro, J. B. (1988). Distribution of C_3 and C_4 grasses at different altitudes in a temperate arid region of Argentina. *Oecologia* (Berl.) **76**, 273–277.

Chapin, F. S., III (1980). The mineral nutrition of wild plants. *Ann. Rev. Ecol. Syst.* **11**, 233–260.

Chazdon, R. L. (1978). Ecological aspects of the distribution of C_4 grasses in selected habitats of Costa Rica. *Biotropica 10*, 265–269.

Christie, E. K. and Detling, J. K. (1982). Interference analysis of C_3 and C_4 grasses in relation to temperature and soil nitrogen supply. *Ecology* **160**, 1277–1284.

Cloud, P. E. (1968). Atmospheric and hydrospheric evolution on the primitive earth. *Science* **160**, 729–736.

Cockburn, W. (1981). The evolutionary relationship between stomatal mechanism, crassulacean acid metabolism and C_4 photosynthesis. *Plant, Cell Environ.* **4**, 417–418.

Cockburn, W. (1983). Stomatal mechanism as the basis of the evolution of CAM and C_4 photosynthesis. *Plant, Cell Environ.* **6**, 275–279.

Cockburn, W. (1985). Variation in photosynthetic acid metabolism in vascular plants: CAM and related phenomena. *New Phytol.* **101**, 3–24.

Cockburn, W., Goh, C. J. and Avadhani, P. N. (1985). Photosynthetic carbon assimilation in a shootless orchid. *Chiloschista usneoides* (DON) LDL. *Plant Physiol.* **77**, 83–86.

Cockburn, W. and McAulay, A.(1977). Changes in metabolite levels in *Kalanchoe daigremontiana* and the regulation of malic acid accumulation in crassulacean acid metabolism. *Plant Physiol.* **59**, 455–458.

Cockburn, W., Ting, I. P. and Sternberg, L. O. (1979). Relationships between stomatal behavior and internal carbon dioxide concentration in crassulacean acid metabolism plants. *Plant Physiol.* **63**, 1029–1032.

Davies, D. D. (1979). The central role of phosphoenolpyruvate in plant metabolism. *Ann. Rev. Plant Physiol.* **30**, 131–158.

Demmig, B., Winter, K., Kruger, A. and Czygan, F.-C. (1987). Photoinhibition and zeaxanthin formation in intact leaves. A possible role of the xanthophyll cycle in the dissipation of excess light energy. *Plant Physiol.* **84**, 218–224.

Dietz, K.-J. and Heber, U. (1984). Rate-limiting factors in leaf photosynthesis. I. Carbon fluxes in the Calvin cycle. *Biochim. Biophys. Acta* **767**, 432–443.

Edwards, G. E. and Huber, S. C. (1981). The C_4 pathway. *In* "The Biochemistry of Plants. A Comprehensive Treatise. Volume 8. Photosynthesis." (Eds M. D. Hatch and N. K. Boardman), Academic Press, New York.

Edwards, G. E. and Ku, M. S. B. (1987). The biochemistry of C_3–C_4 intermediates. *In* "The Biochemistry of Plants. Volume 10. Photosynthesis." (Eds M. D. Hatch and N. K. Boardman), pp. 276–327. Academic Press, New York.

Edwards, G. E., Ku, M. S. B. and Hatch, M. D. (1982). Photosynthesis in *Panicum milioides*, a species with reduced photorespiration. *Plant Cell Physiol.* **23**, 1185–1195.

Edwards, G. E., Ku, M. S. B. and Monson, R. K. (1985). C_4 photosynthesis and its regulation. *In* "Topics in Photosynthesis. Volume 6. Photosynthetic Mechanisms and the Environment." (Eds J. Barber and N. R. Baker), Elsevier Press, The Netherlands.

Ehleringer, J. R. (1978). Implications of quantum yield differences on the distribution of C_3 and C_4 grasses. *Oecologia* (Berl.) **31**, 255–267.

Ehleringer, J. R. and Pearcy, R. W. (1983). Variation in quantum yield for CO_2 uptake among C_3 and C_4 plants. *Plant Physiol.* **73**, 555–559.

Evans, J. R. (1986). The relationship between CO_2-limited photosynthetic rate and RuBP carboxylase content in two nuclear-cytoplasm substitutions lines of wheat and the coordination of RuBP carboxylation and electron transport capacities. *Planta* (Berl.) **167**, 351–358.

Evans, L. (1971). Evolutionary, adaptive, and environmental aspects of the C_4 photosynthetic pathway. *In* "Photosynthesis and Photorespiration" (Eds M. D. Hatch, C. B. Osmond and R. O. Slatyer), Wiley Interscience, New York.

Farquhar, G. D., O'Leary, M. H. and Berry, J. A. (1982). On the relationship between carbon isotope discrimination and the intercellular carbon dioxide concentration in leaves. *Aust. J. Plant Physiol.* **9**, 121–137.

Fischer, R. A. and Turner, N. C. (1978). Plant productivity in the arid and semiarid zones. *Ann. Rev. Plant Physiol.* **29**, 277–317.

Gebauer, G., Schuhmacher, M. I., Krstic, B., Rehder, H. and Ziegler, H. (1987). Biomass production and nitrate metabolism of *Atriplex hortensis* L. (C_3 plant) and *Amaranthus retroflexus* L. (C_4 plant) in cultures at different levels of nitrogen supply. *Oecologia* (Berl.) **72**, 303–314.

Gil, F. (1986). Origin of CAM as an alternative photosynthetic carbon fixation pathway. *Photosynthetica* **20**, 494–507.

Goh, C. J., Arditti, J. and Avadhani, P. N. (1983). Carbon fixation in orchid aerial roots. *New Phytol.* **95**, 367–374.

Grime, J. P. (1979). "Plant Strategies and Vegetation Processes." Wiley, Chichester.

Hatch, M. D. and Slack, C. R. (1970). The C_4 dicarboxylic-acid pathway. *In*

"Progress in Phytochemistry. Volume 2." (Eds L. Reinhold and Y. Liwschitz). Interscience Publishers, London.

Hattersley, P. W. (1983). The distribution of C_3 and C_4 grasses in Australia in relation to climate. *Oecologia* (Berl.) **57**, 113–128.

Hattersley, P. W. and Stone, N. E. (1986). Photosynthetic enzyme activities in the C_3–C_4 intermediate *Neurachne minor* S. T. Blake (Poaceae). *Aust. J. Plant Physiol.* **13**, 399–408.

Herold, A. (1980). Regulation of photosynthesis by sink activity—the missing link. *New Phytol.* **86**, 134–144.

Hew, S. C. and Wong, Y. S. (1974). Photosynthesis and respiration in ferns in relation to their habitats. *Am. Fern J.* **64**, 40–48.

Holaday, A. S. and Black, C. C. (1981). Comparative characterization of phosphoenolpyruvate carboxylase in C_3, C_4, and C_3–C_4 intermediate *Panicum* species. *Plant Physiol.* **67**, 330–334.

Holaday, A. S. and Bowes, G. (1980). C_4 acid metabolism and dark CO_2 fixation in a submerged aquatic macrophyte (*Hydrilla verticillata*). *Plant Physiol.* **65**, 331–335.

Holaday, A. S., Brown, R. H., Bartlett, J. M., Sandlin, E. A. and Jackson, R. C. (1988). Enzymic and photosynthetic characteristics of reciprocal F_1 hybrids of *Flaveria pringlei* (C_3) and *Flaveria brownii* (C_4-like species). *Plant Physiol.* **87**, 484–490.

Holaday, A. S. and Chollet, R. (1983). Photosynthetic/photorespiratory carbon metabolism in the C_3–C_4 intermediate species, *Moricandia arvensis* and *Panicum milioides*. *Plant Physiol.* **73**, 740–745.

Holaday, A. S., Lee, K. W. and Chollet, R. (1984). C_3–C_4 intermediate species in the genus *Flaveria*: leaf anatomy, ultrastructure, and the effect of O_2 on the CO_2 compensation concentration. *Planta* (Berl.) **160**, 25–32.

Holaday, A. S., Shieh, Y.-S., Lee, K. W. and Chollet, R. (1981). Anatomical, ultrastructural, and enzymic studies of leaves of *Moricandia arvensis*, a C_3–C_4 intermediate species. *Biochim. Biophys. Acta* **637**, 334–341.

Holaday, A. S., Talkmitt, S. and Doohan, M. E. (1985). Anatomical and enzymic studies of leaves of a $C_3 \times C_4$ *Flaveria* F_1 hybrid exhibiting reduced photorespiration. *Plant Sci.* **41**, 31–39.

Holbrook, G. P. and Chollet, R. (1986). Photorespiratory properties of protoplasts from C_3–C_4 intermediate species of *Moricandia*. *Plant Sci.* **46**, 87–96.

Holbrook, G. P., Jordan, D. B. and Chollet, R. (1985). Reduced apparent photorespiration by the C_3–C_4 intermediate species, *Moricandia arvensis* and *Panicum milioides*. *Plant Physiol.* **77**, 578–583.

Holthe, P. A., Sternberg, L. S. L. and Ting, I. P. (1987). Developmental control of CAM in *Peperomia scandens*. *Plant Physiol.* **84**, 743–747.

Jain, S. K. (1976). The evolution of inbreeding plants. *Ann. Rev. Ecol. Sys.* **7**, 469–495.

Jordan, D. B. and Ogren, W. L. (1981). Species variation in the specificity of ribulose bisphosphate carboxylase/oxygenase. *Nature* **291**, 513–515.

Jordan, D. B. and Ogren, W. L. (1983). Species variation in kinetic properties of ribulose 1,5-bisphosphate carboxylase/oxygenase. *Arch. Biochem. Biophys.* **227**, 425–433.

Kanai, R. and Kashiwagi, M. (1975). *Panicum milioides*, a Graminae plant having Kranz leaf anatomy without C_4 photosynthesis. *Plant and Cell Physiol.* **16**, 669–679.

Kalisz, S. and Teeri, J. (1986). Population-level variation in photosynthetic metabolism and growth in *Sedum wrightii*. *Ecology* **67**, 20–26.

Keeley, J. E. (1981). *Isoetes howellii*, a submerged CAM plant. *Am. J. Bot.* **68**, 420–429.

Keeley, J. E. (1983). Crassulacean acid metabolism in the seasonally submerged aquatic *Isoetes howellii. Oecologia* (Berl.) **58**, 57–62.

Keeley, J. E., Osmond, C. B. and Raven, J. A. (1984). *Stylites*, a vascular land plant without stomata absorbs CO_2 via its roots. *Nature* **310**, 684–685.

Kluge, M. and Ting, I. P. (1978). "Crassulacean acid metabolism." Springer-Verlag, Berlin.

Kennedy, R. A. and Laetsch, W. M. (1974). Plant species intermediate for C_3, C_4 photosynthesis. *Science* **184**, 1087–1089.

Kestler, D. P., Mayne, B. C., Ray, T. B., Goldstein, L. D., Brown, R. H. and Black, C. C. (1975). Biochemical components of the photosynthetic CO_2 compensation point of higher plants. *Biochem. Biophys. Res. Comm.* **66**, 1439–1446.

Ku, M. S. B., Monson, R. K., Littlejohn, R. O., Nakamota, H., Fisher, D. B. and Edwards, G. E. (1983). Photosynthetic characteristics of C_3–C_4 intermediate *Flaveria* species. I. Leaf anatomy, photosynthetic responses to CO_2 and O_2, and activities of key enzymes in the C_3 and C_4 pathways. *Plant Physiol.* **71**, 944–948.

Laetsch, W. M. (1971). Chloroplast structural relationships in leaves of C_4 plants. *In "Photosynthesis and Photorespiration"*. (Eds M. D. Hatch, C. B. Osmond and R. O. Slatyer). pp. 323–352, Wiley-Interscience, New York.

Latzko, E. and Kelly, G. J. (1983). The many-faceted function of phosphoenolpyruvate carboxylase in C_4 plants. *Physiol. Veg.* **21**, 805–815.

Lee, H. S. J. and Griffiths, H. (1987). Induction and repression of CAM in *Sedum telephium* L. in response to photoperiod and water stress. *J. Exp. Bot.* **38**, 834–341.

Leegood, R. and Furbank, R. T. (1986). Stimulation of photosynthesis by 2% oxygen at low temperatures is restored by phosphate. *Planta* (Berl.) **168**, 84–93.

Long, S. P. (1983). C_4 photosynthesis at low temperatures. *Plant, Cell Environ.* **6**, 345–363.

Lorimer, G. H. and Andrews. T. J. (1981). The C_2 chemo- and photorespiratory carbon oxidation cycle. *In "The Biochemistry of Plants. Volume 8. Photosynthesis"*. (Eds. M. D. Hatch and N. K. Boardman), Academic Press, New York.

Makino, A., Mae, T. and Ohira, K. (1985). Photosynthesis and ribulose 1,5-bisphosphate carboxylase/oxygenase in rice leaves from emergence through senescence. Quantitative analysis by carboxylation/oxygenation and regeneration of ribulose 1,5-bisphosphate. *Planta* (Berl.) **166**, 414–420.

Martin, C. E., Higley, M. and Wang, W.-Z. (1988). Ecophysiological significance of CO_2-recycling via Crassulacean acid metabolism in *Talinum calycinum* Engelm. (Portulacaceae). *Plant Physiol.* **86**, 562–568.

Martin, C. E. and Jackson, J. L. (1986). Photosynthetic pathways in a midwestern rock outcrop succulent, *Sedum nuttallianum* Raf. (Crassulaceae). *Photosyn. Res.* **8**, 17–29.

Martin, C. E. and Zee, A. K. (1983). C_3 photosynthesis and Crassulacean acid metabolism in a Kansas rock outcrop succulent, *Talinum calycinum* Engelm. (Portulacaceae). *Plant Physiol.* **73**, 718–723.

Monson, R. K. (1989). The relative contributions of reduced photorespiration, and improved water- and nitrogen-use, to the advantages of C_3–C_4 intermediate photosynthesis in *Flaveria. Oecologia* (Berl.) (in press).

Monson, R. K., Edwards, G. E. and Ku, M. S. B. (1984). C_3–C_4 intermediate photosynthesis in plants. *BioScience* **34**, 563–574.

Monson, R. K., Littlejohn, R. O. and Williams, G. J. (1982a). The quantum yield for CO_2 uptake in C_3 and C_4 grasses. *Photosyn. Res.* **3**, 153–159.

Monson, R. K. and Moore, B. d. (1989). On the significance of C_3–C_4 intermediate photosynthesis to the evolution of C_4 photosynthesis. *Plant, Cell Environ.* (in press).

Monson, R. K., Moore, B. d., Ku, M. S. B. and Edwards, G. E. (1986). Co-function of C_3- and C_4 photosynthetic pathways in C_3, C_4, and C_3–C_4 intermediate *Flaveria* species. *Planta* (Berl.) **168**, 493–502.

Monson, R. K., Schuster, W. and Ku, M. S. B. (1987). Photosynthesis in *Flaveria brownii* Powell. A C_4-like C_3–C_4 intermediate. *Plant Physiol.* **85**, 1063–1067.

Monson, R. K., Stidham, M. A., Williams, G. J., Edwards, G. E. and Uribe, E. G. (1982b). Temperature dependence of photosynthesis in *Agropyron smithii* Rybd. I. Factors affecting net CO_2 uptake in intact leaves and contribution from ribulose 1,5-bisphosphate carboxylase measured *in vivo* and *in vitro*. *Plant Physiol.* **69**, 921–928.

Monson, R. K., Teeri, J., Ku, M. S. B., Gurevitch, J., Mets, L. and Dudley, S. (1988). Carbon-isotope discrimination by leaves of *Flaveria* species exhibiting different amounts of C_3- and C_4-cycle cofunction. *Planta* (Berl.) **174**, 145–151.

Moore, B. d., Franceschi, V. R., Cheng, S.-H., Wu, J. and Ku, M. S. B. (1987a). Photosynthetic characteristics of the C_3–C_4 intermediate *Parthenium hysterophorus*. *Plant Physiol.* **85**, 984–989.

Moore, B. d., Ku, M. S. B. and Edwards, G. E. (1987b). C_4 photosynthesis and light-dependent accumulation of inorganic carbon in leaves of C_3–C_4 and C_4 *Flaveria* species. *Aust. J. Plant Physiol.* **14**, 657–668.

Moore, B. d., Monson, R. K., Ku, M. S B. and Edwards, G. E. (1988). Activities of principal photosynthetic and photorespiratory enzymes in leaf mesophyll and bundle-sheath protoplasts from the C_3–C_4 intermediate *Flaveria ramosissima*. *Plant and Cell Physiol.* **29**, 999–1006.

Morison, J. I. L. and Gifford, R. M. (1983). Stomatal sensitivity to carbon dioxide and humidity. A comparison of two C_3 and two C_4 grass species. *Plant Physiol.* **71**, 789–796.

Mott, K. A., Jensen, R. G., O'Leary, J. W. and Berry, J. A. (1984). Photosynthesis and ribulose 1,5-bisphosphate concentrations in intact leaves of *Xanthium strumarium* L. *Plant Physiol.* **76**, 968–971.

Nakamoto, H., Ku, M. S. B. and Edwards, G. E. (1983). Photosynthetic characteristics of C_3–C_4 intermediate *Flaveria* species. II. Kinetic properties of phosphoenolpyruvate carboxylase from C_3, C_4, and C_3–C_4 intermediate species. *Plant Cell Physiol.* **24**, 1387–1393.

Nishio, J. N. and Ting, I. P. (1987). Carbon flow and metabolic specialization in the tissue layers of the Crassulacean acid metabolism plant, *Peperomia camptotricha*. *Plant Physiol.* **84**, 600–604.

Nobel, P. S. and Hartstock, T. L. (1986). Leaf and stem CO_2 uptake in the three subfamilies of the Cactaceae. *Plant Physiol.* **80**, 913–917.

Nobs, M. A., Pearcy, R. W., Berry, J. A. and Nicholson, F. (1972). Reciprocal transplant responses of C_3 and C_4 *Atriplexes*. *Carnegie Inst. Wash. Yearbook* **71**, 164–169.

O'Leary, M. (1981). Carbon isotope fractionation in plants. *Phytochemistry* **20**, 553–567.

O'Leary, M. H. (1982). Phosphoenolpyruvate carboxylase: an enzymologists view. *Ann. Rev. Plant Physiol.* **33**, 297–315.

O'Leary, M. (1988). Carbon isotopes in photosynthesis. *BioScience* **38**, 328–336.

Ogren, W. L. (1984). Photorespiration: pathways, regulation and modification. *Ann. Rev. Plant Physiol.* **35**, 415–442.

Osmond, C. B. (1978). Crassulacean acid metabolism: a curiosity in context. *Ann. Rev. Plant Physiol.* **29**, 379–510.

Osmond, C. B. (1982). Carbon cycling and stability of the photosynthetic apparatus in CAM. *In* "Crassulacean Acid Metabolism." (Eds I. P. Ting and M. Gibbs), pp. 112 127, American Society of Plant Physiol., Waverly Press, Baltimore.

Osmond, C. B. (1984). CAM. Regulated photosynthetic metabolism for all seasons. *In* "Advances in Photosynthesis Research" (Ed. C. Sybesma), pp. 557–563, W. Junk, The Hague.

Osmond, C. B. and J. A. M. Holtum. (1981). Crassulacean acid metabolism. *In* "The Biochemistry of Plants. Volume 8. Photosynthesis." (Eds M. D. Hatch and N. K. Boardman), pp. 283–329, Academic Press, New York.

Osmond, C. B., Winter, K. and Ziegler, H. (1982). Functional significance of different pathways of CO_2 fixation in photosynthesis. *In* "Encyclopedia of Plant Physiology, Volume 12B, Physiological Plant Ecology II." (Eds. O. L. Lange, P. S. Nobel, C. B. Osmond and H. Ziegler), pp. 479–548, Springer-Verlag, Berlin.

Patel, A. and Ting, I. P. (1987). The relationship between respiration and CAM-cycling in *Peperomia camptotricha*. *Plant Physiol.* **84**, 640–642.

Pearcy, R. W., Tumosa, N. and Williams, K. (1981). Relationships between growth, photosynthesis and competitive interactions for a C_3 and a C_4 plant. *Oecologia* (Berl.) **48**, 371–376.

Pierce, J. (1986). Determinants of substrate specificity and the role of metal in the reactions of ribulose bisphosphate carboxylase/oxygenase. *Plant Physiol.* **81**, 943–945.

Pierce, J., Lorimer, G. H. and Reddy, G. S. (1986). Kinetic mechanism of ribulosebisphosphate carboxylase. Evidence for an ordered, sequential reaction. *Biochemistry* **25**, 1636–1644.

Powell, A. M. (1978). Systematics of *Flaveria* (Flaverinae-Asteraceae) *Ann. Missouri Bot. Gard.* **65**, 590–636.

Powles, S. B. (1984). Photoinhibition of photosynthesis induced by visible light. *Ann. Rev. Plant Physiol.* **35**, 15–44.

Ramos, C. and Hall, A. E. (1983). Effects of photon fluence rate and intercellular CO_2 partial pressure on leaf conductance and CO_2 uptake rate in *Capsicum* and *Amaranthus*. *Photosynthetica* **17**, 34–42.

Raschke, K. (1975). Stomatal action. *Ann. Rev. Plant Physiol.* **26**, 309–340.

Raven, J. A. (1984). *Energetics and Transport in Aquatic Plants.* A. R. Liss Co., New York.

Raven, J. A. (1985). The CO_2 concentrating mechanism. *In* "*Inorganic carbon uptake by aquatic photosynthetic organisms*". (Eds W. J. Lucas and J. A. Berry), pp. 67–82, Waverly Press, Baltimore.

Raven, J. A., Handley, L. L., MacFarlane, J. J., McInroy, S., McKenzie, L., Richards, J. H. and Samuelsson, G. (1988). The role of CO_2 uptake by roots and CAM in acquisition of inorganic C by plants of the isoetid life-form: a review, with new data on *Eriocaulon decangulare* L. *New Phytologist* **108**, 125–148.

Rawsthorne, S., Hylton, C. M., Smith, A. M. and Woolhouse, H. W. (1988). Photorespiratory metabolism and immunogold localization of photorespiratory enzymes in leaves of C_3 and C_3–C_4 intermediate species of *Moricandia*. *Planta* (Berl.) **173**, 298–308.

Rayder, L. and Ting, I. P. (1981). Carbon metabolism in two species of *Pereskia* (Cactaceae). *Plant Physiol.* **18**, 139–142.

Reed, J. E. and Chollet, R. (1985). Immunofluorescent localization of phosphoenolpyruvate carboxylase and ribulose 1,5-bisphosphate carboxylase/oxygenase proteins in leaves of C_3, C_4 and C_3–C_4 intermediate *Flaveria* species. *Planta* (Berl.) **165**, 439–445.

Robichaux, R. and Pearcy, R. W. (1980). Photosynthetic responses of C_3 and C_4 species from cool shaded habitats in Hawaii. *Oecologia* (Berl.) **47**, 106–109.

Rumpho, M., Ku, M. S. B., Cheng, S.-H. and Edwards, G. E. (1984). Photosynthetic characteristics of C_3–C_4 intermediate *Flaveria* species. III. Reduction of photorespiration by a limited C_4 pathway of photosynthesis in *Flaveria ramosissima. Plant Physiol.* **75**, 993–996.

Rundel, P. (1980). The ecological distribution of C_4 and C_3 grasses in the Hawaiian Islands. *Oecologia* (Berl.) **45**, 354–359.

Sage, R. F. and Pearcy, R. W. (1987). The nitrogen-use efficiency of C_3 and C_4 plants. Leaf nitrogen, growth, and biomass partitioning in *Chenopodium album* (L.) and *Amaranthus retroflexus* (L.). *Plant Physiol.* **84**, 954–958.

Sage, R. F. and Sharkey, T. D. (1987). The effect of temperature on the occurrence of O_2 and CO_2 insensitive photosynthesis in field grown plants. *Plant Physiol.* **84**, 658–664.

Sayre, R. T. and Kennedy, R. A. (1977). Ecotypic differences in the C_3 and C_4 photosynthetic activity in *Mollugo verticillata*, a C_3–C_4 intermediate. *Planta* (Berl.) **134**, 257–262.

Sayre, R. T., Kennedy, R. A. and Pringnitz, D. J. (1979). Photosynthetic enzyme activities and localization in *Mollugo verticillata* populations differing in the levels of C_3 and C_4 cycle operation. *Plant Physiol.* **64**, 293–299.

Sharkey, T. D. (1985). Photosynthesis in intact leaves of C_3 plants: physics, physiology and rate limitations. *Bot. Rev.* **51**, 53–105.

Sharkey, T. D. and Raschke, K. (1981). Separation and measurement of direct and indirect effects of light on stomata. *Plant Physiol.* **68**, 33–40.

Sipes, D. and Ting, I. P. (1985). CAM and CAM-modifications in *Peperomia camptotricha. Plant Physiol.* **77**, 59–63.

Smith, B. N., Martin, G. E. and Boutton, T. W. (1979). Carbon isotopic evidence for the evolution of C_4 photosynthesis. *In* "Stable Isotopes: Proceedings of the Third International Conference." (Eds E. R. Klein and P. D. Klein), pp. 231–237, Academic Press, New York.

Smith, B. N. and Powell, A. M. (1984). C_4-like F_1-hybrid of $C_3 \times C_4$ *Flaveria* species. *Naturwissenschaften* **71**, 217.

Smith, F. A. and Raven, J. A. (1979). Intracellular pH and its regulation. *Ann. Rev. Plant Physiol.* **30**, 289–311.

Smith, J. A. C., Griffiths, H. and Lüttge, U. (1986a). Comparative ecophysiology of CAM and C_3 bromeliads. I. The ecology of the Bromeliaceae in Trinidad. *Plant, Cell Environ.* **9**, 359–376.

Smith, J. A. C., Griffiths, H., Lüttge, U., Crooks, C. E., Griffiths, N. M. and Stimmel, K.-H. (1986b). Comparative ecophysiology of CAM and C_3 bromeliads. IV. Plant water relations. *Plant, Cell Environ.* **9**, 395–410.

Smith, M. and Martin, C. E. (1987). Photosynthetic responses to irradiance in three forest understory species of the C_4 genus. *Muhlenbergia. Bot. Gaz.* **148**, 275–282.

Smith, S. D. and Nobel, P. (1987). Deserts. *In* "*Topics in Photosynthesis. Volume 7. Photosynthesis in Contrasting Environments*". (Eds N. R. Baker and S. P. Long), pp. 13–62, Elsevier Press, Amsterdam.

Stebbins, G. L. (1974). "*Flowering Plants: Evolution Above the Species Level*". Belknap Press, Cambridge.

Stitt, M. (1986). Limitation of photosynthesis by carbon metabolism. I. Evidence for excess electron transport capacity in leaves carrying out photosynthesis in saturating light and CO_2. *Plant Physiol.* **81**, 1115–1122.

Stowe, L. G. and Teeri, J. A. (1978). The geographic distribution of C_4 species of the dicotyledonae in relation to climate. *Am. Nat.* **112**, 609–623.

Szarek, S. R., Johnson, H. B. and Ting, I. P. (1973). Drought adaptation in *Opuntia basilaris*. Significance of recycling carbon through Crassulacean acid metabolism. *Plant Physiol.* **52**, 539–541.

Szarek, S. R. and Ting, I. P. (1975). Physiological responses to rainfall in *Opuntia basilaris* (Cactaceae). *Am. J. Bot.* **62**, 602–609.

Teeri, J. A. (1982a). Carbon isotopes and the evolution of C_4 photosynthesis and Crassulacean acid metabolism. *In* "*Biochemical Aspects of Evolutionary Biology*". (Ed. H. Nitecki), pp. 93–130, The University of Chicago Press, Chicago.

Teeri, J. A. (1982b). Photosynthetic variation in the Crassulaceae. *In* "*Crassulacean Acid Metabolism*". (Eds I. P. Ting and M. Gibbs), pp. 244–259, American Socity of Plant Physiol., Waverly Press, Baltimore.

Teeri, J. A. and Gurevitch, J. (1984). Environmental and genetic control of crassulacean acid metabolism in two crassulacean species and an F_1 hybrid with differing biomass $\delta^{13}C$ values. *Plant, Cell Environ.* **7**, 589–596.

Teeri, J. A. and Stowe, L. G. (1976). Climatic patterns and the distribution of C_4 grasses in North America. *Oecologia* (Berl.) **23**, 1–12.

Teeri, J. A., Stowe, L. G. and Livingstone, D. A. (1980). The distribution of C_4 species of the Cyperaceae in relation to climate. *Oecologia* (Berl.) **47**, 303–310.

Teeri, J. A., Stowe, L. G. and Murawski, D. A. (1978). The climatology of two succulent plant families: Cactaceae and Crassulaceae. *Can. J. Bot.* **56**, 1750–1758.

Teeri, J. A., Tonsor, S. J. and Turner, M. (1981). Leaf thickness and carbon isotope composition in the Crassulaceae. *Oecologia* (Berl.) **50**, 367–369.

Tieszen, L. L., Senyimba, M. M., Imbamba, S. K. and Troughton, J. H. (1979). The distribution of C_3 and C_4 grasses and carbon isotope discrimination along an altitudinal and moisture gradient in Kenya *Oecologia* (Berl.) **37**, 337–350.

Ting, I. P. (1985). Crassulacean acid metabolism. *Ann. Rev. Plant Physiol.* **36**, 595–662.

Ting, I. P. and Burk, J. H. (1983). Aspects of carbon metabolism in *Welwitschia. Plant Sci. Lett.* **32**, 279–285.

Ting, I. P. and Osmond, C. B. (1973). Photosynthetic phosphoenolpyruvate carboxylases. Characteristics of alloenzymes from leaves of C_3 and C_4 plants. *Plant Physiol.* **51**, 439–447.

Ting, I. P. and Rayder, L. (1982). Regulation of C_3 to CAM shifts. *In* "*Crassulacean Acid Metabolism*". (Eds I. P. Ting and M. Gibbs), pp. 193–207, American Soc. Plant Physiologists, Waverly Press, Baltimore.

Walker, D. A. and Herold, A. (1977). Can the chloroplast support photosynthesis unaided? *In* "*Photosynthetic Organelles: Structure and Function*". (Eds Y. Fujita, S. Katoh, and K. Shinata), pp. 295–310. Spec. Iss. *Plant and Cell Physiol.*

Ward, D. A. and Woolhouse, H. W. (1986). Comparative effects of light during growth on the photosynthetic properties of NADP-ME type C_4 grasses from open and shaded habitats. I. Gas exchange, leaf anatomy and ultrastructure. *Plant, Cell Environ.* **9**, 261–270.

Winter, K. (1985). Crassulacean acid metabolism. *In* "*Topics in Photosynthesis. Volume 6. Photosynthetic Mechanisms and the Environment*". (Eds J. Barber and N. R. Baker), Elsevier, The Netherlands.

Winter, K., Lüttge, U. and Winter, E. (1978). Seasonal shift from C_3 photosynthesis to Crassulacean acid metabolism in *Mesembryanthemum crystallinum* growing in its natural environment. *Oecologia* (Berl.) **34**, 225–237.

Winter, K., Schmitt, M. R. and Edwards, G. E. (1982a). *Microstegium vimineum*, a shade adapted C_4 grass. *Plant Sci. Lett.* **24**, 311–318.

Winter, K., Schröppel-Meier, G. and Caldwell, M. M. (1986). Respiratory CO_2 as carbon source for nocturnal acid synthesis at high temperatures in three species exhibiting Crassulacean acid metabolism. *Plant Physiol.* **81**, 390–394.

Winter, K., Usuda, H., Tsuzuki, M., Schmitt, M. R., Edwards, G. E., Thomas, R. J. and Evert, R. F. (1982b). Influence of nitrate and ammonium on photosynthetic characteristics and leaf anatomy of *Moricandia arvensis*. *Plant Physiol.* **70**, 616–625.

Winter, K., Wallace, B. J., Stocker, G. and Roksandic, Z. (1983). Crassulacean acid metabolism in Australian vascular epiphytes and some related species. *Oecologia* (Berl.) **57**, 129–141.

Wong, S. C., Cowan, I. and Farquhar, G. D. (1985). Leaf conductance in relation to rate of CO_2 assimilation. I. Influence of nitrogen nutrition, phosphorus nutrition, photon flux density, and ambient partial pressure of CO_2 during ontogeny. *Plant Physiol.* **78**, 821–825.

Wong, S. C. and Hew, C. S. (1976). Diffusive resistance, titratable acidity and CO_2 fixation in two tropical epiphytic ferns. *Am. Fern J.* **66**, 121–123.

Woodrow, I. and Berry, J. A. (1988). Enzymatic regulation of photosynthetic CO_2 fixation in C_3 plants. *Ann. Rev. Plant Physiol. Plant Mol. Biol.* **39**, 533–594.

Woodrow, I. (1986). Control of the rate of photosynthetic carbon dioxide fixation. *Biochim. Biophys. Acta* **851**, 181–192.

Yoshie, F. (1986). Intercellular CO_2 concentration and water-use efficiency of temperate plants with different life-forms and from different microhabitats. *Oecologia* (Berl.) **68**, 370–374.

Zalenskii, O. V. and Glagoleva, T.(1981). Pathway of carbon metabolism in halophytic desert species from Chenopodiaceae. *Photosynthetica* **15**, 244–255.

Note Added in Proof

Since this review was written, enhanced activities of glycine decarboxylase have been demonstrated in bundle-sheath cells, compared to mesophyll cells, of C_3–C_4 intermediates in the genera *Panicum*, *Molungo* and *Flaveria* using an immunogold labelling technique (Hylton, C. M., Rawsthorne, S., Smith, A. M., Jones, D. A. and Woolhouse, D. W. (1988). Glycine decarboxylase is confined to the bundle-sheath cells of C_3–C_4 intermediate species *Planta* (Berlin), **175**, 452–459).

Dendroecology: A Tool for Evaluating Variations in Past and Present Forest Environments

H. C. FRITTS and T. W. SWETNAM

I. INTRODUCTION

It is well known that yearly tree-ring width sequences, called chronologies, have been used to date structures, such as archaeological ruins, historic buildings and early Dutch paintings (Anon. 1977; Baillie, 1982; Trefil, 1985). A. E. Douglass, an astronomer working in Arizona, is credited with

ADVANCES IN ECOLOGICAL RESEARCH Vol. 19
ISBN 0-12-013919-7

developing tree-ring dating (1919, 1928, 1936) and is considered the founder of the discipline of "dendrochronology" (Webb, 1983). *Dendro* refers to the Greek root word meaning tree and *chronology* to the study of time. The discipline is most appropriately characterized as the systematic use of tree-ring "cross-dating", a procedure that uses variability of ring characteristics to establish the exact year in which each ring was formed. Cross-dating was first used to date beams or charcoal fragments from archaeological and historical structures in the southwestern United States, and the technique provides archaeologists with the most precise time control ever devised spanning the last two thousand years (Douglass, 1935, 1937; Dean, 1986).

Similar tree-ring width sequences have also been used to reconstruct records of past climatic changes (Douglass, 1914; Schulman, 1947, 1951, 1956; Fritts, 1976; Hughes *et al.*, 1982) and to study past hydrologic history (Cook and Jacoby, 1983; Stockton *et al.*, 1985). The terms *dendroclimatology* and *dendrohydrology* are commonly used to refer to dendrochronological studies of climatic and hydrologic phenomena. Dendrochronology can also be used to study various ecological problems. The term, *dendroecology*, refers to applications of dendrochronological techniques to problems in ecology. The first conference devoted entirely to this subject was held in August, 1986 in Tarrytown, New York (Jacoby and Hornbeck, 1987).

There are four basic ways that dated tree-ring information can be applied to ecological studies: (1) specific ecological events can be dated by their association with dated ring structures or injuries; (2) past forest disturbances can be dated and their importance evaluated by distinctive changes in ring widths or other ring features; (3) climatic or hydrologic conditions can be calibrated and reconstructed by using the variations in ring structure; and (4) climatically related variations in animal populations and behavior can be identified and reconstructed.

In the first application, all associated tree-ring materials are carefully dated to place all growth rings in their correct time sequence (Douglass, 1941, 1946; Stokes and Smiley, 1968; Baillie, 1982; Holmes, 1983). Unusual ring features or evidence of injury are dated by their association with rings formed in a particular growing season. Thus, Sigafoos (1964) used the rings in flood-damaged trees to date scars and to deduce past flooding history. Shroder (1978, 1980), Alestalo (1971) and Giardino *et al.* (1984) dated the scars of trees growing on steep slopes to study rock slides and other types of geomorphic changes.

Dieterich (1980), Ahlstrand (1980) and Swetnam and Dieterich (1985) used tree-ring dated fire scars to reconstruct fire history. Unfortunately, many fire histories have been based upon simple ring counting techniques that can lead to large dating inaccuracies and uncertain conclusions. Madany *et al.* (1982) and Dieterich and Swetnam (1984) demonstrated that dendrochronological dating is considerably more reliable and can be used to establish the exact fire-history sequences. The high precision of dendrochronology also allows comparisons of fire histories among trees within study areas and between study areas (Swetnam and Dieterich, 1985) as well as between fire occurrence and variations in climate.

Dendrochronology can also be used to date fire scars in standing dead trees, logs or stumps, while this cannot be accomplished with simple counting of the rings. The use of tree-ring dating techniques in the study of tree death and dynamics of woody debris in forest ecosystems also has considerable potential (e.g. Gore *et al.*, 1985), but has been infrequently applied in this important and developing field of forest ecology (Harmon *et al.*, 1986; Franklin *et al.*, 1987).

The second general application of dendroecological analysis deals with past forest disturbances that may leave no scar but affect the ring character by influencing the tree's productivity and growth (Fritts, 1976; Lorimer, 1985). For example, Marchand (1984) dated rings and analyzed their characteristics to evaluate wood production efficiency in wave regenerated fir (*Abies balsamea* [L.] Mill.) forests. LaMarche (1966) identified periods of accelerated stream erosion by observing growth reduction effects in trees that have sustained a rapid uncovering of their roots. LaMarche (1968) dated exposed roots and used their age and depth to estimate the rate of slope degradation. Smiley (1958) and Yamaguchi (1983, 1985) dated eruptions of volcanoes by observing distinctive ring patterns of trees that had grown within the ash-fall zone. Brubaker and Greene (1979), Ferrel (1980) and Swetnam *et al.* (1985) have studied the effects of insect defoliation on tree growth by dendrochronological dating of the annual rings, and by comparing the differences in ring growth of host trees and non-host trees. Accurate dating is especially important in studies of insect defoliation and related forest disturbances, because many rings can be locally absent or missing in severely defoliated or suppressed trees (Keen, 1937; Evenden, 1940; Wagener, 1961), and this condition would be undetected by simple ring counting.

In the third general application of dated tree-ring information to ecological problems, the characteristics of the dated rings are used to reconstruct past variations in drought (Stockton and Meko, 1975, 1983; Cook and Jacoby, 1977), temperature and precipitation (Fritts *et al.*, 1979), stream flow (Phipps, 1972; Stockton and Jacoby, 1976; Holmes *et al.*, 1979; Cook and Jacoby, 1983) and water levels (Stockton and Fritts, 1973; Phipps *et al.*, 1979). Taylor (1981) and Graumlich and Brubaker, (1987) have also used dendrochronological methods in studies of forest productivity.

This third type of dendroecological application is usually needed because the existing climatological or hydrological records are too short to detect long-term climatic variability and changes (Hecht, 1985; Gates and Mintz,

1975), and because understanding and resolving the ecological problems at hand require this information. The tree-ring information on climate from the oldest trees is calibrated and verified with the existing records of climate and the tree-ring variations in the past are converted to estimates of past variations in climate (Fritts, 1976, Stockton *et al.*, 1985). Consider how tree-ring analysis of past hydrologic variability was used to help resolve the following environmental problem.

A newly constructed dam, which was completed in 1967, reduced the peak spring discharge of the Peace River, Canada that usually flooded the rich natural wildlife habitat of the 1·5 million-acre Peace–Athabasca delta. Measurements of levels in Lake Athabasca showed a marked decline after the gates in the dam had been closed. A public outcry, as early as 1972, expressed fears that this valuable natural habitat would be destroyed by these abnormal changes. It was argued by the proponents of the dam that the 33 years of water-level measurements made before dam construction were anomalous because of recent climatic changes, and the recent low water levels after dam closure were simply a return to natural conditions.

It was hypothesized that rings of white spruce (*Picea glauca* [Moench] Voss) growing along the flood plains and levees in the delta region might be influenced by unusually high or low water levels and that a dendroclimatological study might provide an objective basis for resolving this issue. Trees with 200 or more rings were found growing on the better-drained sites throughout the delta area. Visible differences in ring-width patterns in widely scattered trees suggested that some macro-scale environmental conditions had been growth limiting.

The rings were sampled by coring 10 or more replicate trees (two cores per tree) growing on six widely scattered sites over the delta system. The locations of these sites were largely determined by availability of old trees, but where possible they were chosen so as to reflect differences in the water levels within the inlets and outlets of the delta hydrologic system. The rings were dendrochronologically dated, the widths measured, and these data analyzed to obtain six different records of growth throughout the region (Stockton and Fritts, 1973).

While 61% of the ring-width variations were common to the six growth chronologies, there were differences that appeared to be associated with site drainage and soil–water relationships. Multivariate techniques were used to calibrate the variance in the six chronologies with the 1935–1967 record of water levels averaged for three 10-day periods (May 21–30, July 11–20, and September 21–30) for the year in which the rings were produced and for the year prior to that. The partial regression coefficients were applied to the tree-growth data to obtain reconstructions (statistical estimates) of average water levels for the six 10-day periods.

The 1810–1967 means of the reconstructions were 686·0, 688·7 and 685·5 ft above mean sea level which differed by only 0·6, 0·6 and −0·1 ft from the means for 1935–1967 gauged records. However, the reconstructed variances for the 1810–1967 time period were estimated to have been 5·62, 7·71 and 3·66, which are 309%, 179% and 89% of the variances for 1935–1967 calibration period. It was concluded that the variances of the May 21–30 and the July 11–20 gauge measurements were substantially below the variances for the longer record estimated from the tree-ring measurements, but the variance of the September 21–30 gauge measurements was substantially higher than for the longer estimates.

A few years were reconstructed when water levels were as low as they were early in the season after impoundment, but the reconstructed low levels were isolated cases with higher water levels in the years both before and after them. No low levels had been reconstructed to persist as long as was observed after impoundment. These results led to the conclusion that the low values after impoundment were extreme but within the range of the earlier estimates and hence within the range of natural variability. However, the persistence of these extremes was unprecedented in the earlier estimates.

These results led to the recommendation that the present-day Lake Athabasca levels might be managed to counteract the effect of the impoundment. The extremely low levels from impoundment could be interrupted frequently, raising water levels to match the higher levels associated with pre-impoundment conditions. For example, this might be accomplished by constructing temporary earthen dams across one or more of the outlets before the spring floods begin. This would temporarily impound the water in the Peace–Athabasca area for the early part of that year until the temporary dams washed out, helping to maintain the habitat conditions nearer to pre-impoundment levels.

The fourth type of dendroecological application involves one or more animal species that are so affected by climate that an aspect of their behavior or population can be related to the climatic information found in tree-ring width variations. For example, Clark *et al.* (1975) were interested in determining whether the large scale changes in the fisheries along the North Pacific coastal waters during the last four decades were unusual compared to earlier time periods. They studied the correlations between dated chronologies from arid-site trees of western North America and yearly "landings" of albacore tuna (*Thunnus alalunga*) which reflected population changes or migrations of tuna into or out of the coastal waters north of San Francisco. These changes were found to reflect variations in sea surface temperatures that were, in turn, related to the atmospheric conditions that influenced the growth of arid-site trees in the West. As in the previous example, a regression was obtained between the fish landing data and the variations in the tree-ring chronologies from 1938 through 1973. The regression was then applied to tree growth before 1938 to reconstruct yearly tuna landings, which were inferred to be surrogates of variations in the population of albacore tuna during the past. It was concluded that the past fluctuations in tuna population were as variable as in the 1938–1973 period, and such variations would be expected to continue in the future.

In other similar applications, Young (1979) studied factors influencing hunting success of mule deer (*Odocoileus hemionus*) in Arizona and used tree-ring widths as indicators of the availability of natural forage. Spencer (1964) used tree-ring chronologies as well as dated scars from trees grazed by porcupine (*Erethizon epixanthum*) to document past changes in porcupine populations at Mesa Verde, Colorado.

Today we are faced with a number of very important ecological problems for which dendroecological techniques are well suited. These include (1) widespread outbreaks of herbivorous insects in forests (Doane and McManus, 1981; Schmitt *et al.*, 1984; Brookes *et al.*, 1987), (2) tree decline that was first observed in forests of central and northern Europe and is now reported for some areas of the United States (Abrahamsen *et al.*, 1976; Binns and Redfern, 1983; Johnson and Siccama, 1983; Bormann, 1985; Hornbeck and Smith, 1985; Smith, 1985) and (3) potential environmental changes brought about by the rising concentration of atmospheric CO_2 and other gases (Lemon, 1983; Woodwell *et al.*, 1983; LaMarche *et al.*, 1984).

Because the relevant tree-ring analysis techniques needed for such research may be unfamiliar to some ecologists or are simply misunderstood, the following discussion will summarize the basic principles and practices of the field, describe some of the most relevant new developments, indicate some of the limitations and suggest how dendrochronology can help to illuminate some of the current day environmental issues. This work updates an earlier article on the subject (Fritts, 1971).

II. PARAMETERS REFLECTING TREE-RING VARIABILITY

A variety of structural characteristics of tree rings, such as width, wood density (Schweingruber, 1983; Schweingruber *et al.*, 1978a) and vessel size (Eckstein and Frisse, 1982), show variability from one ring to the next. The variations in ring width have been studied most often (Fritts, 1976; Baillie, 1982), because width can be observed and measured easily from a finely sanded or microtomed surface by using a hand lens or dissecting microscope.

Early in the 1960s, Polge (1963, 1966, and 1970) discovered that a thin section of wood can be X-rayed, and the image on exposed film scanned by an optical densitometer to obtain detailed ring density measurements. Parameters such as minimum density, maximum density, earlywood width and latewood width are now derived from these types of measurements

(Schweingruber, 1983). These can be correlated with climatic variations as well as various physical, chemical and biological features of the environment (Keller, 1968; Parker and Henoch, 1971; Fritts, 1976; Huber, 1976; Schweingruber *et al.*, 1978a, 1978b; Conkey, 1982a, 1982b, 1986; Cleaveland, 1986).

Stable and unstable isotopes of oxygen, hydrogen and carbon in tree rings, as well as accumulations of heavy metals, are other promising sources of information (Lepp, 1975; Jacoby, 1980; Long, 1982; Wigley, 1982; Brubaker and Cook, 1983; Guyette and McGinnes, 1987). While these new sources of tree-ring data may become viable alternatives to studies of ring width, they often require custom-made equipment, relatively complicated procedures and some of the methods are not fully worked out.

Densitometric analysis appears to offer considerable promise for environmental and climatic analysis of moderately moist forest habitats such as the central deciduous forest and southern boreal forest of North America, as well as in the forests of western Europe. Here the rings may be wide and the widths may exhibit much less variability than wood density characteristics. In areas of little width variation, dendroecology may be possible only because of the greater variability in density measurements.

The implementation of densitometry for studies in more extreme sites may not be practical, because the rings from many old trees growing on stressed sites are sometimes too small to be resolved by the densitometer and the additional information provided by densitometric measurements may not justify the high cost of measuring these variations. These problems may be partly circumvented with further development of image analysis techniques (Telewski *et al.*, 1983; Jagels and Telewski, in press), where digitized measurements of reflected light from surfaces of tree-ring specimens may provide estimates of wood density as well as other detailed information on structural variations within annual rings.

In the meantime, some collection teams sample and mount cores in a manner that allows for both ring-width and densitometric work (Holmes *et al.*, 1986). The ring widths are processed and when better equipment and more reliable densitometric procedures become available and financially supportable, these cores can be X-rayed and the film scanned for wood density analysis. With a few modifications, the principles and practices described for ring-width analysis also apply to densitometric analysis. The information that can be obtained from both types of measurements are complementary. Most laboratories are experimenting with densitometric analysis, and the procedures of the Radiodensitometric Laboratory at the Eidgenössische Anstalt für das forstliche Versuchswesen, Birmensdorf, Switzerland are most often emulated.

III. GENERAL PRINCIPLES AND PRACTICES OF DENDROECOLOGY

The "principles" referred to in the following text are essentially the fundamental framework of understanding including the operational procedures that are followed in the discipline of dendroecology. These "principles" are not laws or even necessarily rules of nature but are rather well-tested best inferences based upon known facts at a particular time. Principles are more or less stable over sustained time periods and are revised or overthrown only during periods of scientific revolutions (Kuhn, 1970). The general principles of one field, such as ecology, frequently apply to subfields, such as dendroecology, but they may be viewed and applied in unique ways. The following is a brief description of what we believe to be the most important principles and practices of dendroecology, particularly those that relate to dendrochronological applications.

A. Uniformitarianism

The principle of "uniformity in the order of nature" was first enunciated for geology by James Hutton in 1785. It is commonly stated as: "The present is the key to the past". Its dendroecological implications are (1) the physical and biological processes that link today's environment with today's variations in tree growth were also operating in the past, (2) tree-ring features in a given tree are related to the same environmental conditions in the present as in the past, and (3) environmental conditions associated with present day ring features may be inferred to have existed when that feature occurred in the past. This principle does not imply that environmental or tree growth conditions were all the same in the past as in the present, only that the relationships that govern them must have been the same.

However, a set of unusual conditions in the past could be missed if that set is completely absent from the modern record. This is referred to as the "no analog problem" because there may be no analog in the present to deduce a similar condition in the past. Similarly, there may be no analog in the past for present day conditions. For example, a new condition, such as pollution or a forest management practice, could create a "no analog condition" in that it does not occur anywhere in records of the past. The lack of an analog in the past could prohibit applying some dendroecological inferences to the past from information collected after that condition was introduced. Sometimes, there may be alternative strategies available that do not violate the uniformitarian principle, or will at least minimize the violation's effects.

For example, Bryson (1985) discusses the "no analog problem" applied to paleoclimatic inference and research. His analysis focuses on interpretations of spatial patterns in pollen assemblages and Holocene climate. He points out: "Climate is multidimensional (a vector), not a *single scalar datum*." He suggests that there are so many independent variables in the climate system that probably no *perfect* analog could be found. Partial analogs (involving a large number of correlated as well as causal relationships) may be all that are necessary; and methods such as canonical analysis that consider a large number of variables may provide a good way to transfer proxy data records into estimates of climate. Increased availability of density as well as ring-width information may help to mitigate some "no analog problems". However, in the case of new environmental conditions created by pollution or other effects of man, the absence of a true analog remains a serious problem.

Since Hutton's day geologists and paleontologists have studied the spatial arrangement of Earth's strata with the perspective of modern processes that are now producing similar strata in order to reconstruct the "deep time" chronologies of Earth's history. This method of inference, utilizing layered records and uniformity of process, has been an extremely important tool for reconstructing past Earth history and thereby building our knowledge of this planet and its lifeforms.

The essential point here is that documentation of history, i.e., the correct ordering and placement of events in time, depends on the understanding of process, and *vice versa*. We believe that understanding and exploiting this duality of history and process (or "time's arrow" and "time's cycle", see Gould, 1987) will continue to be a fruitful approach in the study of the geosphere and biosphere. In this chapter we advocate the use of dendroecology techniques where possible for obtaining many new chronologies (histories) of disturbances, climatic change and other environmental changes in forest ecosystems of the world. We also recognize the need for more experimental studies of the ecophysiological linkages between trees as organisms and the environment. This effort will provide the necessary documentation of history that must go in hand with improved understanding of processes and mechanisms toward a clearer picture of our world.

B. Limiting Factors

The well known biological principle of "limiting factors" also leads to sampling procedures important to dendroecology. It is not our intent here to weigh the relative merits or weaknesses of the arguments concerning additive versus interactive effects or whether the concept of single factors acting in successive limitation is a profitable paradigm. We recognize that a variety of factors can influence the complex process involved in ring growth, and each factor may operate in a variety of ways and at different times. For example, the number of tracheid cells within annual rings, their size and wall thickness are often controlled by a variety of limiting factors. However, if one examines individual years in which growth is minimal or sites where a particular ring feature is evident, some kind of limiting condition can usually be found to dominate. For example, the narrowest rings in arid-site and low-elevation trees from southwestern North America can be attributed to climatic patterns in years when drought had been the most probable but not the only limiting condition. For years in which the rings are wider, drought-related conditions are less likely to have occurred and other factors were probably more limiting. The rings with the lowest maximum density are related to climate through the predominance of limiting (low) temperatures (Schweingruber *et al.*, 1978a; Conkey, 1986; Kienast and Schweingruber, 1986).

The relative importance of different limiting factors also varies over space and time owing to micro-environmental changes as well as changes in trees as they grow and occupy different positions in the forest. In spite of these complexities, simplification of the problem is possible. Ecological insights and field observations can be tested by sampling along particular ecological gradients and measuring any changes in ring characteristics. This assumes that there is adequate replication to minimize the confounding effects of factors not associated with the gradient that was sampled. It is also helpful to measure tree growth and other related factors throughout the year, or to manipulate, control or modify the environment to study these relationships. Once relationships are documented in this manner it is possible to make ecological inferences from tree-ring information (Fritts *et al.*, 1965b; Fritts, 1969, 1976; Norton, 1983; and Kienast and Schweingruber, 1986).

Fox *et al.* (1986) use this principle to study pollution effects on ring-width chronologies. They report an increase in mean width and an increase in variance for trees growing downwind from the point pollution source after emission controls had been installed. Presumably, growth increased as pollution levels became less limiting and became more variable as the variations in climatic conditions became more critical.

The specific growth response to environmental factors can vary markedly from one kind of habitat to another. For example, the ring widths of northern hemisphere trees growing at their species' northern or high altitude limit are likely to be affected adversely by unusually cold growing conditions which slow down the growth processes (LaMarche, 1974; Schweingruber *et al.*, 1979; Garfinkle and Brubaker, 1980; Jacoby *et al.*, 1985; Graumlich and Brubaker, 1986). The ring-width growth of such trees located at the southern or low-elevation limits of the species are more likely to be adversely affected by lack of moisture or unusually warm conditions in summer (LaMarche, 1974; Fritts, 1976). Temperature is most likely to become limiting through its effect on evapo-transpiration, tree water potential and water stress. Different species of trees, growing at widely different altitudes, may respond directly to temperature in spring and autumn if the individuals are equally close to the upper margin of their species' habitat range. In addition, one might also expect trees growing at increasing elevations to be decreasingly responsive to low moisture, because evapo-transpiration may become less limiting as the ambient temperature declines with the higher altitude (LaMarche, 1974, 1982; Fritts, 1976; Phipps, 1982). These relationships are by no means simple ones. The Kienast and Schweingruber (1986) paper is a landmark that documents the tremendous variability in growth that can be noted in both Europe and North America at different elevations and exposure associated with variations in macroclimate. More discussion of limiting factors is included under the topic Tree and Site Selection (p. 134).

C. Crossdating

Crossdating refers to both a principle and a practice. The principle refers to the general year-to-year agreement or synchrony between variations in ring characteristics of different trees. This synchrony can be shown to be the limiting effects of variations in climate on tree growth (Fritts, 1976). The practice involves detecting and correcting for any lack of synchrony in ring features when the patterns are out of sequence. This occurs when the ring for a particular year is missing from a sampled radius, when intra-annual latewood bands are indistinguishable from the true boundary of the annual ring, or when there are simply mistakes in counting or in growth-layer identification. Practiced correctly, crossdating assures that all ring features are placed properly in the correct time sequence.

All the various approaches to crossdating involve the following six basic steps:

(1) Visual and statistical comparisons are made of the ring sequence through time involving features such as total ring width, earlywood/latewood width, color, wood density or other differences in structure. It is sometimes useful to plot one or more features, or to use symbols to represent the varying ring characteristics.

(2) If one and only one ring is formed in a year, the chronology of these features is synchronous among trees as long as the count is begun with the same ring. If a ring is absent or two layers of latewood are formed in one year, the chronology from that tree will match the chronology in other trees up to the location of the discrepancy. Beyond that location the chronologies will appear to be one year out of phase. This pattern usually becomes complicated by shifts in the matching of patterns at a number of locations.

(3) Crossdating utilizes the presence and absence of synchrony from different cores and trees to identify the growth rings that may be misinterpreted.

(4) Knowledge of ring structure and chronology is used to correct the sequence by entering zero values for locally absent rings, combining the widths of double rings, and adjusting for miscounting or any other problem with the ring sequence.

(5) The comparison, checking and correcting are replicated many times using different trees. The time period spanned by the rings must be long enough to include some years when climate had been limiting to growth. Crossdating of samples is repeatedly checked, so any prior interpretations of the chronology are verified by the successful matching of all independent ring information from additional samples. Computer programs have been developed that analyze the statistical crossdating of tree-ring series (Baillie, 1982; Holmes, 1983). These programs are most useful for assisting in the dating of samples that have a relatively small number of errors, and they are also valuable as a check of dating accuracy and measurements. No computer program has been able to substitute for the thought processes and judgment required for successful crossdating.

(6) The final step in crossdating is checking the site dating sequence against the independent chronology sequences from trees in other more distant sites. This last step may not be absolutely necessary because of the internal consistency and rechecking that has already been built into the dating of trees within the sample. If enough trees including various age classes are properly crossdated, the likelihood of a mistaken date is so small that for all practical purposes it is zero. If step 6 is included, it serves as an additional confirmation that the chronology is accurate to the exact year. Step 6 should always be used if there is a likelihood that other chronologies have not been dated accurately.

The synchrony in ring features that is used in crossdating is relatively easy to observe in the ring-width patterns of certain conifers growing on arid sites. For example, in south-western United States, approximately 60 to 80% of the ring-width variation or variance is coincidental in most trees of a given species and region if they are growing on similar sites. Even in trees of different species there may be as much as 50 to 70% of the variance in common as long as the trees are subjected to aridity. The similarity in the ring-width pattern is correlated with seasonal variations in macroclimatic factors that are closely coupled with local environmental conditions controlling physiological processes important to ring growth. Thus, the synchrony used in crossdating is evidence that a factor like macroclimate, which varies from one year to the next throughout the region, has affected ring growth. In fact, one can infer how often climatic factors are more or less limiting to ring-width variation from the amount of crossdatable variance (Fritts, 1976).

After the growth rings have been crossdated, the measurements for rings produced in each year can be standardized into chronology values (discussed in the next section), and averaged to show the yearly ring-width variations that are common to the trees of a stand. These measurements also can be related to the environmental or climatic conditions that had originally affected tree growth on the site.

Unusual conditions may sometimes produce characteristic ring features that may not be due exclusively to climate but will enhance the crossdating between trees or sites. Abnormalities of the ring structure caused by early or late frosts (Glock, 1951; Glock *et al.*, 1960) can be recognized and have been used as an independent test of the accuracy of crossdating (LaMarche, 1970; LaMarche and Harlan, 1973). Filion *et al.* (1986) demonstrated that "light rings", characterized by one or a few latewood cell layers, were an indication of shortened growing seasons in black spruce (*Picea mariana* [Mill.] BSP) at the tree-line in Quebec. They found that light rings were useful for crossdating because there was little ring-width variation for that species on those sites, and that low temperatures occurring at the end of the growing season seemed to initiate the light ring formation.

Sudden removal of the overstory will usually be reflected as a sudden increase in growth of all surviving trees (Fig. 1a), while a stand-wide disturbance such as a volcanic ash fall may result in the sudden decrease in growth of surviving trees (Fig. 1b). Such growth effects are usually distinguishable from the effects of climate by the sudden onset of the growth change, the persistence of the effect and the spatial variability in the response of trees. Some caution should be exercised in the use of disturbance events reflected in tree rings for crossdating, because the stressful effects of the disturbance on the tree increases the likelihood of ring anomalies such as ring absence.

Many cases of absent rings have been found in fire history studies (Craighead, 1927; Zackrisson, 1980; Dieterich and Swetnam, 1984). Madany *et al.* (1982) compared fire scar dates that were estimated only by ring counting and matching of fire scars between samples with fire scar dates determined by crossdating the rings of the same samples. Crossdating revealed numerous absent rings, and as a consequence, the dates estimated by ring counting were found to be accurate only 26% of the time.

Other dendroecology studies have clearly demonstrated the necessity of crossdating, especially when studying stressed trees. Marchand (1984) found many cases of absent rings in a study of wave-regenerated fir forests. Evenden (1940), O'Neill (1963) and Swetnam (1987) noted many rings were missing in trees that had been defoliated by insects. Athari (1981) emphasized that precise ring dating is essential in tree-ring/air pollution studies, especially where increment losses are to be calculated. He reported 1,879 missing rings in samples from 328 emission-damaged Norway spruce trees and 119 missing rings from 148 trees from undamaged check stands. There

(a)

(b)

Fig. 1a. A rapid increase in the ring widths in this ponderosa pine follows the removal of competing trees in a 1966 timber harvest (Dieterich and Swetnam, 1984).

Fig. 1b. A rapid decrease in the ring widths of this Douglas-fir specimen from a northern Arizona archaeological site appears to be the result of a volcanic eruption which may have damaged the crown or roots of the tree in 1064 or 1065 AD (Smiley, 1958).

was a maximum of 19 missing growth rings per radii in the damaged trees, while the undamaged trees had a maximum of 6 missing rings per radii.

D. Standardization

The growth potential of the seedling and its capacity to respond to climate change slowly as the seedling grows, matures and attains a dominant position in the canopy of the forest. These changes affect the character of the rings in young trees, creating the well-known ring-width sequences of Duff and Nolan (1953). Some of the techniques of Duff and Nolan were applied to older trees used for dendrochronology by comparing 20-year means of ring characteristics rather than comparing the yearly ring-width values (Fritts *et al.*, 1965a). They show that the rings are widest near the base and central portions of the stem. The ring width decreases with increasing age of the cambium, with increasing height in the young stem, and with decreasing amounts of apical growth. These changes produce a downward trend in ring width and variance that are due to intrinsic factors such as aging and changes in bole geometry. To study changes in the extrinsic environment of trees, the time series of these measurements must be transformed before applying most statistical analyses.

The procedure of standardization is designed to make this transformation. It usually involves the fitting of a curve or straight line to correspond to the average growth potential as it changes over time (Fritts *et al.*, 1969) (Fig. 2a). The curve is data dependent in that its values vary for each year, measured radius, tree, species and site. To correct the series for the intrinsically related decline in mean and variance the ring width is divided by the value of the curve to express width as an index or percent (× 100) of the potential average growth for that year (Fig. 2b). The mean or expected value of the index is 1·0, and the variance of the standardized index is generally stationary through time. With a stationary variance, the indices can be averaged with the indices from other cores and trees to obtain a chronology for the site (Figs. 2c,2d) and the data can be easily analyzed using techniques of time series analysis (Jenkins and Watts, 1968; Meko, 1981; Monserud, 1986).

The lowest possible index value is zero, when the ring is absent, but there is no upper limit. Even with this constraint, approximately 90% of the standardized chronologies from drought-subjected sites in western North America are normally distributed (Fritts, 1976). The normality of indexed chronologies from other types of sites and for other regions has not been examined extensively.

The standardization curve for ring widths is not a simple linear function of the volume growth and changing geometry of the growth layer throughout the tree. Models based exclusively on geometric considerations rarely

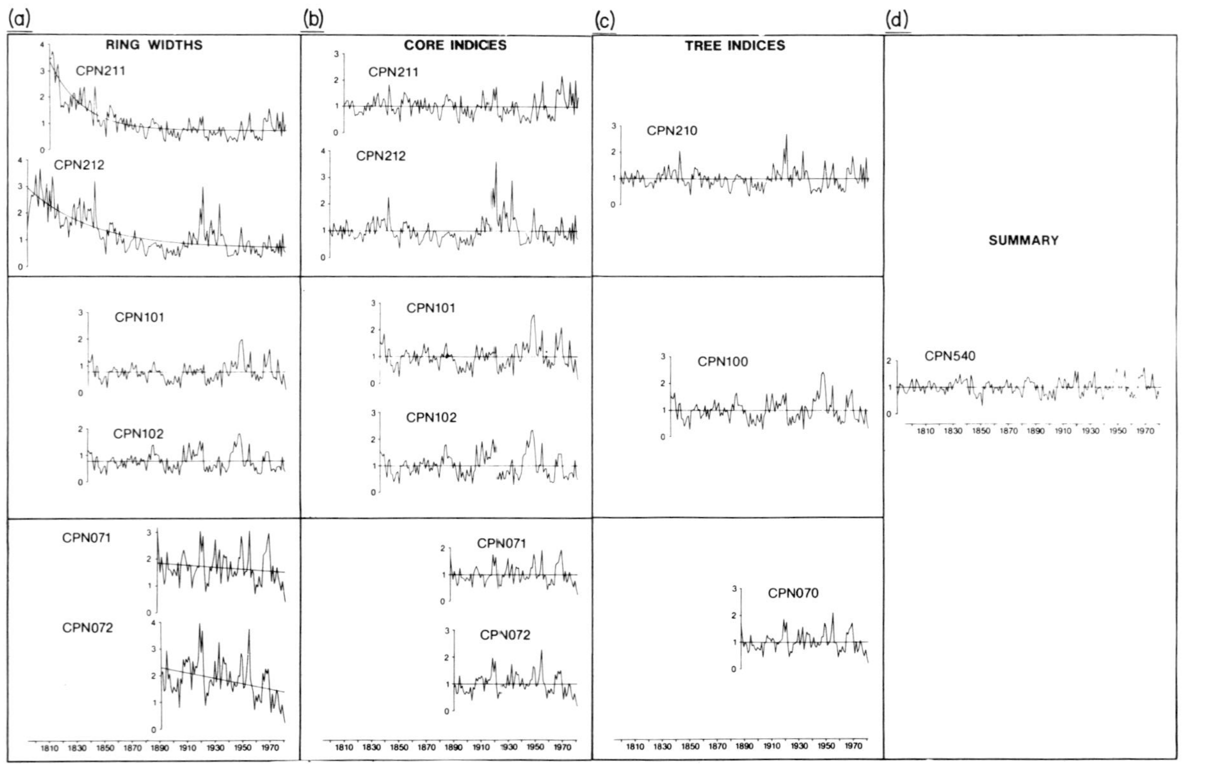

Fig. 2. The dated ring widths are transformed into a standardized chronology by: a. Fitting a curve or straight line to the ring widths from each core, b. Dividing by the values of the fitted curve to obtain the indices, c. Averaging the cores for each tree to obtain the tree indices, and d. Averaging the tree indices to obtain the summary or the site chronology.

linearize the observed changes in ring width over time (Fritts, 1976). In addition, a variety of intrinsic and extrinsic factors associated with increasing tree height, changes in the overstory, the varying proximity of neighbors and other within stand conditions are added to the geometric effects. Flexible empirical models, rather than inflexible physical models, appear to be needed to express these complex changes (Fritts, 1976; Graybill, 1979; Cook and Peters, 1981; Blasing *et al.*, 1983; Cook, 1985). Graybill (1982) and Cook (1987) have constructed what Cook calls a linear aggregate model that distinguishes between the major sources of ring-width variation by considering their inter-relationships and statistical differences. The model suggests several statistical approaches to the study of these sources of variation. A brief description of this model is presented in Section IV with a discussion of its implication in the study of "forest decline" and pollution.

In the past, dendrochronologists have been more interested in studying climatic factors than in studying age or stand-related changes in tree growth, so the changes in growth potential have been estimated primarily to remove them from the ring-width measurements. Only the indices are retained for further analysis. The dendroecologist, however, may have different objectives in mind. Standardization may still be advisable to remove age-related growth changes that contribute unwanted variance. However, other problems and questions may require measurement and assessment of some or all of this variation. In these cases standardization techniques may be modified or an alternative analysis used to preserve the age or growth variations related to stand dynamics for subsequent analysis.

E. Variance of the Mean and the Signal-to-Noise Ratio

After standardization, the indices for all trees from a site are averaged for the cores from each tree and then averaged for all trees in the site (Figs. 2c, 2d) to obtain the mean indices or yearly chronology values. The variance of these values is always less than the pooled population variance of the individual standardized indices making up each set.

The variance of the yearly chronology value can be considered the result of two kinds of influences: large-scale factors operating on the forest as a whole and small-scale factors that act on the individual trees or part of the tree. The first is represented by the variance in common among all trees on a site; the second is the error variance which can be associated with measurement error, growth differences from one side of the tree to the next and growth differences from tree to tree in the same site (Fritts, 1976; Cook, 1987).

Assuming that the ring series are relatively free of the effects of large-scale pollution and exogenous disturbances (see Section IV), the variance in common can be considered to represent the effects of macroclimate or the "climate signal" (s) because it seems to result primarily from large scale climatic variations that have been limiting either directly or indirectly to growth in all trees (Fritts, 1976). Using this same terminology, the remaining more-or-less random error can be considered as "noise" (*n*) which dilutes and perturbs the expression of the "signal" in the chronology for a given site. For example, conifers in semi-arid sites commonly have about 60% or more of the variance in common among trees. This leaves 40% of the variance as noise, so that an individual tree has a signal-to-noise ratio of 60/40 with a decimal value of 1·5.

The amount of variance attributed to the signal can be estimated from the components of standard analysis of variance (Fritts, 1976), but autocorrelation must be considered when estimating the degrees of freedom. Also, the variance of the signal can be approximated by calculating and averaging the correlation coefficients between all possible combinations of trees on the site, excluding the correlations between cores from the same tree (Wigley *et al.*, 1984). An adjustment for degrees of freedom is made by multiplying the average correlation by $(n-1)/n$, where n is equal to number of years in the analysis period. The percentage of signal variance is subtracted from 100 to estimate the average error variance (i.e., information that is not common among trees). An s/n ratio for a tree is converted to the s/n ratio of a chronology by multiplying by the number of trees included in the chronology. In the above example, a chronology made up 10 trees with a 1·5 s/n ratio would have a chronology s/n ratio of $1{\cdot}5 \times 10$ or 15:1. This is a typical s/n ratio for chronologies from western United States semi-arid sites.

For ring widths in the more moist deciduous forests of the eastern United States or western Europe, the percentage of signal in a single radius index series can be as low as 20 to 40%, although judicious sampling and standardization can be used to minimize some of the noise and to increase the percentage of variance in the signal. The s/n ratios for a single radius from the deciduous forest trees would then range from 20/80 to 40/60, which are decimal values of 0·25 and 0·67. However, if the indices from 10 trees are averaged the chronologies would have s/n ratios ranging from 2·5:1 to 6·67:1.

The s/n ratio in a standardized ring-width chronology is dependent in part upon how limiting climatic factors and any other stand-wide conditions have been to growth of the trees during the years that were sampled, and in part upon the number of cores and trees that were averaged to obtain the chronology for a site. Figure 3 illustrates how this relationship can vary in a chronology with changes in sample size over time (Holmes *et al.*, 1986). Four separate 10-year segments have been selected from a well-replicated ring-width chronology from a site named Hager Basin, California. The chronology plots over the four segments are enlarged and shading added to show the standard error around each chronology value. The signal is the variance of the 10 average index values shown in each 10-year segment. The error is calculated by squaring each standard error, multiplying by the sample size,

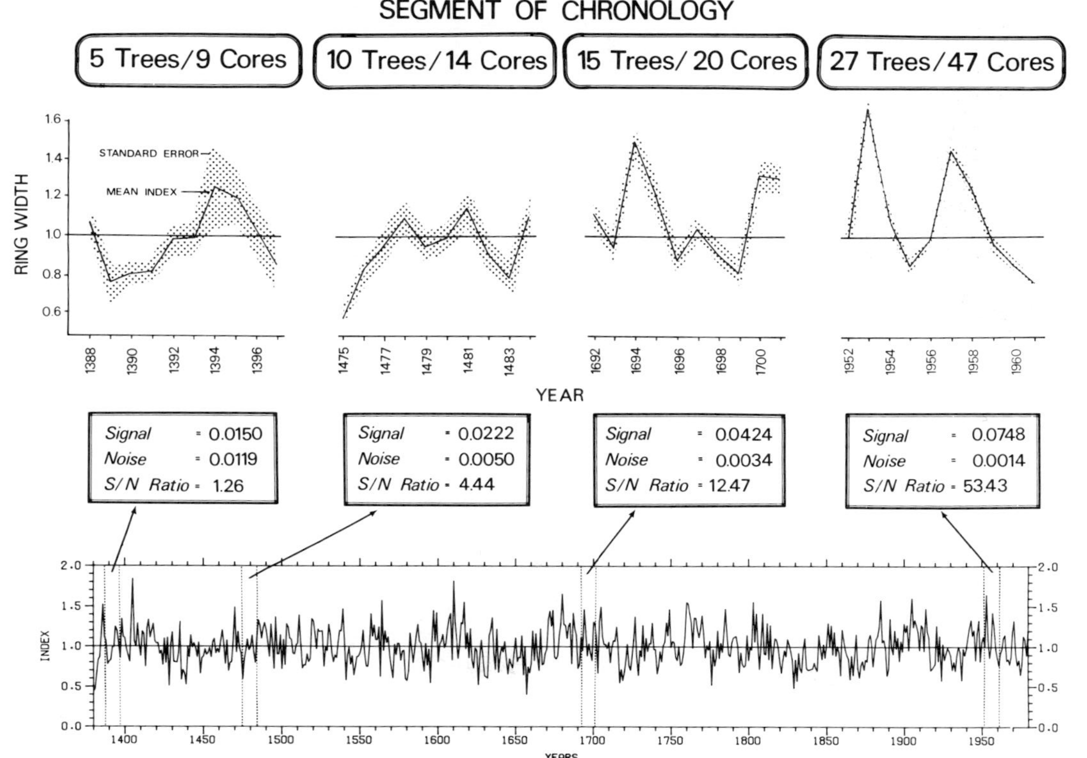

Fig. 3. Index plots for four 10-year segments of a dated and standardized chronology from western juniper in Hager Basin, California. The shading shows the standard error as a departure from each mean index value. The number of trees and cores available for each segment, and the signal, noise and *s*/*n* ratio are shown in boxes.

summing the results over the decade and dividing by 10, the number of years in each segment. The s/n ratios are calculated from these two values using the numbers of trees and cores shown above each segment. To simplify the calculation for the purpose of this illustration, the variance in common within trees is assumed to be identical to the variance in common between trees (Fritts, 1976; Cropper, 1982).

The first segment began in 1380, and measurements from nine cores were standardized and averaged to obtain the chronology. The s/n ratio, estimated in the above fashion, amounts to only 1·26. In the second segment which began in 1475, the indices from 14 different cores were averaged. The error, shown by shading, was reduced and the s/n ratio rose to 4·44. By the year 1692, the beginning of the third segment, 20 ring measurements were available for each chronology value. The error was reduced further with an s/n ratio of 12·47. The last segment began in 1952 and was the average of 47 different measurements. The error estimate is so small as to be barely visible and the s/n ratio is 53·43.

Techniques such as crossdating and standardization influence the s/n ratio, because they reduce the error resulting from incorrect dating and nonclimatic growth features of ring widths (Fritts, 1982). Figures 4a,b serve as an example. Figure 4a shows the standardized mean ring-width chronology for a group of dated cores sampled from trees growing in California. The dating indicated that about 1% of the rings were locally absent from the cores that were sampled. A chronology made up entirely of undated ring widths was simulated by simply deleting all zero values representing missing rings from the dated ring-width measurements and reassigning dates by counting from the outside to the inside rings. These data were restandardized and the indices pooled to simulate an "undated" chronology series.

The dated chronology (with points) and the "undated" series (without points) are superimposed and shown on the same scale in the lower plot of Figure 4a. The uppermost plot is the total number of samples. The next plot is the number of incorrectly dated samples, so the difference between the two is the number of correctly dated samples. The values from the correctly dated set are subtracted from the values of the incorrectly dated set and the differences are plotted in Figure 4b. The chronologies of the counted and dated sets from 1976 to 1980 were almost the same, because no ring absence (missing ring) was encountered over this time period. The small differences that do occur reflect the higher average values of the curves fitted to the incorrectly dated series.

However, the 1975 ring was missing in one out of 27 cases and the 1972 ring was missing in 11 out of 27 cases. The result of these discrepancies is a reduction of the deviations from the overall series average value of 1·00 and an increase in the magnitude of the differences between the two series. As one moves further back in time, more and more absences are encountered, the

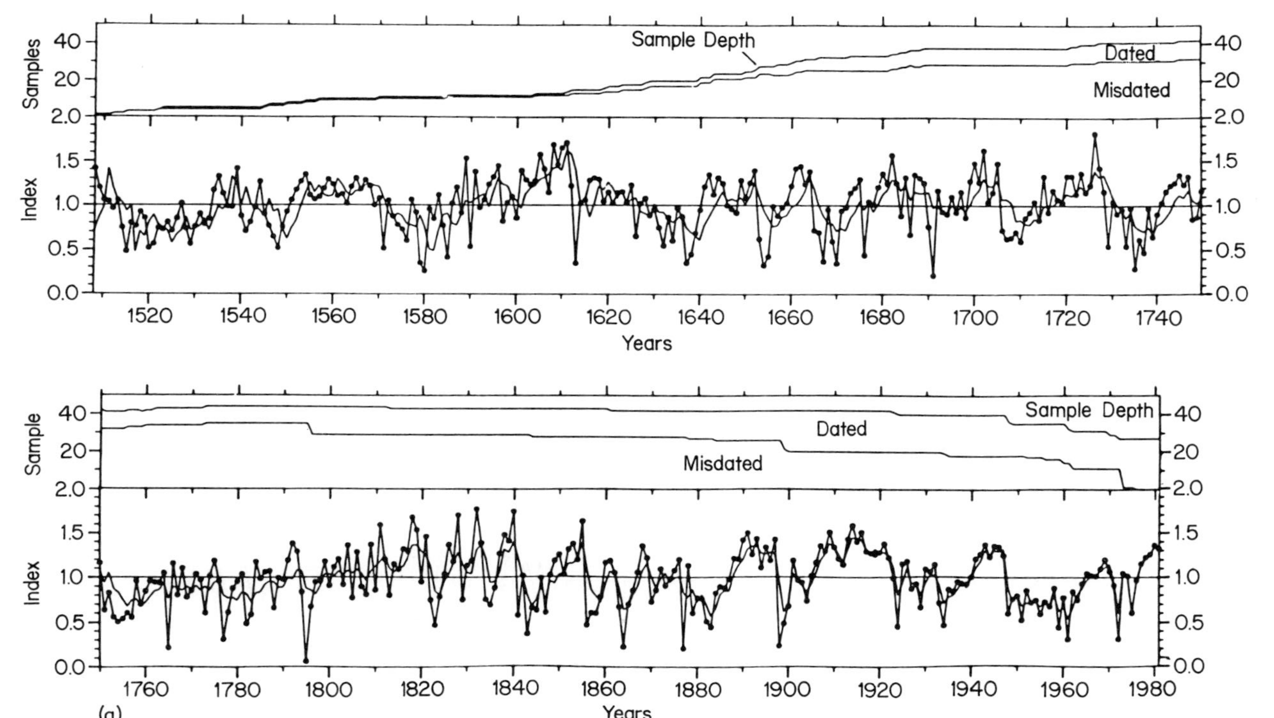

Fig. 4a. A Jeffrey pine chronology from Sorrel Peak, California with ring widths correctly dated (with dots) and incorrectly dated using only ring counts (without dots). The total numbers of samples and the number of misdated samples in the counted set for each year are shown in the uppermost plots. The difference between these two values is the number correctly dated in the counted set.

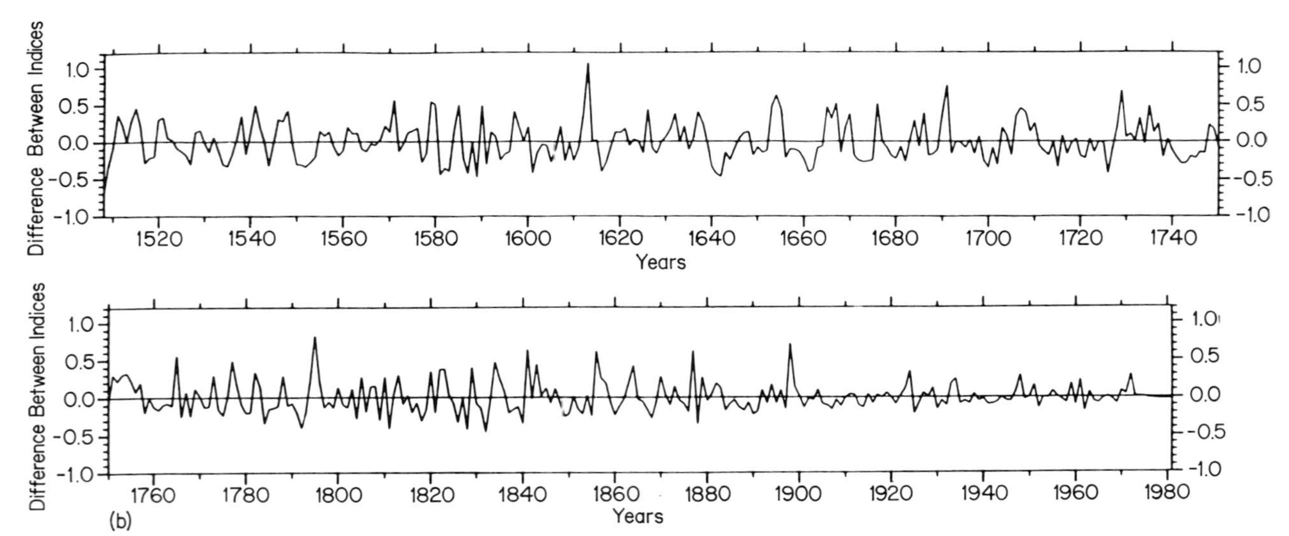

Fig. 4b. The chronology value of the counted set minus the chronology value of the dated set. This difference increases from right to left as the dates on more and more samples become incorrect.

differences become larger, and the ring index average values of the counted series include increasing numbers of rings formed in other years. The net effect is to reduce the magnitudes of chronology variations, especially the very low values where there were missing rings. The peaks and troughs of the undated chronology become displaced several years forward in time (to the right) of the series with the correct dating.

The percent variance in common to all trees after standardization is 57·8 in the dated set and only 27·6 in the counted set. The *s*/*n* ratio for a chronology from 10 trees was estimated to be 13·70 for the dated and 3·81 for the counted ring series. The dated set is not only more variable, but its variations are clearly better related to causal limiting conditions. The variance of the dated series is stable with the passing of time, and can be used to relate 20th-century conditions to those in earlier times (Fritts, 1976; Stockton and Jacoby, 1976; LaMarche, 1978; Hughes *et al.*, 1982; Cook and Jacoby, 1983). On the other hand, the variance characteristics of the counted set are unstable. The variance decreases and the first order autocorrelation increases as more and more rings become mismatched going backwards in time. With this instability comparisons from one time period to the next cannot be made.

F. Sample Replication

Sample replication and the law of large numbers are included here as their consequences in dendrochronological procedures are frequently overlooked. If a large number of trees and cores are crossdated and analyzed in each site so all years have been identified on all specimens (i.e. no rings are missing from the final dated series), it is highly likely that the chronology dating is correct. This is easily checked by comparing two independently dated chronologies from the same geographic area (LaMarche and Harlan, 1973).

The actual number of cores, trees and sites to be sampled for a particular investigation will depend in part upon the strength of the signal in the individual core samples and the strength of the *s*/*n* ratio desired in an analysis. The *s*/*n* ratio and other related statistics have been described in Section E. These statistics can be routinely calculated (Fritts, 1976; Wigley *et al.*, 1984) to help one decide on the adequacy of a particular sample.

Although it is difficult and probably unwise to argue that all studies must attain a given *s*/*n* ratio, this ratio can help one to evaluate data comparability and to decide upon future collection strategies. For example, in the early 1960s before the *s*/*n* statistics had been applied to dendrochronological work, the faculty of the Laboratory of Tree Ring Research launched a program to update and expand their arid-site conifer chronologies. They arbitrarily decided that a collection strategy of two cores from approximately 10 trees in a site appeared to provide adequate replication.

Later, when analysis of variance techniques were applied to the new collections, it was noted that chronologies made up of 10 trees often had a *s*/*n* of 15:1 or more (Fritts, 1976; DeWitt and Ames, 1978). When the same calculations were applied to chronologies from more mesic sites, such as in the eastern United States, it was noted that the signal in the final chronology was weaker. It was estimated that more than 30 or 40 trees would have to be collected and analyzed to obtain an *s*/*n* ratio as high as 15:1. Although they did not recommend that an *s*/*n* of 15:1 was necessary for all applications, the result suggests that the signal should be estimated at an early stage in an investigation, a decision made as to what *s*/*n* was desirable and a sampling strategy developed which would achieve the desired objective.

Cook (personal communications) suggests an alternative sampling strategy for dendroclimatic reconstruction that makes sense in the temperate forests of eastern North America. Instead of increasing the size of each sample to enhance the *s*/*n* ratio, he recommends keeping the same sample size but increasing the total number and density of the sites that were collected. He argues that in eastern North America the effects of local stand competition and other nonclimatic factors are more important than those in western sites. These factors are more likely to be retained as part of the signal in a given eastern North American site chronology, even if a large number of trees are sampled, because they arise from stand-wide disturbances. Assuming that these effects are more likely to be random between sites than within sites, an increase in the replication of sites, rather than trees within sites, would provide the best estimate of macroclimatic conditions. The same strategy seems reasonable for dendroecological studies. A certain amount of replication is needed not only for enhancing the signal of the chronology but also to provide an adequate number of specimens for crossdating.

G. Tree and Site Selection

Tree and site selection is an extension of the principles of limiting factors and sample replication. The choice of trees and sites for a dendroecological problem requires an understanding of the micro-environmental relationships affecting growth in the landscape being investigated in order to visualize and hypothesize how the rings may vary in response to changes in limiting factors (LaMarche, 1982). Then the effects of a particular control factor can be investigated by selecting trees from a number of "target sites" (LaMarche, 1982) along an environmental gradient upon which that control factor is thought to vary (Fritts *et al.*, 1965b; Fritts, 1969; LaMarche, 1974; Norton, 1979, 1983; Kienast and Schweingruber, 1986). However, the replicated trees within each target site must be sampled very carefully to minimize any differences in the ring patterns due to the control factor. An attempt should also be made to minimize the effects of as many other factors as possible by using within-site replicates from trees of the same species, age and stand history.

The selection and replication of trees and sites are simply a stratification of the sample to minimize the noise and maximize the common signal in the site chronology, so that subtle differences in the chronologies can be statistically tested. Based on these selection criteria, sampling can be considered essentially "random" within the bounds of the target population representing a specific set of environmental conditions (LaMarche, 1982).

Studies addressing different ecological issues may require different stratifications of sites and a more formal sampling design. However, if it is a dendroecological problem, attention must be given to the age of the trees, the importance of limiting conditions that provide crossdating and the homogeneity of forest stands that are to be sampled. A dendrochronologist working in semi-arid southwestern North America who is interested in climatic reconstruction, can recognize the oldest climate-stressed trees at a distance by the flat or spike-topped shape of the crown and the large size of the lateral branches, and he would sample the trees accordingly. A similar strategy may or may not be required in an ecological or climatological investigation elsewhere, depending upon the question to be investigated. The following studies provided new information and perspective on different ecological problems by examining trees along particular gradients using a variety of sampling strategies.

Morrow and LaMarche (1978) studied the effect of insect herbivory on ring-width and related phenomena. They stratified ring series from insect-infested trees before and after treatment with insecticides, and they also compared ring series of treated and untreated trees and examined the differences in ring growth.

Brubaker and Greene (1979), Ferrel (1980) and Swetnam *et al.* (1985) investigated the effects of insect infestation on forest growth. They sampled species that were susceptible and unsusceptible to the insect and looked for differences in ring character associated with periods of known insect outbreaks.

Fox *et al.* (1986) studied the effects of a point pollution source on ring widths by obtaining replicate samples of trees at increasing distances down wind from the pollution source. Several stands with no obvious pollution source were sampled for controls. The study also involved differences in the ring-width chronologies for years before pollution, after pollution and after pollution controls were installed.

Schweingruber *et al.* (1983) studied ring characteristics of trees growing at increasing distances from a pollution source. By studying a large number of trees and noting site characteristics and related conditions, they were able to evaluate differences in the pollution effect associated with tree age, with soil acidity and with elevation of the tree sites as well as with distance from the source.

Marchand (1984) compared replicated samples of older balsam fir stands (with an average age 74 years) to replicated samples of younger stands (with an average age 35 years) to determine whether wave-mortality (dieback) was age related.

Payette *et al.* (1985) compared the rings from living trees to those from trees that had died at some earlier time period. They inferred from differences in ring structure, as well as from differences in age distributions, that climate had been different in the earlier time period.

H. Calibration and Verification

If the *s*/*n* ratios in tree-ring chronologies are large enough, and the chronology values are linearly related and well correlated with one or more environmental factors, it is possible to derive a relatively simple equation using least-squares techniques that rescales the chronologies in terms of the correlated environmental factors (Lofgren and Hunt, 1982; Blasing and Duvick, 1984). The variations of past growth can be substituted in the equation to estimate statistically the variations in past environmental factors.

The statistical procedure of deriving the coefficients of the equation is called "calibration". Tree-ring data and climatic data are compared over an interval of time called the calibration period. The coefficients are unique to the data in the calibration period and independent of any observations not included in the period (Gordon, 1982). Therefore, it is possible that even a well-calibrated equation may produce unreliable reconstructions when it is applied to data independent of the calibration.

The reliability of a calibrated equation can be tested by withholding some of the observations on the variable to be reconstructed when making the calibration, and then using these independent observations to test whether the reconstructions for those particular years are correct. This procedure is called "verification". An array of verification statistics (Gordon and Leduc, 1981; Gordon, 1982; Fritts *et al.*, in the press) can be calculated to estimate the similarities between the independent estimates and the observations that were withheld from the calibration. These statistics are tested for significance to ascertain whether the independent reconstruction is better than would be expected solely by chance variations.

A simple correlation, however, is usually an inadequate model for calibration because the tree-ring chronologies are noisy records of a complex set of interacting climatic and environmental factors (Fritts, 1976). One solution is to calibrate a chronology using least-squares techniques in a multivariate regression model that statistically estimates the chronology value from a number of predictor climatic variables such as monthly precipitation and temperature. The sign, size and significance of the partial regression coefficients obtained in the calibration are examined and interpreted as a tree-ring response to the monthly climatic factors (Fritts *et al.*, 1971; Fritts, 1974; Guiot, 1986). This solution is called a response function.

Another solution is to use a number of chronologies in a multivariate regression model to predict statistically one or more climatic factors. In this solution the values of the chronologies are calibrated with the climatic factors, and the resulting equation is used to reconstruct past climatic variations from the variations in past tree growth (Fritts *et al.*, 1971; Blasing, 1978; Lofgren and Hunt, 1982). This solution is called a transfer function, which Bryson (1985) defines as "an empirically determined quantitative relationship between a proxy 'data vector' and a climatic 'data vector'". In the case of tree rings, the variance of the chronologies is transferred to statistical estimates of climatic variance.

Several attributes of tree-ring chronologies and climatic data complicate the form of the equation used in calibration. The predictor variables may be so colinear that the reliability of the coefficients may be underestimated. This and related problems with the response function can be partly resolved by using special types of multiple regression (Cropper 1985; Marquardt and Snee, 1985). ARMA modeling techniques (Jenkins and Watts, 1968; Meko, 1981; Richards, 1981; Monserud, 1986; Cook, 1985; Graumlich and Brubaker, 1986) also provide alternative procedures. Guiot (1986) and Fritts (in press) use both univariate and multivariate ARMA modeling along with canonical analysis to accomplish calibration (Fritts *et al.*, in the press).

Often a certain amount of tree-growth variance may be attributed to a lag in growth of one or more years behind the occurrence of climate because of autocorrelation and other types of persistence in the tree-ring chronologies. These effects can be modeled by (1) lagging the tree-ring series one or more years behind the climate occurrence in a transfer function (Fritts, 1976), (2) considering the size of the prior ring as well as climate as a predictor factor in the response function (Fritts *et al.*, 1979), (3) using variable transformations (Fritts and Gordon, 1982), and (4) using techniques of time series analysis (Meko, 1981; Jones, 1985; Guiot, 1986). The last techniques efficiently handle time-series properties and can be combined with various multivariate models to obtain reconstructions of both single points representing climatic stations or grid points and arrays of points portraying spatial surfaces (Briffa *et al.*, 1988b).

Calibration and verification procedures can be applied whenever tree-ring or climate relationships are important contributors to the tree-ring variance. They can be used to derive an equation for extending a relatively short record of instrumental climatic data, drought severity indices or hydrologic information, from a study area, backward in time (Fritts, 1976; Hughes *et al.*, 1982; Brubaker and Cook, 1983; Stockton *et al.*, 1985). Observations on environmental and other data, which may be correlated with the relationships related to ring growth, can also be correlated with and reconstructed from the chronology variations (Stockton and Fritts, 1973; Clark *et al.*, 1975). Separating and removing the effects of climatic factors on ring growth from the effects of other factors can also be done (Nash *et al.*, 1975; Swetnam *et al.*, 1985).

It is not possible to recommend a particular technique for all calibrations. Time series analysis is currently in vogue but its application to reconstructing large spatial grids has yet to be tested and evaluated. Simulation techniques like those of Cropper (1985) can be used to generate time series with a known signal of climate. The various calibration procedures can then be applied to these data to see how well they can recover the climatic information.

I. Modeling

As in other disciplines, it is helpful to construct models that depict hypothesized physical or physiological interrelationships. Models not only provide hypotheses to be tested; they can be used to make projections or calculations of conditions resulting from changes in the model inputs (Cooper *et al.*, 1974). Horton and Bicak (1987) add, "Models provide a framework for organizing knowledge as it is acquired. This framework is particularly convenient since, as knowledge expands, the number of attributes in a model can be increased without a change in the fundamental form."

Early attempts at dendrochronological modeling took the form of diagrams. Figure 5 includes examples that model some of the interacting relationships linking temperature and precipitation with ring-width variations (Fritts, 1976; Koerber and Wickman, 1970; Stockton, 1971; Young, 1979). Other models are more quantitative ranging from energy balance relationships within the tree (Brown, 1968; Gates, 1980) to simulations of the response function relationships (Cropper, 1982).

Modeling of cambial activity for conifers, which was introduced by Wilson (1964) and Wilson and Howard (1968), was adapted to dendroecology by Stevens (1975). The model simulates daily cell growth and differentiation in a single radial file of cells which ultimately forms a tree ring. The growth processes are controlled by 22 model parameters which are the only inputs to the program. Variations in limiting environmental factors must act on these inputs to affect ring structure. Fritts (in press) is attempting to model these linkages, and also has developed a stochastic model that generates a number of time series representing limiting conditions in different seasons. The tree-ring responses to these time-series are weighted averages, and random variance is added to simulate the error component. ARMA processes, the growth trend, and other features can then be introduced to simulate the ring response to various forest influences.

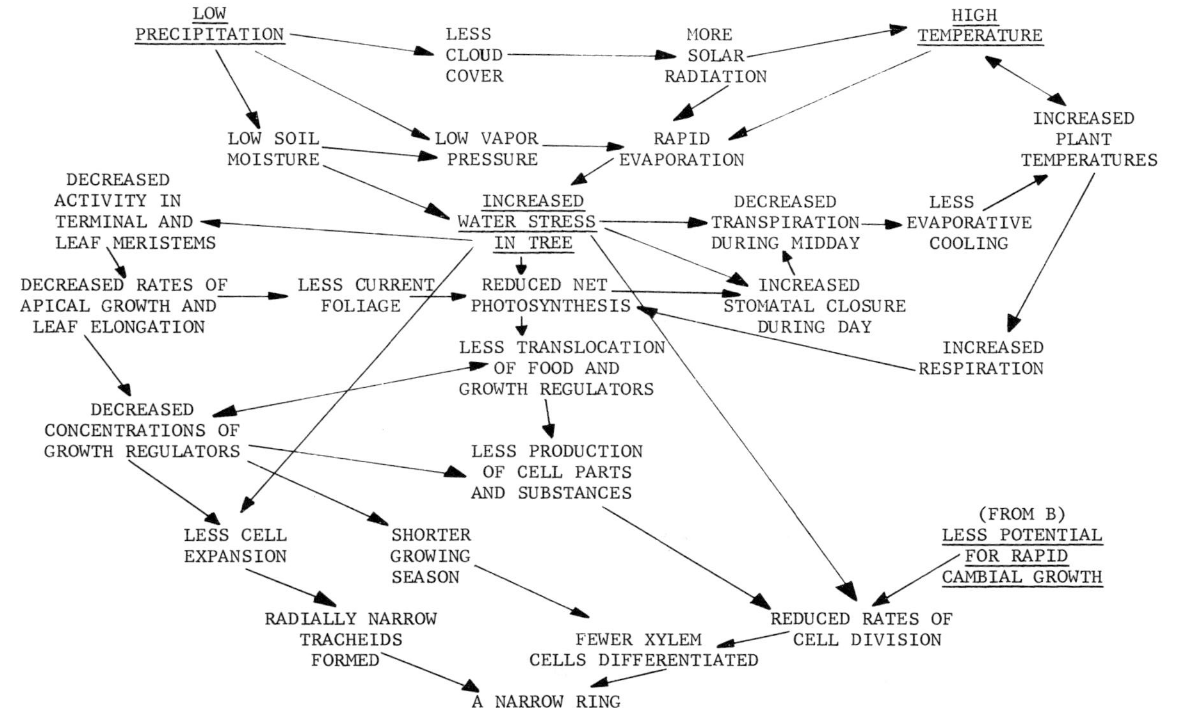

Fig. 5a. Model diagram representing some of the relationships that cause climatic factors of low precipitation and high temperature during the growing season to lead to the formation of a narrow ring in arid-site trees. Arrows indicate the net effects and include various processes and their interactions. It is implied that the effects of high precipitation and low temperature are the opposite; that is, ring width will increase (from Fritts, 1976).

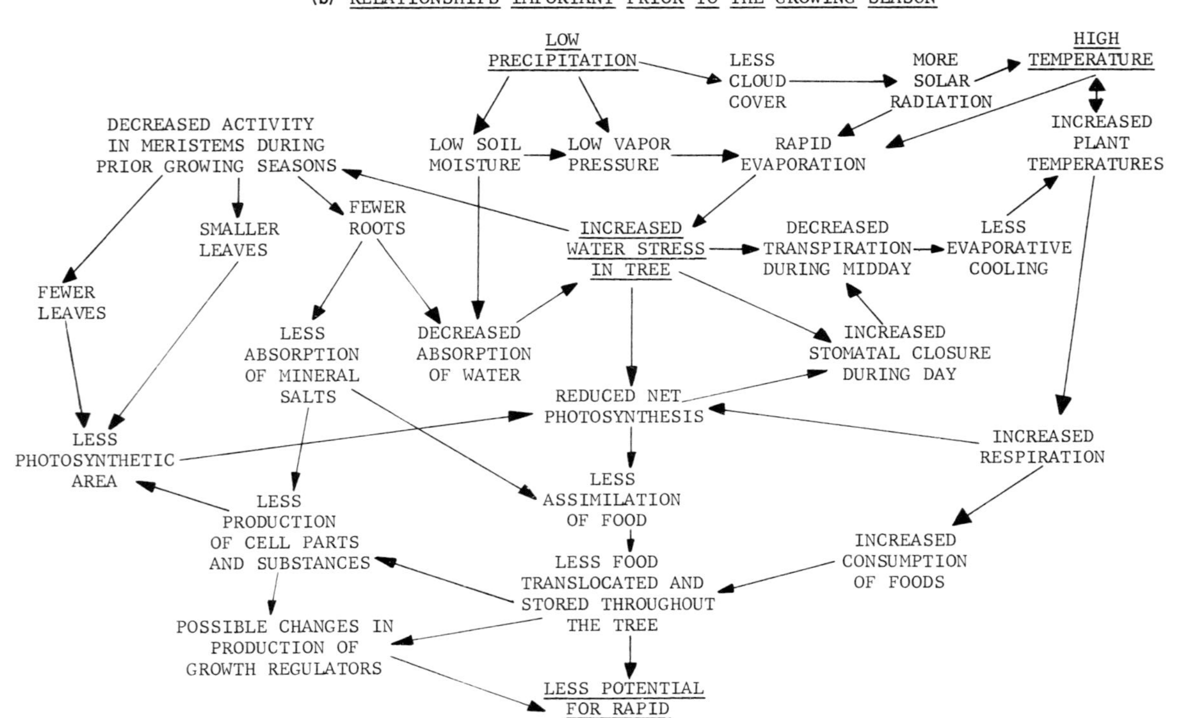

Fig. 5b. Model diagram representing some of the relationships that cause climatic factors of low precipitation and high temperature occurring prior to the growing season to lead to the formation of a narrow ring in arid-site trees (from Fritts, 1976).

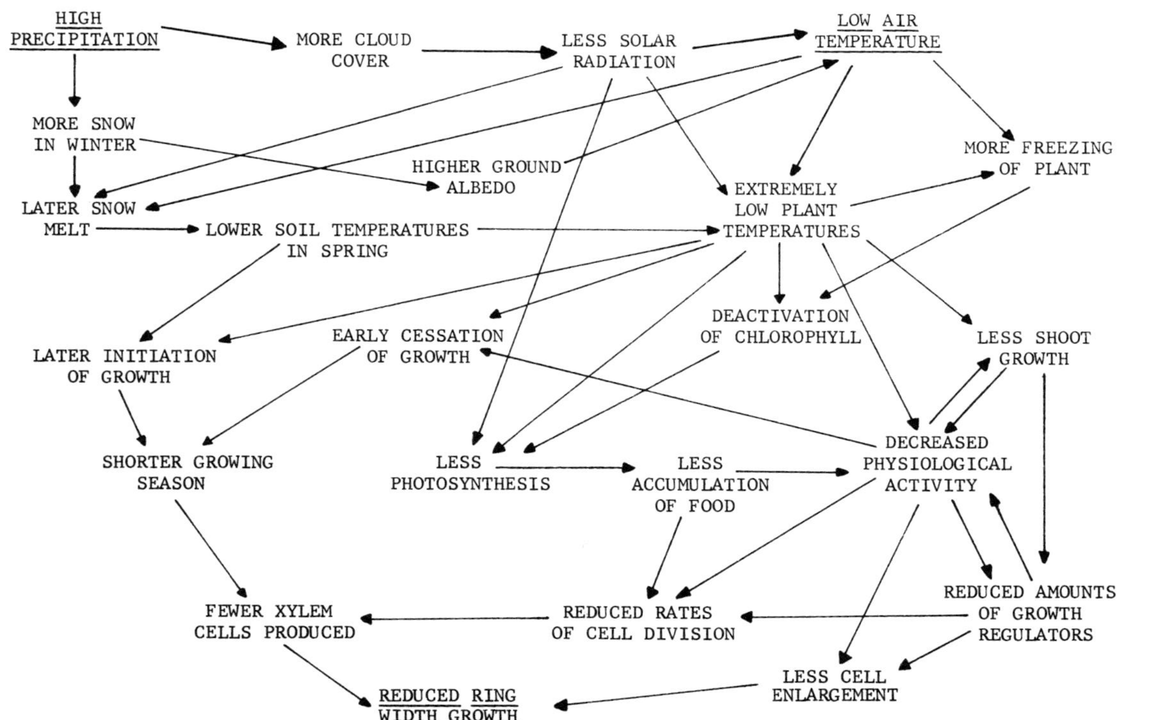

Fig. 5c. Model diagram representing some of the relationships that cause the climatic factors of high precipitation and low temperature to lead to the formation of a narrow ring (from Fritts, 1976).

Another model, developed by Zahner and Myers (1986a, 1986b), calculates the impact of soil water deficits on the radial growth of southern pines in the Piedmont region of the southeastern United States. Daily water deficits affect daily radial growth, and thus account for the major environmental stress acting on southern pine growth. The model adjusts annual ring widths upwards to the long-term mean during periods of severe water stress and downwards during years of mild water stress. Zahner and Myers (1987) applied this model to adjust ring widths measured on over 2000 increment cores and were able to use the adjustments to test whether there has been a significant decline in ring width over time independent of soil water deficits in the Piedmont region. They found a decline in ring width of approximately two-thirds of the annual growth increment for equivalent trees 36 years earlier that was associated with an undetermined stress factor.

A model with four different controls of ring-width growth was proposed by Graybill (1982). The controls include (1) a macroclimatic signal common to the trees on a site, (2) a biological growth curve of variable form, (3) tree disturbance signals including disturbance resulting in growth differences among the sampled trees as well as disturbance resulting in similar features of growth in all trees, and (4) a randomly varying component analogous to error. Section IV describes a modification of this model to include six controls (Cook, 1987) which was applied to the forest decline problem.

IV. A LINEAR AGGREGATE MODEL OF RING-WIDTH MEASUREMENT

The dependent variable in this model (R_t) represents a ring-width measurement in the radial direction along a cross-section of a tree stem. Each ring is assumed to have been accurately dated in year t. Ring width is expressed as an aggregate of 6 basic components of the equation:

$R_t = G_t + C_t + aD1_t + aD2_t + aP_t + E_t$ where

G = the age-related growth trend value in year t that is shared by that species growing on that type of site;

C = the climatically related growth variations common to a stand of trees in year t including the mean persistence of these variations due to physiological preconditioning and interaction of climate with site factors

$D1$ = the endogenous disturbance pulse originating from competition, changing tree stature and other forces acting on specific trees in year t

$D2$ = the exogenous disturbance pulse originating from forces outside the forest community such as that due to an ice storm, a late frost or an insect infestation in year t

P = the variance due to anthropogenic pollutants that have stand-wide impact on radial growth in year t

E = the series of more-or-less random variations representing growth influencing factors in year t unique to each tree or radius, including error in the ring-width measurement.

The a associated with $D1_t$, $D2_t$ and P_t is a binary indicator of the presence ($a=1$) or absence ($a=0$) of a subseries in R for some year or a group of years. There is an assumption of linearity and independence among variables that is not strictly true, as most environmental variables are highly interrelated. In spite of this oversimplification, the model does provide a useful framework for identifying and separating the different types of influences.

Component G_t corresponds to the slowly changing ring-width variations that are commonly modeled and removed by standardization. Cook (1987) states, "The growth trend, G_t, is a non-stationary process that arises, in part, from the geometric constraint of adding a volume of wood each year to a stem of increasing radius." For many open grown trees, where competition is a minor factor, the trend in radial growth associated with G_t can be adequately modeled and removed by deterministic mathematical models fitting a variety of curves (Fritts *et al.*, 1969; Fritts, 1976; Graybill, 1979, 1982). Cook continues, "Unfortunately, the growth trends of trees growing in closed-canopy forests are usually very complex and stochastic because of disturbances and competitive interactions within the forest. Therefore, G_t must be generalized to allow for a variety of linear and curvilinear growth trends of arbitrary slope and shape."

According to Cook (1987), the input that produces C_t reflects certain broad-scale meteorological variables that directly or indirectly limit the growth processes of trees in a stand. These variables are assumed to be uniformly important for all trees of a given species and on similar sites. "As a common signal in the ring widths of all trees, C_t could be mistakenly identified as a pollution signal if the recent behavior of C_t mimics the expected pollution effect on ring widths. Thus, the effects of climate on ring width must be carefully modeled and removed before a pollution effect can be inferred" (see Zahner and Myers, 1987).

Cook (1987) goes on to explain that $D1_t$, the endogenous disturbances,

> ... are caused by factors related to characteristics of the vegetation that are independent of the environment (White, 1979). Such disturbances occur when dominant overstory trees senesce, die and topple as a natural consequence of competition, aging and stand succession. In the context of searching for a pollution signal in tree rings, endogenous disturbances can be expected to occur randomly in space and time in forest communities. Thus, the loss of a dominant tree in one section of a stand is not likely to be related temporally or spatially to similar losses at other locations in the stand. This property suggests that endogenous disturbance pulses in tree rings will rarely be synchronous among separated trees in a stand except by chance alone. Thus, the lack of synchrony

in ringwidth fluctuations between trees during a hypothesized pollution effect period may be used as evidence for rejecting the presence of a pollution signal.

Exogenous disturbances are caused by natural environmental forces that lie external to and are independent of the vegetation (White, 1979). Unlike endogenous disturbances, these disturbances can affect large areas of forest. Some of the important causal agents are fire, windstorms, ice storms, frost damage, disease and insect infestation. Since the areal extent of an exogenous disturbance can be great, the resultant disturbance pulse, $D2_t$, may occur contemporaneously in virtually all trees in a stand. This property presents obvious difficulties for differentiating a pollution-caused ringwidth decline from that caused by a natural exogenous disturbance. Historical documentation of exogenous disturbances in forests may be needed to determine the presence or absence of this confounding source of variance.

The pollution signal, P_t, is assumed to be common to all ringwidth series in the sampled trees of a stand. This assumption could be criticized for being too restrictive in requiring a pollution effect on all sampled trees. It could be argued, for example, that the crowns of canopy trees "scrub out" wet and dry atmospheric pollutants before they reach understory trees. If this were the case, then the understory trees might not show a pollution effect. This possibility indicates the need for the stratified sampling of trees based on *a priori* criteria such as crown class or canopy position.

Hopefully $D2_t$ and P_t can be distinguished from C_t because they have more persistence. If their time-series characteristics are similar to those for climate, it may be necessary to model and remove as many climatic effects as possible before study of these non-climatic factors. It is self-evident that methods of standardization that might remove some of the variance of $D2_t$ and P_t could not be used if these factors were the object of the investigation.

Component E_t is the random variance in the ring-width series due to such variables as localized responses to micro-environmental factors, variations around the circuit of a ring and measurement errors, and is assumed to be unrelated to the variance accounted for by the other components. In addition, it is assumed to be serially uncorrelated within each tree and spatially uncorrelated within the stand of trees. The usual way to reduce this random variance is through replicate sampling and averaging after standardization (Fritts, 1976).

The above model is only descriptive at present. It uses statistical characteristics of ring widths or standardized time series to identify the source of the variation and does not consider features other than ring-width variation. As our understanding of ring features including cell size, density of the wood and chemical composition improves, it may be possible to make the model more quantitative and to add features other than ring width, which should increase its capability of discriminating between sources of variation.

V. DENDROCHRONOLOGICAL APPLICATIONS TO SPECIFIC ENVIRONMENTAL ISSUES

We have described the field of dendroecology by listing the nine most important principles along with illustrations of various applications. The most distinctive feature of all dendrochronological studies is the requirement that the tree-ring sequences must be dendrochronologically crossdated. This is the only method by which the integrity of the dendrochronological time series can be assured because the procedure automatically validates the results whenever two time series that were independently dated are compared. Most dendroecological methods assume that the annual measurements are correctly identified as to the year in which they belong. Thus, dendroecology provides an exact time control as well as a historical perspective to ecological investigations.

Assuming the above, we will direct our attention to three environmental issues that have been given much public attention: insect infestations and forest growth; the forest decline problem; and ecological questions of climatic variability and change over periods of years to centuries.

A. Spruce Budworm Effects on Forest Growth

Spruce budworms (*Choristoneura* spp.) are considered to be the most destructive forest pests in North America (Sanders *et al.*, 1985). Forestry records indicate that outbreaks lasting 10 to 20 years have recurred several times in the last century in North American forests. Substantial losses to timber resources have stimulated intense research efforts in the last decade to understand better the ecology and impacts of these insects (Sanders *et al.*, 1985). Tree-ring studies have played an important role by providing a long-term historical perspective of this episodic phenomenon. In addition to documenting growth impacts, which are necessary for the formulation and justification of forestry management strategies, tree-ring reconstructions of budworm history have been useful for improving our understanding of the dynamics of insect populations and forest stands, including the interactions of climate and human activities. The following dendroecological investigation provides an example of one approach for reconstructing past budworm history and impacts and the ecological implications of such tree-ring derived disturbance chronologies.

Eleven recently defoliated mixed conifer stands in the southern Rocky Mountains were sampled along a north–south transect from northern Colorado to northern New Mexico (Swetnam, 1987). The working hypothesis was that a recognizable exogenous disturbance component in host trees (Douglas-fir, *Pseudotsuga menziesii* [Mirb.] Franco., and white fir, *Abies concolor* [Gord. and Glend.] Lindl.) could be used to infer past insect outbreaks. Response function analysis had shown that a large and similar climatic component can be found in Douglas-fir and ponderosa pine (*Pinus ponderosa* Laws.) growing on similar semi-arid sites (Fritts, 1976). Therefore, the standardized ring-width chronology in ponderosa pine, which is a non-host species, could show the variations in the climatic component without the variations due to spruce budworm infestation. If graphical and statistical comparisons of the host and non-host series, in conjunction with forestry records of budworm outbreaks, showed that the above hypothesis was reasonable, then variations in the difference between the host and non-host chronologies could be used as an index of budworm outbreaks.

The eleven stands were originally chosen by the United States Forest Service for monitoring budworm populations and impacts and were generally considered to be representative of areas currently defoliated by budworms. These stands included a fairly broad range of topographic positions, elevations (2400 to 3000 meters), soil moisture conditions, and mixtures of conifer tree species (Swetnam, 1987).

Cores were obtained from at least 15 randomly selected mature host trees in each of the eleven stands (more than 40 trees were sampled in five of the stands). Two replicate samples were extracted from each tree. Five to ten of the oldest host trees within the stands were also sampled to ensure that the collection included the maximum possible record length. An additional 15 to 20 non-host trees (ponderosa pine) were collected within and near the study plots for comparative purposes.

All of the Douglas-fir and ponderosa pine ring-width series were cross-dated. Crossdating revealed that absent rings were relatively common in the cores from host trees, with a maximum of 30% of the cores from a stand having one or more absent rings. In nearly all cases the absence of annual rings in host trees were observed to be associated with low growth during periods of known or inferred budworm outbreaks. Absent rings were also detected in non-host trees, but they were more evenly distributed throughout the ring series and were often associated with drought years.

The Douglas-fir and ponderosa pine chronologies from one of the 11 stands are shown together in Figure 6a. The general correspondence between the year-to-year values in the plot of the host chronology (line with triangles) and the non-host chronology (line without triangles) was characteristic of most samples.

The average correlation between the host and non-host chronologies was 0·65. Additional statistical comparisons of the host and non-host chronologies included climatic response function analysis and spectral analysis (LaMarche, 1974; Fritts, 1976). These analyses generally confirmed the working hypothesis that the chronologies from both species had similar climatic signals, especially in the higher frequencies.

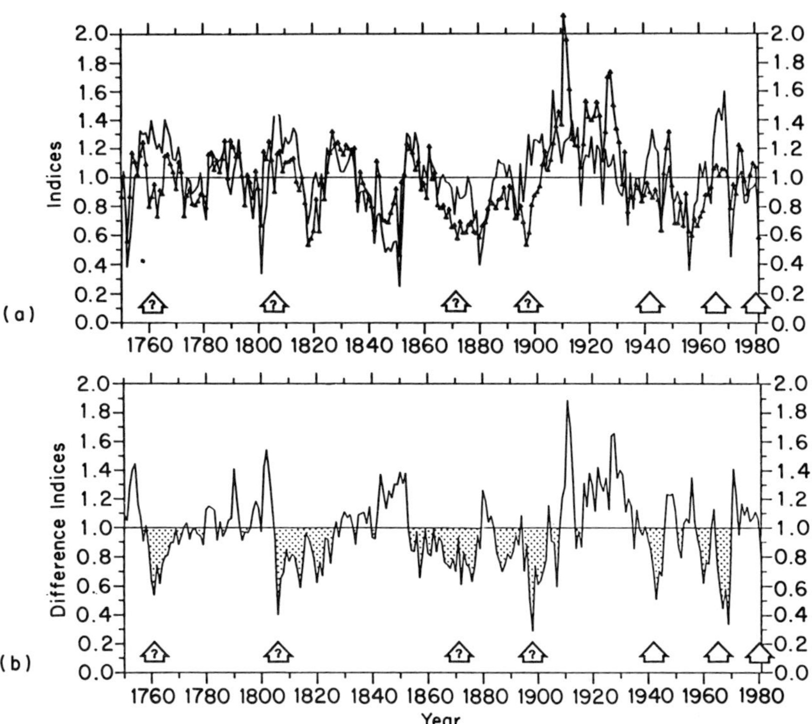

Fig. 6. Comparisons of ponderosa pine (lines without triangles) and Douglas-fir (lines with triangles) chronologies from a northern New Mexico spruce budworm study plot. (a) shows the standardized chronologies and (b) shows the difference series derived by rescaling the ponderosa pine chronology (non-host series) and subtracting it from the Douglas-fir chronology (host series). The arrows with question marks indicate periods of low growth in the Douglas-fir that are inferred to be records of past budworm outbreaks. The open arrows are known periods of budworm outbreak.

The climatic signal in the host chronology was then removed, or at least reduced, by computing the "difference" chronology (Figure 6b). This chronology was computed by subtracting a rescaled version of the non-host chronology from the host chronology (Nash *et al.*, 1975; Swetnam *et al.*, 1985). There are three periods of known budworm outbreak after the 1900s (shown by open arrows in Figure 6). The first two outbreaks are visible as low growth periods (difference indices less than 1·0) during the early 1940s and 1960s. The duration of these low growth periods, and the years of lowest growth, correspond very well with historic documentation of budworm defoliation in this area. During the most recent outbreak, defoliation of trees did not exceed 10% of current years foliage until 1980, and a growth reduction (index value less than 1·0) does not appear until 1981, the last year sampled.

The positive differences following several of the inferred outbreaks reflects an increased growth in the host trees that is not matched in the non-host trees. This phenomenon is probably due to opening of the stand through budworm induced mortality or thinning of crowns of competing trees. The non-host trees were sampled from a pure ponderosa pine stand growing nearby, so these trees would not have benefited from mortality or thinning of host trees. The positive differences may also be a result of other unidentified systematic differences in the growth response of the host and non-host trees. However, analysis of the ring-width series and age class data from this stand provides strong evidence that the inferred outbreak in the 1890s resulted in considerable mortality, followed by growth release of survivors (thus, the large positive differences in the 1910s and 1920s) and establishment of a younger age class of trees (Swetnam, 1987).

The results of this study indicate that as many as nine spruce budworm outbreaks have occurred within southern Rocky Mountain mixed conifer forests since 1700. The average duration of budworm induced low growth periods was about 13 years and the period between outbreaks (period between initial outbreak years) was about 35 years. Although no apparent change in the frequency of outbreaks was observed, the pattern of outbreaks was noticeably different in the twentieth century compared with earlier periods. When chronologies were compared among all sampled stands the timing of initial and maximum growth reduction years of outbreaks was relatively non-synchronous prior to 1900, but they were markedly more synchronous during the twentieth century. This evidence suggests that spatial and temporal pattern of pre-1900 budworm outbreaks was relatively patchy, while post-1900 outbreaks were more widespread or coincident in time among the stands.

One explanation for these results may be derived from the perspective of "patch dynamics" (Pickett and White, 1985) and the recognition that disturbance regimes, forest structure, and forest dynamics are intimately linked. For example, consider the following historical observations: the structure (age distribution, and species composition) of mixed conifer forests in many areas of the southern Rocky Mountains has been altered by human activities, especially through fire control and timber harvesting (Peet, 1981; Veblen and Lorenz, 1986). Elimination of periodic surface fires has led to increased stand density and multi-level canopies through establishment of tree seedlings that otherwise would have been killed by surface fires.

Timber harvesting has primarily involved removal of the non-host pine species, which has led to stands composed of a larger host component than was present in pre-1900 forests. Dense stands dominated by host species are known to be highly susceptible and vulnerable to budworm infestation. Thus, the general pattern of development of these forests has been toward greater homogeneity across the landscape. This is not to say that the age structure or species composition is necessarily less heterogeneous within stands, but that stands throughout the region are more similar in structure to each other than they were before the settlement era. Owing to widespread tree establishment, stands are now also more closed and continuous. In contrast, the structure and spatial distribution of pre-1900 forests was almost certainly more patchy. Indeed, the patchiness of these stands was created by, and interacted with, the natural disturbance regimes of these ecosystems, including fire and insects. The basic interpretation is that a patchy pre-1900 forest structure favored a patchy disturbance regime, and vice versa, but now forests are less patchy and less open with larger areas covered by contiguous, dense host stands and this has led to a less patchy disturbance regime (i.e. larger outbreaks).

There are many additional potential uses of disturbance chronologies derived by tree-ring analysis, such as the budworm history described above, which have not yet been explored in any depth. For example, studies of the interactions of climate, and disturbances (especially fire and insect outbreaks), and the influence of disturbances on the dynamics of forest populations could greatly benefit from the long historical perspective provided by dendroecology. Accurate long-term disturbance chronologies in combination with forest structure data will also be most useful in development and/or testing of forest dynamics models (e.g. Shugart, 1984, 1987).

B. Studies of Forest Decline

Dead and dying conifers have been observed in scattered high elevation locations in the eastern United States and in more widespread areas in northern and central Europe (Johnson and Siccama, 1983; Tomlinson, 1983; Bormann, 1985; Blank, 1985; Hornbeck and Smith, 1985; Zedaker *et al.*, 1987). This phenomenon has generated considerable concern among the general public, as well as forest managers and scientists. It has stimulated interest in research to determine the probable causes and consequences of the problem (Morrison, 1984; McLaughlin, 1985; Smith, 1985).

Acid deposition has been suggested to be a causal agent of the observed forest decline in some areas. The first tree-ring evidence suggesting that there may be a link was reported by Jonsson and Sundberg (1972) for areas in Sweden, but they were careful to state that their results were not conclusive. In a follow-up investigation Jonsson and Svenssen (1982) found that their tree-ring evidence linking acid deposition and forest decline was even more tenuous than originally thought. There was no indication of either positive or negative growth due to acid deposition; nevertheless, this work did focus world-wide attention on forest decline and tree-ring analysis.

Several other tree-ring studies have attempted to identify the timing, severity and causes of observed forest decline (Cogbill, 1977; Johnson *et al.*,

1981; Johnson and Siccama, 1983; McLaughlin, 1984), but the tree-ring evidence presented in these studies lacks the precision expected from dendroecological analysis and was therefore unconvincing or inconclusive. Some of the difficulties with these studies is due partly to the regional nature of forest decline, which poses the problem of finding adequate control tree-ring data to compare to the air pollution data set. It also appears that a number of the principles and practices of dendroecology were ignored or improperly applied. Often no mention is made of crossdating, and many of the illustrations show the inconsistent variations that would be expected to result from improperly dated samples (see Figure 4).

McLaughlin *et al.* (1983) acknowledge the importance of dendroecology in their review of the forest decline problem. However, they presented data consisting of averaged ring-width series uncorrected for age, competition and stand history variations, and it was not clearly stated that the rings were crossdated. Similarly, photographs of tree cores and cross sections showing growth increment reduction from the pith to the bark have been presented as suggestive evidence of declines in growth of pines in the southeastern United States (Knight 1987; Sheffield and Cost, 1987). Though some of these data may be based on large numbers of measurements, they can not be considered evidence of forest decline until the effects of tree age, past forest history and climatic variations on growth have been dealt with (Sheppard and Jacoby, 1987; also see Zahner and Myers, 1987 for one approach to this problem).

1. Point Source Pollution

Dendroecological investigations of point source pollution and its effects on tree rings usually have resulted in more definitive and statistically supportable results than the above mentioned studies. Significant declines in ring width that are unrelated to age and climatic factors have been associated with rising emission levels and the increasing proximity of smelters or factories (Fox, 1980; Heikkinen and Tikkanen, 1981; Fox *et al.*, 1986). One study on arid-site trees from Nevada reported an increase rather than decrease in growth associated with smelter emissions (Thompson, 1981). This could have been a fertilizing effect of acidic deposition through nutrient release in the alkaline desert soils of the region, but this possibility was not tested or confirmed by any follow-up investigation. Other studies have related ring density changes to emission levels (Keller, 1980; Kienast, 1982; Kienast *et al.*, 1981; Schweingruber *et al.*, 1983; Yokobori and Ohta, 1983).

A strategy employed in point source pollution studies involves the sampling of forest stands at varying distances from a known emission source but otherwise affected by similar site, stand-history, age and climatic variations. For example, Fox *et al.* (1986) sampled the ring-width variations in trees from five stands of western larch (*Larix occidentalis* Nutt.) chosen at different distances down wind from the lead/zinc smelter at Trail, BC,

Canada. Three control stands outside the area of known pollution were also sampled and studied. Two cores were extracted from a number of trees in each stand.

All materials were crossdated, the ring widths were measured, and these values were standardized by dividing by the sample mean so that any information in the growth trends was not removed by standardization. The chronology values of the control trees, and two lagged values from them, were used in regression models to estimate and remove the climate component from the pollution effects. Separate analyses were performed for years before and after installation of two tall stacks, for drought and non-drought years and for years prior to initiation of smelting.

For the period after smelting began, but before stack installation, the growth variation in the affected trees explained by the pollution decreased with the increasing distance from the smelter. Concomitantly, the variation explained by the climatic controls increased with distance. As pollution levels became more and more limiting to ring growth with the increasing proximity of the smelter, there appeared to be less opportunity for climate to be limiting. After pollution abatement procedures were installed, this pattern was reversed, with the greatest recovery observed in the trees that were nearest the smelter. No other environmental changes or stand conditions could be found to explain the large and systematic growth changes that were measured.

It is often difficult to ascertain what portions of the trend in a tree-ring chronology can be attributed to stand history, aging of the tree, long-term climatic variation or atmospheric pollution. When this happens, some of the variance caused by pollution may be removed in the process of standardization.

Nash *et al.* (1975) developed a procedure that can identify the signal due largely to climatic variation, remove it from the variance due to other effects and then restore the original ring-width variations without the effects of climate. The procedure begins with the usual standardization, which fits a growth curve and divides the width by the growth curve estimate to obtain the index. The climatic signal is estimated from the average yearly value of a number of chronologies from trees of the same species, age, and site characteristics, but from sites outside and surrounding the area that had the pollution effect. The normalized values of the averaged chronology are subtracted from the normalized indexed series. This difference is then multiplied by the value of the growth curve that had been fitted in the original standardization to produce the original ring widths minus the effects of climatic variation.

2. Regional Forest Decline in North America

Peterson (1985) and Peterson *et al.* (1987) report dendroecological studies of

Jeffrey pine (*Pinus jeffreyi* Grev. & Balf.) in Sequoia and Kings Canyon National Parks, California, that were exposed to moderate concentrations of ambient ozone (mean hourly concentrations of 6 to 9 ppm, and maximum hourly concentrations of 15 ppm during the late afternoon). Ozone is usually regarded as a regional air pollution problem (Peterson *et al.*, 1987). Dated ring-width chronologies from exposed trees were compared to dated chronologies from control trees growing in more remote sites where ozone exposure had not been reported. No significant difference between the chronology values was noted before 1965. After that date, the ozone exposed chronology values were approximately 10% lower, and the difference was statistically significant. Even though the ozone damage is classified as a regional pollution problem (Peterson *et al.*, 1987) in that it occurs over large areas, mountain terrain and wind patterns appeared to protect some trees, which Peterson and his colleagues used as control chronologies for the ozone effect.

Ozone damage to conifers in southern California has been extensively studied (Miller, 1973; McBride *et al.*, 1975; Miller, 1985). However, dendroecological techniques have not been applied in that area with the exception of one study (Gemmill *et al.*, 1982) reporting on crossdated tree-ring chronologies from an ozone-stressed forest.

In the eastern United States and Europe it may be more difficult to find control sites for the study of forest decline than it was for the study of ozone effects in California. If the hypothesized air pollution effects are widespread, trees free of the effects that could serve as controls might simply be unavailable. In addition, few, if any, long-term records of emissions on such a large scale are available. This limits the possibility of comparing rings in affected and unaffected trees during the same time period, and it limits the use of calibration. Differences in species or individual tree tolerance to various emissions may offer a possibility, but such an approach has not been reported yet.

The species and habitats that are available from the forest decline areas in the eastern United States and Europe may have a number of less desirable dendroecological characteristics than those in the more arid North American West. Some of the problems that may be encountered: (1) the climatic signal in the ring-width variations may be weaker and the noise stronger (Cropper, 1982), which would make crossdating more difficult to detect and apply (wood density variations exhibit a stronger climatic signal (Schweingruber *et al.*, 1978a) and this is one reason why densitometric analysis is considered so important in these areas); (2) the stands are more densely stocked, with more possible interactions between stand dynamics, aging of trees and pollution; and (3) the growth response to the various controlling factors may involve lags lasting for several years, resulting in autoregression and possible nonlinear or synergistic influences.

The comparison of growth before and after the onset of pollution is one promising research strategy for the study of regional forest decline. However, the diminishing ring width associated with increasing age of a tree could resemble the hypothesized forest decline effect. In addition, the ring-width changes associated with increasing age may vary greatly from tree to tree and from one stand to the next in these dense forests (Cook, 1985, 1987). Flexible standardization curves using cubic splines may help to resolve some of this difficulty, but if too much flexibility is used, one runs the risk of removing the pollution signal along with the age-related effects.

Sample stratification to include only the oldest trees, even though they are widely scattered throughout the forest, might be preferable to using many younger individuals from the same local habitat (Ashby and Fritts, 1972). The age-related growth changes in old trees are more consistent, smaller and less likely to resemble the hypothesized growth decline.

The primary objective of dendroecological studies, in the eastern United States (Cook, 1987; Cook *et al.*, 1987), in Germany (Eckstein *et al.*, 1984; Greve *et al.*, 1986) and in Switzerland (Kienast, 1982) was to determine if an exogenous growth decline could be measured. Comparisons were made of the chronologies between what were thought to be pre- and post-decline periods.

The growth-climatic variations in the chronology were calibrated using the pre-decline chronology and climatic data for the corresponding period. Climatic information from the post-decline period was applied to the post-decline chronology to estimate what the chronology would have been if it was affected only by climate. If the chronology had been adequately replicated, dated and standardized, then a departure of this estimate from the actual chronology should not be the result of aging, stand changes or climatic variation. If there is no evidence that nonclimatic factors such as fire, insect infestation, cutting history or some other factor can explain the departure, then pollution damage can be inferred to be the probable cause of the effect. Also, if the growth–climate relationship is significantly stronger before the decline than after it, the reduced strength may be considered as evidence for an increasing influence of nonclimatic factors, such as pollution, on ring growth. Some of these studies are described in detail to illustrate this type of application.

Cook (1987) applied dendroecological techniques to the problem of forest decline in the Adirondack Mountains of northern New York State. He obtained a stratified sample from 20 dominant and co-dominant trees, two cores per tree, of red spruce (*Picea rubens* Sarg.) growing at an elevation of 1150 meters and only 27·5 km from Whiteface Mountain where symptoms of red spruce decline were reported by Scott *et al.* (1984), and Johnson and Siccama (1983). Three cores were eliminated as they had unusual growth distortions (reaction wood) due to changes in the direction of stem growth that were unique to those individuals. The remaining 37 cores were crossdated, the ring-widths measured, the measurements converted to standardized indices and the indices averaged to obtain a chronology.

Cook applied his linear aggregate model (Section IV) to this growth decline problem. He described experiments with different statistical techniques to estimate the growth curve and apply it to standardization (Cook, 1985). Two estimates of the chronology were obtained. One was equivalent to using standardization techniques described in Section IIIE. For the other estimate, time-series techniques (Box and Jenkins, 1976; Cook, 1985; Holmes *et al.*, 1986) were applied to the ring-width data to remove excessive low frequency variations and autoregressive relationships. The terminology of time-series modeling calls this procedure "prewhitening", i.e. unusually large amounts of long-wave variations ("red noise") are removed to produce a "white noise" time series with equal amounts of variations at all wave lengths. Therefore, Cook applies the terms "unwhitened" and "prewhitened" to the first and second estimates of the chronologies. Cook reported that the mean chronology for the period common to all trees, 1837–1964, had an s/n ratio of 21·7:1.

Calibration equations for predicting the yearly chronology values were then developed using monthly divisional climatic averages from the prepollution period of 1890–1950. These data were the 19 candidate predictor variables. They were the monthly mean temperature from March one year before the annual ring growing season through September at the end of the ring growing season. Stepwise multiple regression analysis was used to select a subset of significant monthly climate predictors of the ring-width chronology. The equations for the unwhitened and prewhitened chronologies accounted for 50·2% and 53·8% of the chronology variance, and the climatic predictors were consistent with other reports in that temperature during and prior to the growing season were associated with growth (Conkey, 1982a; Cook, 1982). The prewhitened calibration gave slightly better growth estimates (see plots in Fig. 7).

The equation was applied to the climate data for 1951–1976 to estimate the ring-width chronology. The 1951–1967 period had no apparent trends that may have been due to the pollution component, so these data provided an independent test of verification. The 1968–1976 interval provided a test for possible effects of pollution.

The results are shown as plots in Fig. 7. The statistical estimates of the two equations mimic the chronology variations in both the calibration and verification periods. The calibration was excellent for this kind of data, and the independent verification statistics were found to be more significant than would be expected by chance variations. However, the amounts of disagreement between the estimates and the chronology values increase from 1968 to 1976, the post-pollution period.

As would be expected from these results, there are certain years in the

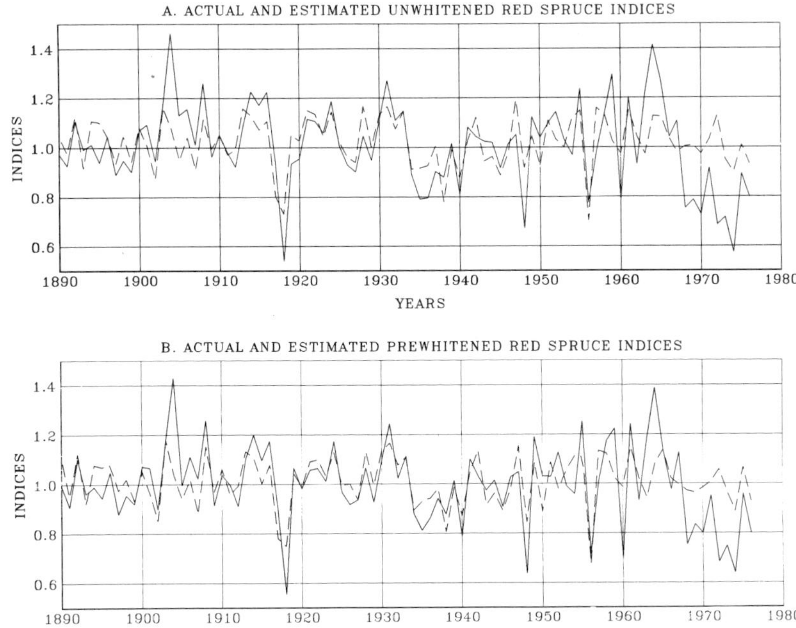

Fig. 7. Actual and estimated red spruce chronology values based on a temperature response model. The model was developed using the 1890–1950 data. Model predictions run from 1951 through 1976 (from Cook, 1987).

calibration period when estimates are substantially different from the chronology (e.g., 1904–1906, 1918, 1935–1936, 1948, 1959, 1963–1964). However, the residuals from the regression are more or less random until the 1960s. (A test of randomness was not reported.) After 1967 the chronology values are always lower than the estimates.

The estimates from a complex nonlinear equation with more variables might have approximated the actual biological relationships more closely. However, fewer degrees of freedom would have remained, thus reducing the reliability of the independent estimates.

The high calibrated variance, the significant independent statistics and the random variation of the residuals before but not after the 1960s, support the conclusion that factors other than the climatic variables that were calibrated, are responsible for the growth decline starting in the 1960s. While several alternative explanations are possible, acid deposition and other forms of pollution must be considered as possibilities. Pollution was also suggested by other investigations (Scott *et al.*, 1984; Johnson and Siccama, 1983).

Cook (1987) acknowledges: "These results indicate that the observed ring-width decline of red spruce in this stand cannot be explained by the verified climatic response models developed here. As a result, a change in the growth environment of these trees has probably occurred, which had a stand wide impact on the sampled trees." He goes on to emphasize, however, that climatic effects (representing variables other than the monthly climatic data used in the calibration) still cannot be excluded as possible causal agents in the observed ring-width decline. One hypothesis is that winter foliar damage might be involved (Friedland *et al.*, 1984). In this case it is possible that a climatic change may have led to a higher frequency of winter freeze damage events since 1967, or it is possible that red spruce have recently developed a heightened sensitivity to winter freeze damage that could be related to a predisposing stress such as acid deposition or nitrate fertilization (Friedland *et al.*, 1984). Cook (1987) suggests that a test of this hypothesis should involve both an examination of recent winter climate in relation to conditions necessary to cause freeze damage and studies of the long-term occurrence of this phenomenon.

Later, Cook *et al.* (1987) took this research one step further. They used ordinary least-squares for estimating climatic response models for forest decline studies in different stands of red spruce throughout the Appalachian Mountains. In northern Appalachian trees, the regression models were significant up to 1976, the last year with no observable pollution effects, but not over the 1968–1976 period when there was a marked growth decline and pollution. More importantly, a test for bias revealed there was a significant overestimation in the forecast of radial growth. This suggested there was a change in the relationship between red spruce tree rings and climate after 1967 which might have been the result of climatic change or a new form of stress, such as pollution.

Results from southern Appalachian trees supported the results from the northern trees, except that the climatic relationship with temperature was weaker, probably because important climatic variables such as precipitation had not been considered. The bias in the decline period was about half that found in the northern Appalachians.

The chronologies from 21 red spruce sites in New York and Vermont were examined very closely to determine whether the relationship between tree rings and climate during the most recent 20-year period examined was anomalous compared to earlier periods or whether the bias could be attributed to the standardization of the tree ring data (Cook *et al.*, 1987). They therefore removed all trends from the tree-ring data and compared these data to a regionally averaged temperature record that had been estimated back to 1820. The climate response model was calibrated using the data from the 1885–1940 time period, leaving ample data for two pre-decline verification periods, 1856–1884 and 1941–1960. The 1961–1981 interval was set aside as the decline period. Between 12 and 55% of the tree-ring variance was calibrated, and the most commonly selected predictors were July and

August temperatures of the previous growing season (negatively) and December and January temperatures prior to the growing season (positively). Verification statistics indicated that the climatic response models were, in general, quite time stable up to 1961 even though the variance calibrated was sometimes low. After that time little or no relationship with climatic data could be found.

They also considered possible effects of extreme temperatures in August and December by using all years when the standard normal deviates exceeded the 0·9 probability level and plotting them as indices of stress. The occurrence of stressful years was notable in the 1870s, the late 1930s and the late 1950s to early 1960s. There were also periods of abnormally high red spruce mortality. However, a period of noted high spruce mortality in the 1840s and 1850s does not correspond with their stress index results.

They believe that their results suggest that "abnormally high climatic stress may be acting as a predisposing factor or cause of red spruce decline, both past and present." They had not ruled out the possibility that some additional stress factor was present, such as air pollution. However, they caution that the question of survivorship bias in the analysis should be investigated. "It is possible that had the red spruce been sampled in 1880s, the relationship to climate would have broken down due to the inclusion of trees that eventually died from the mortality episode." Thus the role of anthropogenic pollution in causing or intensifying the present decline of red spruce remains uncertain. Nevertheless, it is clear that dendroecology has provided new insights and understanding of this problem, and like most seminal research, it has suggested new questions and areas of investigation.

3. Forest Decline in Europe

Extensive dendroecological investigations of regional forest decline problems have been conducted by dendrochronologists in West Germany (Eckstein *et al.*, 1984; Eckstein, 1985; Greve *et al.*, 1986). Their work has involved tree-ring sampling along suspected pollution gradients, study of sulfur and fluoride content in spruce needles as indicators of such gradients, and dendroclimatic analysis to determine if climatic changes such as drought could explain the observed growth reductions. Their findings have been similar to those of Cook and his colleagues in the eastern United States, in that growth reductions observed during modern portions of tree-ring chronologies (generally post-1940) cannot be explained by the climatic response models calibrated for earlier periods. Evidence of greater growth reductions in areas of West Germany with higher levels of suspected pollution than in areas with lower suspected levels seems to support, but does not prove, a cause and effect relationship between air pollution and tree growth reduction.

As an example we will describe dendroecological studies of forest decline in Switzerland. Kienast (1982) describes investigations of possible fluoride emission damage in the Rhône Valley in Switzerland. He used annual rings to ascertain in which years damage occurred and to estimate the radial growth loss. X-ray techniques were used to obtain a continuous wood density profile (Lenz *et al.*, 1976), and five parameters per ring, including width and density measurements, were derived for subsequent analysis. Instead of fitting the standardization age curves to the entire length of record, the age trends were estimated using the longest possible emission-free period (1874–1940). The trends in these curves were extrapolated through the period affected by pollution (1941–1979) (Pollanschütz, 1971). The indices were obtained and the standardized chronology calculated. The calibration between the chronology values and climate factors of temperature and precipitation used regression methods described by Kienast (1982) and response function techniques described earlier in this paper. The calibration was applied to the pre-pollution period and the equation was then applied to climatic data for the pollution period to estimate the expected growth for 1941–1979 due solely to climate. No mention was made of independent verification tests.

Comparisons among the densitometric measurements largely confirmed the visual dating and observations. On average, damage was most clearly established for latewood width, followed by total ring width, then by earlywood width and then maximum latewood density. The loss in growth over the pollution period was 20–30% of the ring width for the calibration period and 30–40% of the latewood width over the same period. No explanation was given for these differences.

Figure 8 includes two plots for maximum latewood density and the corresponding estimates obtained from climate. One plot is for an undamaged tree and the other for a damaged tree. The data were calibrated over the first period and the calibration applied in the extrapolation period to estimate density variations due solely to climate. The difference between the actual measurement and the estimated value is shaded to emphasize values indicating a probable pollution effect.

Kienast (1982) summarizes his results from the Rhône Valley as follows: "The pine forests of the Rhône Valley are severely damaged. Possible causative agents are fluoride and other harmful gases such as SO_2, HCl and NO_x in the emissions of nearby aluminum smelters and chemical plants. Droughts and aging may also have contributed to the damage ... Growth disturbances occurred most commonly during the droughts that began in 1938 when a new aluminum smelter was built in the main valley. It stood in the mainstream of the prevailing winds and was without precipitators until 1965. The harmful emissions of the late thirties may have damaged the trees to such an extent that they could no longer withstand natural stresses and finally died 30–40 years later. Although the fluoride emissions are not

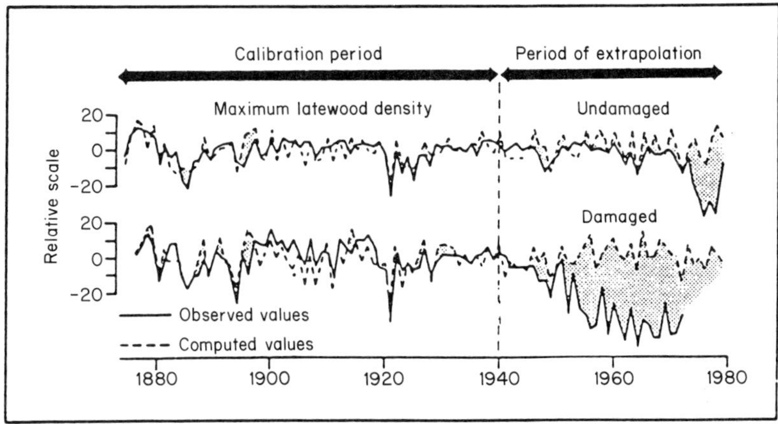

Fig. 8. Estimate of damage from indexed maximum density plots of two pine trees from Valais, Switzerland. The computed curves which are superimposed on the measured values were calculated from meteorological data and the difference which is shaded shows a possible pollution effect (from Kienast, 1985).

causally related to the forest damage, they provide a valid explanation of the results."

Kienast's study was an intensive analysis directed at point-source air pollution. It was a part of a larger effort concerned with a survey of possible regional decline in Swiss forests (Schweingruber *et al.*, 1983). This work is described below.

Dendroecological techniques were used to help answer the following questions. How has the vitality of tree growth changed from earlier times? Where and when did it occur? How extensive are the damaged areas? New rapid sampling and analysis techniques were needed to survey and map the ring changes over such a large area. They included a simple visual dating method that used the matching of "pointer" rings with narrow or reduced latewood bands that were known to occur in particular years. Cores were sampled and ring sequences dated in the field using the "pointer" rings for time control. Major changes in ring structure were visually identified from the dated series, and simple diagrams and other graphical techniques were used to record the information while at the collection sites.

The data from some 3800 cores and stem disks from firs and pines in northern Switzerland were recorded in this manner. From 75 to 86% of the materials were successfully dated by the "pointer" method. This is an acceptable margin for temperate forests in Europe. The "pointer" rings in subalpine trees often were associated with cold, moist and cloudy summers. For trees at lower elevations and dryer sites, the pointer rings were frequently associated with warm and dry summers. Only in the extremely dry years of 1921 and 1976 could pointer rings be identified in all stands that were sampled (Schweingruber *et al.*, 1983; Schweingruber, personal communications).

These well-dated materials established that disturbances increased markedly in the 1940s, especially in 1942, 1944 and 1947 (Kontic *et al.*, 1986). Mortality of trees in the study areas peaked several years later. They found that diseased trees exhibited an abrupt reduction in annual growth increments that was more than 50% in most cases. It was usually easy to recognize the first year in which the injury occurred.

The decline in ring-width or latewood density, however, begins at different times for different species and for different locations in Switzerland. It begins approximately at the turn of the century in the alpine region of central Valais with reduced growth of Scots pine (*Pinus sylvestris* L.) reflecting local pollution caused by industrialization. This general pattern is overlaid by the effects of climatic variation. More trees showed growth reductions apparently due to pollution in drought periods than in moist periods. The growth decline does not appear in silver fir (*Abies alba* Mill.) growing in the Central Plateau, until the 1940s (Schweingruber *et al.*, 1983). In the heavily populated areas north of the Alps, extensive damage was present in all the sites they investigated (Schweingruber, 1986). Firs that were examined in the southern Alps were undamaged (Schweingruber, personal communication).

In the Valais area of Switzerland, the decline in latewood density and ring width were observed from 3 to 6 years earlier than the appearance of unhealthy crowns (Kontic *et al.*, 1986). The possible causes of these differences are now being investigated. The initial changes in the crown may begin with increased shedding of the needles at the same time as the change in ring structure, but the crown may not appear unhealthy until several years later. These results suggest that changes in ring structure may be a more reliable indicator of the beginning of forest decline than the crown appearance (Schweingruber, personal communication).

The general forest decline as proposed by foresters could not be fully confirmed because the density of the crown does not always relate to abrupt growth changes. Schweingruber and his colleagues observed a combination of climatic effects and apparent disease, but they were uncertain as to whether pollution was the cause of abrupt growth changes, except in heavily polluted areas (Schweingruber, 1986).

C. Climatic Variability and Change

It was shown in Section I that a tree ring reconstruction of past hydrologic variability can provide a longer time perspective for viewing ecological problems than is generally available from short instrumental measurements.

In like manner, dendroclimatology also can provide a time perspective that generally focuses on environmental variations related to climatic variations from one year to the next. Thus, tree ring data provide paleoclimatic information over time scales representing years to centuries (Fritts, 1971). Most other paleoclimatic data can not resolve annual variations, and many respond to variations no shorter than a century.

In some situations and with some species such as high altitude *Pinus longaeva* Bailey and *P. flexilis* James, the tree-ring chronologies can span more than 1000 years, providing long records of yearly climatic variations (Ferguson, 1968; LaMarche, 1974, 1978; Graybill, 1987). A thorough discussion of dendroclimatology is beyond the scope of this chapter. However, we will cite some important reviews, comment on a few aspects relevant to dendroecological applications, and conclude with concerns about the adequacies of climatic models, including possible contributions of dendroclimatic studies in a systems approach to the problems of climatic change.

A general treatment of dendroclimatology can be found in Fritts (1976). Later developments and discussions on various methodologies are reported by Hughes *et al.* (1982). Schweingruber (1983) summarizes the field from the European point of view and describes new opportunities provided by wood density measurements. Brubaker and Cook (1983) and Stockton *et al.* (1985) summarize the field in a geological and meteorological context with considerable attention given to the world-wide distribution of climatically responsive tree-ring chronologies. Kairiukstis and Cook (in the press) have collected a series of papers which summarize some current methodologies. This could be a landmark volume, however, caution about the validity of some methods may be warranted. For example, some Soviet dendrochronologists apply Fourier analysis to existing tree-ring chronologies to forecast future variations but it is not clear from their papers whether such projections can be validated by independent verification.

Tree-ring chronologies from all continents except Antarctica are available for modern dendroclimatic analysis. ARMA modeling and both simple and multivariate regression techniques have been used for reconstructing past environmental variations (Fritts *et al.*, 1971, 1979; Fritts, unpublished; Stockton *et al.*, 1985; Guiot, 1986; Briffa *et al.*, 1983, 1986). Many involve the climate at one location or an average over a region (Briffa *et al.*, in the press a) and are based upon ring-width variations (Hughes *et al.*, 1982). However, analysts are beginning to deal with spatial variations in both tree-ring chronologies and climatic variables (Fritts *et al.*, 1971, 1979; Fritts, in press; Schweingruber *et al.*, 1978a; Briffa *et al.*, 1988a). Wood density is also being exploited for reconstructing past climatic variations (Briffa *et al.*, 1988b).

Sometimes dendroclimatological inferences based solely upon simple observation of marked anatomical features can lead to valuable insights about past ecological conditions. For example, LaMarche and Hirschboeck

(1984) note that frost injuries within annual rings of high altitude bristle cone pine in southwestern North America are significantly associated with years of major volcanic eruptions. They have now demonstrated that the rings with frost damage (called frost rings) represent new, independent proxy records of climatically important eruptions occurring during the past several thousand years. They report a notable occurrence of frost rings in 1626 BC which they attribute to the dust veil and the associated widespread cooling from the eruption of Santorini in the Aegean Sea (Lamb, 1977). This date was at first highly criticized by archeologists, who preferred the more conventional 1500–1450 BC date for the eruption based upon ceramic and artefact chronologies. However, LaMarche and Hirschboeck argued that their frost-ring date falls well within the error range of radiocarbon dating of organic artifacts and that the archaeological dates were more likely to be the problem. Hammer *et al.* (1987) recently announced an ice-core date of 1645 BC for the eruption; and this was followed by a review of the archeological evidence by Betancourt (1987) with the conclusion that the radiocarbon and associated tree-ring dates appeared to have been correct all along.

These arguments stimulated Baillie and Munro (1988) to examine more closely the crossdated ring-width data from bog oaks (*Quercus* sp.) growing under poorly drained conditions in Northern Ireland. They report that extremely narrow rings can be noted in these oak trees for the same dates suggested for major volcanic eruptions already demonstrated by other methods. They examined all of their data using an index of ring narrowness in bog oak chronologies from 5289–116 BC. A number of periods with narrow rings were identified, including a conspicuous interval with narrow rings in the 1620s BC coinciding with the frost ring date for Santorini. Baillie and Munro point out that the more precise tree-ring dates should take precedence over the ice-core date because the tree-ring dates can be shown to be internally more consistent.

Baillie and Munro conclude, "These results may have implications for interpreting the effects of volcanic dust veils. Even if volcanic ash clears from the stratosphere after two–three years, the Belfast tree-rings show effects which last much longer (*ca.* 10 years), suggesting than an initial trigger event (flooding?) caused severe long-term problems for trees growing on bogs. Other biological systems may show similarly extended responses. There are possible implications for the impact on human societies, which might suffer the effects of runs of bad harvests, poor pasturage and impeded communications."

Dendroclimatic reconstructions can be obtained for a single climatic record or for multiple records. Some of these multiple records can provide information on spatial variations in climate from spatial variations in tree-ring chronologies (Fritts, 1976; Hughes *et al.*, 1982; Stockton *et al.*, 1985).

Seasonal to century-long climatic variations can be reconstructed and mapped over both space and time using various dendroclimatic methods. Dendroclimatic reconstruction of these kinds can reveal synoptic scale climatic features and suggest certain ecological consequences of these variations (Fritts, in the press).

However, there is no simple solution to this complex ecological problem. Tree growth from a variety of sites representing different topographic localities, as well as from different species, must be related to a geographic array of climatic data. Fritts *et al.* (1979) and Fritts (unpublished) use stepwise canonical regression to deal with this problem. The tree-ring width variations from 65 arid-site chronologies throughout western North America are calibrated with 20th century temperature and precipitation records from the United States and southwestern Canada. A systematic analysis and elimination procedure was used to select the optimum calibration models from a variety of models of different structure. For example, the models differed as to whether or not ARMA modeling was applied to the tree-ring data before calibration. Different-sized grids of climatic data were calibrated to evaluate the effects of distance from the trees on the climatic reconstructions. Principal component analysis of tree-ring and climatic data was used to reduce the number of variables and to orthogonalize the variations; and different numbers of principal components were used to vary the size of the calibration equation. Different lags and different numbers of predicted principal components were entered into regression, a stepwise selection of variables was used to retain only the significant canonical variates, autocorrelation of the residuals was computed and used to adjust the degrees of freedom in the statistical analyses, and most important, verification statistics using all available independent climatic data from the 19th century were used to select the optimum model structure.

At each step in the analysis, only model structures that resulted in reconstructions with significant calibration and verification statistics were retained for the next step of the analysis. The choice depended first upon which models had the most significant statistics and second upon the ecological reasonableness of the model (Fritts *et al.*, 1979; Fritts, in the press) particularly regarding the lag of growth behind the climate input as opposed to preceding it. The simpler of two models with comparable statistics was selected. The results are available from the author on floppy disk.

The first calibrations used monthly, seasonal or annual climatic data to evaluate the response structure. Only seasonally averaged climatic data could be successfully calibrated. The annual climatic data appeared to contain too little meaningful information and the monthly intervals were too short a time span with too many predictors for the spatial analyses. Thus, the climate was calibrated one season at a time with the ring-width variance. At this stage in the analysis, unmodeled variance associated with the effects of climate in other seasons was simply carried along as part of the error variance.

Often several verified models of quite different structure produced seasonal reconstructions of comparable quality. This suggested that the reconstructions from several well verified models (Bates and Granger, 1969) might be combined to improve the statistical estimates. Thus the most promising combinations of two to three well-verified models were tested using the same calibration and verification climatic data used in the original analysis. Combinations of more than three models failed to show significant improvements in verification statistics. For each climatic variable, the combination of two to three models giving the best calibration and verification statistics were selected for the seasonal estimates. The annual values were obtained by summing or averaging the selected combined reconstructions for the four seasons. Calibration and verification statistics that were computed for these data using annually averaged climatic data were far superior to the statistics for the models that were directly calibrated with annual climatic data.

These results are consistent with the idea that the ring-width response of arid-site trees integrates the limiting conditions of climate over the four seasons. The importance of different variables in different seasons can change substantially from site to site and from one species to the next. In addition there were significant lags in growth of one year or more behind the occurrence of climate particularly for summer (July to August), which begins too late in the growing period for there to be a major effect of climate on total width of rings formed in that season.

The annual and seasonal reconstructed temperatures provided more reliable estimates of climate than precipitation. The independent temperature reconstructions span the interval 1602–1900 and extend the 20th century instrumental record for the United States and southwestern Canada to the beginning of the 17th century. This provided a unique opportunity to investigate the effects of volcanic dust-veils on the spatial variations in North American temperature (Lough and Fritts, 1987). Such a study was not possible before the reconstructions had become available because there were too few great eruptions coincident with the period of instrumental data to conduct a spatial analysis.

The beginning (key) dates of the 14 largest eruptions during 1602–1900 were subjected to superimposed epoch analysis to look for a climatic signal associated with the injection of aerosols into the stratosphere and subsequent cooling (Mass and Schneider, 1977; Taylor *et al.*, 1980; Self *et al.*, 1981; Kelly and Sear, 1984). Temperatures for the five years following the key dates were compared to the temperatures reconstructed for the five years prior to the key dates. The differences were calculated and then averaged for all the selected key dates. Monte Carlo techniques were used to assess the statistical significance of these averages. The differences for each data point were mapped to reveal any spatial patterns.

All 24 eruptions were analyzed without regard to the latitude of the eruption, and no significant pattern was noted. The volcanic events were then separated into seven high-latitude eruptions, eight mid-latitude eruptions, and eight low-latitude eruptions. Figure 9 shows the results from these analyses. After high-latitude eruptions there may have been some warming in the central northern United States and cooling in the southeast (Figure 9a), but only three grid-points were statistically significant. For the low-latitude events, the reconstructed warming was significant in the Pacific Northwest and southwestern Canada, and cooling was observed east of the Rocky Mountains to the Atlantic and Gulf coasts. Cooling was most pronounced and significant in the central and some eastern states (Figure 9c). It was concluded, that while annual temperatures decline over much of the United States following low-latitude eruption events, significant and marked warming occurs in many western states.

The pattern of anomalies varied among the seasons. The temperature anomalies reconstructed for winter, spring and summer are shown in Figure 10. Cooling was reconstructed in spring for the central United States (Figure 10b) with 38% of the stations significant. In summer, cooling was pronounced from the Rocky Mountains to the East Coast and significant warming was reconstructed primarily in the western coastal region and adjacent states (Figure 10c). Extensive warming was reconstructed in the West in winter (Figure 10a).

Lough and Fritts (1987) checked other tree-ring data and some climatic reconstructions including high-altitude sites where temperature was expected to be growth-limiting. They also examined some long temperature records and found general confirmation of cooling in the mid-continent and warming in the west, particularly in summer.

The strength of this analysis suggests that the reconstructions provide meaningful climatic information in these temperature reconstructions. They also confirm to some extent the work of LaMarche and Hirschboeck (1984) except that their work dealt with short-lived cold outbreaks at the end of the growing season that froze and injured the cells perhaps during one night. In contrast, Lough and Fritts (1987) reconstructed temperature conditions averaged over seasons to the entire year including a three-year lag after the eruption began.

All of these studies illustrate that the rings from many trees distributed over a wide spatial area can contain information on a variety of climatic variables. They also suggest that there is a detectable vegetational response in North America to low temperatures associated with dust veils from large low-latitude volcanic events. Cooling appears to occur over large areas in the

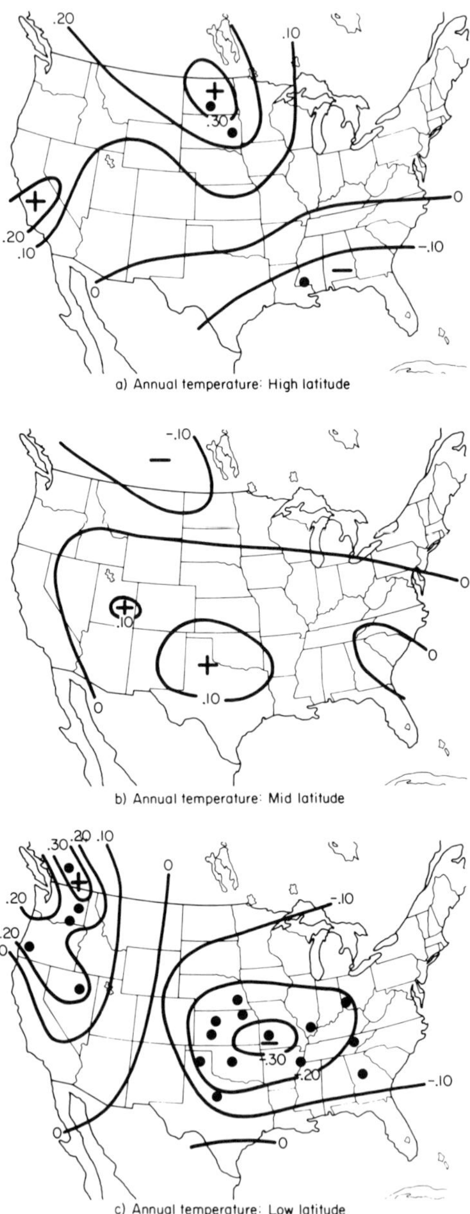

Fig. 9. Average reconstructed annual temperature differences (centigrade degrees) of the average of years 0 to 2 after key dates minus the average of years 1 to 5 prior to key dates for (*a*) high latitude, (*b*) mid latitude and (*c*) low latitude volcanic events. Heavy dots denote stations at which the temperature difference is significant at the 0·95 confidence level (from Lough and Fritts, 1987).

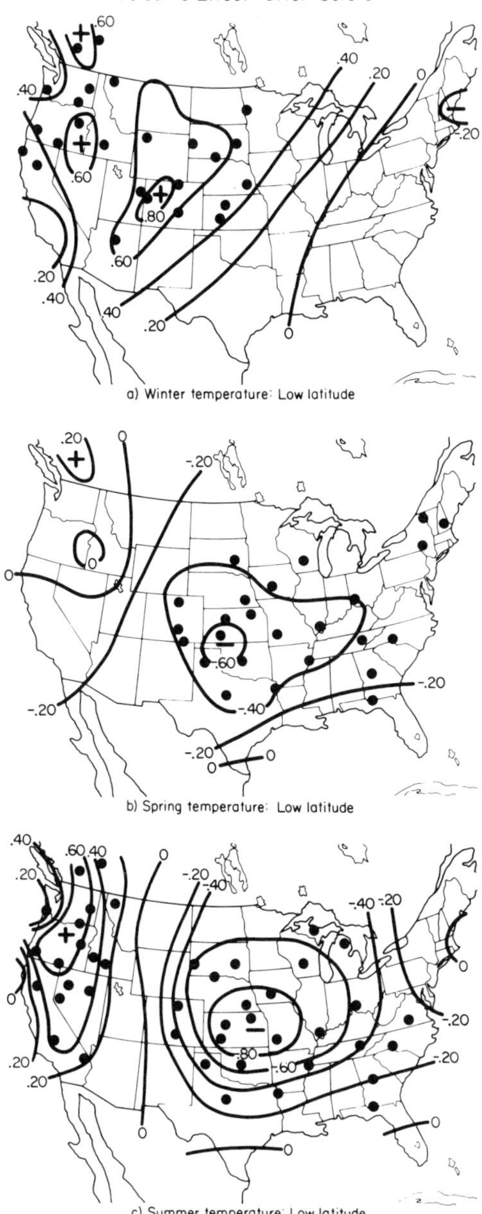

Fig. 10. Average reconstructed temperature differences (centigrade degrees) of the average of years 0 to 2 after key dates minus the average of year 1 to 5 prior to key dates for low-latitude eruptions for (*a*) winter, (*b*) spring and (*c*) summer. Bold dots denote stations at which the temperature difference is significant at the 0·95 confidence level (from Lough and Fritts, 1987).

United States in spring and summer for 0 to 2 years after a great eruption at low-latitudes. However, in winter, an equally marked warming appears to occur over large areas of the western United States, but the warming becomes more restricted in spring and summer including only areas along the Pacific coast and adjacent inland states. Volcanic eruptions at mid, and to some extent at high latitudes, do not appear to produce as marked a change in temperatures. This does not imply that there is no effect but rather that there is no unique response pattern detected through the verified climate–growth transfer functions developed for arid-site trees from western North America.

In previous sections we described the use of tree rings to detect pollution signals in the growth ring record. Now we turn to the problem of detecting possible changes in the amounts of atmospheric gases such as CO_2. LaMarche *et al.* (1984) report rising trends in the ring widths of trees growing at high elevations in the western United States that exceeded the increase in growth expected from known climatic trends in the same region. They proposed that rising levels of atmospheric CO_2 could be the causal factor. It was hypothesized that "subalpine vegetation generally, and upper treeline conifers in particular, could now be exhibiting enhanced growth due directly to rising levels of atmospheric CO_2."

Graybill (1987), who is a dendroclimatologist investigating this subject, explains: "One of the primary physiological bases for this hypothesis is that CO_2 becomes more limiting to photosynthesis as elevation increases because the concentration of CO_2 per unit volume is decreased from that nearer sea level. With substantially increasing CO_2 since the mid 1800s one might then expect to first see improved photosynthetic performance in trees growing under ambient conditions at relatively high elevations." (See also Gale, 1986; LaMarche *et al.*, 1986). These workers suggest that more effort should be directed to evaluating these growth changes at high altitudes, because of the ecological importance of these kinds of vegetational changes, if they have begun, as well as their implications for the global carbon budget.

Graybill (1987) extended the existing collections of very old high altitude *P. longaeva* and *P. flexilis* growing in the Great Basin and Rocky Mountain regions. He continues to find that trees at high altitudes (*ca.* 3400 m) near upper treeline show major and relatively continuous increases in annual growth rate since about 1850. He notes that in almost all cases the particular trees were growing on south exposures or on windswept crests where drought as well as temperature can be limiting. He suggests that water use efficiency of trees in these settings also may have increased with corresponding changes in CO_2, leading to more efficient utilization of site resources and growth enhancement.

Graybill used ARMA modeling techniques to examine the growth trends. He fitted an ARMA model to the 1380–1860 period and applied that model to the post-1859 period, but this failed to remove the trend in these data. Moreover, the mean of the post-1859 residual series was five times greater than the mean of the earlier values. He also examined temperature and precipitation records throughout the region and found that variation in the prewhitened ring-width indices could only be partially explained by those variables. However, he did not find a trend in either of these climatic variables that could account for the rising growth at high altitudes, and was unable to rule out CO_2 as an agent for the growth change.

He also developed several chronologies at the drier and lower limits of growth (*ca.* 2900 m) for Bristlecone pine in the Great Basin. In some cases these are from the same mountain ranges as the upper treeline chronologies. There is no overall upward trend in growth during the past 130 years, but instead these chronologies from lower elevations track the available moisture records of this century with reasonable fidelity (Graybill, personal communication, 1988).

In this section we have given only a few examples of dendroclimatic applications that can help to identify ecological problems and expand our knowledge of past environmental variations and changes. There are other major issues and problems dealing with climatic variation and changes to which dendroclimatology or dendroecology could make unique and significant contributions.

For example, Wood (1988) points out that the global climate system is immensely complex and that existing climate models make assumptions leading to important uncertainties in the model results. One important uncertainty involves how much of the hemispheric warming in the instrumental data analyzed by Jones *et al.* (1986) are really the effects of urban warming. He proposes that a variety of independent data, including the information from proxy records of past climate such as tree-ring measurements and alpine glacier changes (Wood, 1988), should be considered to validate this time series.

Wood also points out that "researchers have looked to the paleoclimatic record as an analogous indication of future climatic change (see Webb *et al.*, 1985; Kutzbach, 1985), especially with respect to the effects of increasing concentration of atmospheric trace gases. An important caveat is that current climatic conditions appear to be unprecedented, and therefore it is not clear whether and to what extent past climate behavior serves as a valid guide to the future."

Wood points out that (1) the atmospheric concentration of CO_2 appears to have already exceeded any level experienced over the last 160 thousand years; (2) that rate of increase in CO_2 concentration appears to be more than an order of magnitude faster than previously experienced; (3) several other trace gas concentrations are increasing almost as fast or faster than CO_2; (4) the present Interglacial is widely estimated to be about 10–11 thousand years old, and would be expected, on the average, to be in its late stages and to be characterized by long-term cooling; (5) the global forest cover has been reduced by a conservatively estimated 1·3 billion hectares since pre-agricultural times. The rate of global deforestation, especially in the tropics, appears to be at least an order of magnitude faster than that previously experienced due to natural causes, over at least the last 10 thousand years. "This suggests an unprecedented occurrence of an extremely rapid increase in trace gases and decreases in forestation during what would in geological terms be the late stage of an interglacial with its expected long-term cooling (Wood, 1988).

Because there are so many uncertainties in our understanding of the climate system and limitations of current climatic modeling efforts, Wood (1988) proposes that an intensive research effort be launched to deal with a wider range of inter-related relationships as a systems problem. An important first-order contribution that such systems research might make is the identification of key factors of climatic change and the integration of knowledge ranging across all of the disciplines that may be potentially relevant. Wood lists 21 relevant disciplines including botany, ecology, forestry and dendrochronology that could make meaningful contributions to such a systems analysis.

A third order contribution that he mentions is the identification of needs for climatic monitoring such as (1) the use of rural temperature networks, (2) a tropospheric temperature network, (3) monitoring alpine glacier trends, (4) permafrost thermal gradients, (5) lake levels and, possibly, lake freeze-up dates, (6) ocean plankton trends, and (7) tree-ring chronologies (as a climate-sensitive indicator of changes in terrestrial biomass, carbon flux, and albedo—perhaps under the auspices of the International Project in Dendroclimatology). This reference was to an organization of collaborating dendrochronologists who are attempting to assemble an international network of tree ring chronologies to check the instrumental record of past climatic variation and to extend that record back to the beginning of the 18th century or earlier where there is adequate dendrochronological coverage (Hughes, 1987).

Wood (1988) highlights various weaknesses of climatic models and stresses the importance of climatic and paleoclimatic information on time scales of years and decades to model improvement. Not only do precisely dated and well-replicated dendrochronological data sets allow us to extend knowledge about environmental variations backward in time; but if the unprecedented changes in the earth atmosphere system are truly induced by man's activities, trees already established around the world will be responding to that change. The ring record from these trees can be subjected to dendroecological analysis to assess the direction, magnitude and date of these changes whether or not sufficient instrumentation is in place to monitor these changes. Both dendroclimatology and dendroecology can provide quantitative and precise information on past environmental conditions associated with past tree-growth activity.

VI. LIMITATIONS OF DENDROECOLOGY

There are some areas of the world where dendroecology is not applicable because there are no suitable trees for the analysis. For example, the rings of many tree species, especially those growing in the tropics, may be indistinct, or bear no relationship with an annual growth cycle, or if they do, there are no visible patterns common between trees that can be used for crossdating (Eckstein *et al.*, 1981; Ogden, 1982). However, there are tropical forest sites with species that have crossdatable rings, which have provided useful chronologies (Ogden, 1982; Villalba *et al.*, 1985). It should also be kept in mind that very mesic and apparently favorable site conditions do not always mean that potentially useful species are not available for tree-ring studies (LaMarche 1982). For example, Stahle *et al.* (1985a,b) have found bald cypress (*Taxodium distichum* L. Rich) growing in swamps in the southeastern United States that are crossdatable, climatically sensitive, and exceeding 1000 years in age.

In other situations the growth of rings may generally follow an annual cycle, but there may be too little variability to distinguish a pattern for dating. In extremely arid regions or where a species is at the limit of its ecological range, so many rings may be missing from a sequence or so many intra-annual growth bands may be produced in any one year that the ring sequence cannot be dated (Glock *et al.*, 1960). Additionally, many trees may not attain sufficient age to be useful. Thus, dendroecology requires a tree species that (1) produces distinguishable rings for most years, (2) possesses ring features that can be dendrochronologically dated, and (3) attains sufficient age to provide the time control required for a particular investigation. Fortunately, there are many suitable species in the temperate forests of the Northern and Southern Hemisphere and crossdating has been observed in many of them (Hughes *et al.*, 1982; Stockton *et al.*, 1985). Many old trees with datable ring patterns can still be found in remote or protected areas of deciduous forests (Hughes *et al.*, 1982; Brubaker and Cook, 1983). Conifers have been investigated most extensively. Angiosperms have also been used, with oaks (*Quercus* spp.) being selected most often for ring-width analysis. The large vessels in oak, however, preclude densitometric analysis of oak species (Leggett *et al.*, 1978).

Even though the rings of many temperate forest species are potentially suitable for dendroecological investigation, the majority of trees, particularly in a second growth forest, may be undesirable for study. For example, youthful trees generally have too few annual rings to be of interest. Their ring patterns may be so dominated by competition between individual trees that no crossdatable pattern can be distinguished. Middle-aged under-story trees may grow so slowly that many rings are missing, or the rings become so compressed that they cannot be readily identified. Later, when these individuals penetrate the forest canopy, the growing conditions may become more favorable and the rings are much wider. Crossdatable ring-width patterns may become more evident, but in many second growth stands the ring-width variations due to nonclimatic factors often are much larger.

In second-growth stands, occasional large trees can be found that were inhabitants of the original forest. Frequently the ring record from these trees is incomplete because the heartwood is rotten or the trunk is hollow. Also, a tree may be large and still not be old. The rings from such large trees are not only wide but there is often insufficient ring-width variation for crossdating. However, density features of the rings from second-growth stands may be more datable (Schweingruber, 1980; Conkey, 1986). The greatest future contribution of densitometric techniques may be in the more productive temperate and boreal forests where there is little ring-width variation in common with other trees that can be used for dating.

The success of a project often depends upon how skilfully the sites and trees have been chosen. A well-trained dendroecologist often spends considerable time in the field examining the various habitats and species available before establishing a sampling strategy. For this reason, the responsibility for developing a strategy and selecting the sites cannot be delegated to technicians or to other professionals with little or no training in dendrochronology.

While it is desirable to remove the age-related trends from ring-width measurements for many applications, in practice it is difficult to be sure that this is all that is removed in the process of standardization. The statistical approaches and time-series analysis of Nash *et al.* (1975) and Cook (1985) are attempts to deal with this problem.

The relationship of indexed tree-ring chronologies to measurements of wood volume and biomass production have not been investigated adequately. Tree-ring indices may be useful proxy records of relative changes in forest growth, but these measurements are necessarily derived from samples of trees that have survived past disturbances. Thus, determination of wood volume lost, for example, due to a past insect outbreak, will not include the wood volume lost through mortality of trees that are no longer present in the stand. The tree-ring chronology does contain information that is useful in the estimation of forest productivity (Taylor, 1981; Graumlich and Brubaker, 1987), and it should be considered seriously as a possible forest mensuration tool.

The correlation statistic is only an empirical measurement of association between variables. It alone cannot prove that a cause and effect relationship exists, such as between acid deposition and forest decline. However, a correlation finding can be considered a conclusive result when the correlation is too high to have occurred solely by chance; it is verified using independent results; it is based upon a well-tested model of cause and effect; and alternative explanations have been examined and ruled out. Carefully conducted dendroecological studies offer a special contribution to the understanding and resolution of environmental problems because they provide a historical perspective and precise time sequence of environmental changes. Development and analysis of networks or grids of tree-ring chronologies may reveal spatial as well as temporal changes in forest decline.

VII. CONCLUSIONS

The basis for some of the fundamental techniques, principles and practices of dendroecology have been reviewed. Much of the evidence presented is drawn heavily from personal as well as cited references. The reference list, however, is far from exhaustive.

More attention should be given to dendroecological investigation of current forest ecology problems. The value of forest inventory collections could be enhanced greatly if crossdating and other dendroecological techniques were added to those procedures that have been adopted. Dendroecology is not in competition with standard forest mensuration approaches nor is it considered a substitute for stem analysis using the three-dimensional approach of Duff and Nolan (1953) (LeBlanc *et al.*, 1987). It provides different kinds of information and requires different kinds of analyses (Fritts *et al.*, 1965a; Swetnam *et al.*, 1985). Furthermore, standardized tree-ring chronologies can be used in combination with other growth projection systems to provide a more sensitive measure of relative changes in forest productivity through time than some of the standard mensurational approaches (Thammincha, 1981). Considering that the forestry values at risk are primarily economic, the measurement of growth reduction or non-reduction would be of considerable value if it could be expressed in terms of some type of timber volume measurement (Morrison, 1984; Peterson, 1985).

One alternative for obtaining such a measurement is to convert ring-width index chronologies, with or without adjustments for climatic effects, back to ring widths (e.g., Nash *et al.*, 1975; Greve *et al.*, 1986) and then these chronologies can be recomputed as basal area increments. Estimates of changes in volume growth from standardized ring-width chronologies would also require additional data on the growth potential of the site and height growth, but it should be possible to relate this index of growth in the lower stem to overall changes in volume growth.

Numerous computer simulation models have been developed to project forest growth and succession, and basic information on disturbance regimes and climatic relationships are often necessary inputs (Cooper *et al.*, 1974; Shugart, 1984, 1987). Computerized growth and yield models for forestry also depend on estimates of growth impacts of insects outbreaks in order to adjust forecasts (Wycoff *et al.*, 1982). We believe that dendroecology studies can provide much of the basic information that is required for calibrating or testing these types of computer models.

Dendroecology studies examining forest decline have a number of advantages over other types of tree-ring investigations. Site and tree selection strategies are designed to minimize sources of variation in the tree-ring series that may have little or no relation to the problem in question. Trees of similar species, age and stature are collected from similar sites, soil types, exposures and elevations. Accurate dating allows for replicate sampling and the averaging of results in a yearly sequence.

The variance due to large scale extrinsic growth controlling factors such as climate or to large scale stand disturbances remains in the stand chronology because the growth in all sampled trees was similarly affected by the same limiting conditions. The small scale variations unique to each tree or measured radius of a tree are minimized by both the sampling strategy and by the averaging process since these variations are approximately random over space and time. Sampling strategies can be designed to vary one factor at a time such as differences in suspected pollution damage (Schweingruber *et al.*, 1983) or differences in distance from a suspected source (Fox *et al.*, 1986; Greve *et al.*, in the press). These allow for a rigorous analysis of variance, regression or time-series analysis (Fritts, 1976; Cook, 1985, 1987).

ACKNOWLEDGEMENTS

We acknowledge L. B. Brubaker, E. R. Cook, J. D. Fay, C. A. Fox, T. H. Nash III, and F. H. Schweingruber for their very helpful reviews of the manuscript. We especially thank F. B. Wood, F. H. Schweingruber, F. Kienast and E. R. Cook for providing published and unpublished materials and for their invaluable suggestions and comments. We also thank J. Mather, B. J. Molloy and R. L. Holmes for their assistance with portions of the manuscript. Preparation of this manuscript was supported by the Utility Air Regulatory Group, Acid Deposition Committee, Washington DC.

REFERENCES

Abrahamsen, G., Horntvedt, R. and Tveite, B. (1976). Impacts of acid precipitation on coniferous forest ecosystems. *General Technical Report* NE-23, US Department of Agriculture, Forest Service, Northeastern Forest Experiment Station, Upper Darby, PA., pp. 991–1009.

Ahlstrand, G. M. (1980). Fire history of a mixed conifer forest in Guadalupe Mountains National Park. In *Proceedings of the Fire History Workshop*, Oct. 20–24, 1980, Tucson, Arizona. *General Technical Report* RM-81, US Department of Agriculture, Forest Service. Rocky Mountain Forest and Range Experiment Station, Fort Collins, CO., pp. 4–7.

Alestalo, J. (1971). Dendrochronological interpretation of geomorphic processes. *Societas Geographic Fenniae* **105**, 1–140.

Anonymous. 1977. Tales the tree-rings tell. *Mosaic* 8(5), National Science Foundation, Washington, DC, 10 pp.

Ashby, W. C. and Fritts H. C. (1972). Tree growth, air pollution, and climate near LaPorte, Indiana. *Bulletin of the American Meteorological Society* **53(3)**, 246–251.

Athari, Said (1981). Missing growth rings, an often disregarded problem in the studies of increment in emission-damaged and healthy Norway spruce stands. *Mitteilungen der forstlichen Bundes-Versuchanst.* (Wien) **139**, 7–25. In German, with English summary.

Baillie, M. G. L. (1982). *Tree-Ring Dating and Archaeology*. University of Chicago Press, Chicago.

Baillie, M. G. L. and Munro, M. A. R. (1988) Irish tree rings, Sanotini and volcanic dust veils. *Nature* **332**, 344–346.

Bates, J. M. and Granger C. W. J. (1969). The combination of forecasts. *Operational Research Quarterly* **20(4)**, 451–468.

Betancourt, P. P. (1987). Dating the Aegean Late Bronze Age with radiocarbon. *Archaeometry* **29**, 45–49.

Binns, W. O. and Redfern D. B. (1983). Acid rain and forest decline in West Germany. *Forestry Commission Research and Development Paper* 131. Edinburgh, Scotland.

Blank, L. W. (1985). A new type of forest decline in Germany. *Nature* **314**, 311–334.

Blasing, T. J. (1978). Time series and multivariate analysis in paleoclimatology. In *Time Series and Ecological Processes*, H. H. Shugart, Jr., ed., pp. 212–226. *SIAM-SIMS Conference Series*, No. 5. Society for Industrial and Applied Mathematics, Philadelphia.

Blasing, T. J. and Duvick, D. N. (1984). Reconstruction of precipitation history in North American corn belt using tree rings. *Nature* **307**, 143–145.

Blasing, T. J., Duvick, D. N. and Cook, E. R. (1983). Filtering the effects of competition from ring-width series. *Tree-Ring Bulletin* **43**, 19–30.

Bormann, F. H. (1985). Air pollution and forests: an ecosystem perspective. *Bioscience* **35**, 434–441.

Box, G. E. P. and Jenkins G. H. (1976). *Time Series Analysis, Forecasting and Control*. Holden-Day; San Francisco.

Briffa, K. R., Jones, P. D., Wigley, T. M. L., Pilcher, J. R. and Baillie, M. G. L. (1983). Climate reconstructions from tree rings: part 1. Basic methodology and preliminary results for England. *Journal of Climatology* **3**, 233–242.

Briffa, K. R., Jones, P. D. Wigley, T. M. L., Pilcher, J. R. and Baillie, M. G. L. (1986). Climate reconstructions from tree rings: part 2, spatial reconstructions of summer mean sea-level pressure patterns over Great Britain. *Journal of Climatology* **6**, 1–15.

Briffa, K. R., Jones, P. D., Pilcher, J. R. and Hughes, M. K. (1988a) Reconstructing summer temperatures in Northern Fennoscandinavia back to 1700 A.D. using tree-ring data from Scots pine. *Arctic and Alpine Research* **20**, 385–394.

Briffa, K. R., Jones, P. D. and Schweingruber, F. H. (1988b) Summer temperature patterns over Europe: a reconstruction to 1750 A.D. based on maximum latewood density indices of conifers. *Quaternary Research* **30**, 34–52.

Brookes, M. H., Campbell, R. W., Colbert, J. J. Mitchell, R. G. and Stark, R. W. technical coordinators. (1987). Western spruce budworm. *Technical Bulletin* No. 1694, US Department of Agriculture, Forest Service, Washington, DC, 198 pp.

Brown, J. M. (1968). The photosynthetic regime of some Southern Arizona ponderosa pine. PhD dissertation, University of Arizona, Tucson.

Brubaker, L. B. and Cook, E. R. (1983). Tree-ring studies of Holocene environments. In *Late Quaternary Environments of the United States*, Vol. 2: *The Holocene*, H. E. Wright, Jr, ed., pp. 222–235, University of Minnesota Press, Minneapolis.

Brubaker, L. B. and Greene, S. K. (1979). Differential effects of Douglas-fir tussock moth and western spruce budworm defoliation on radial growth of grand fir and Douglas-fir. *Canadian Journal of Forest Research* **9**, 95–105.

Bryson, R. A. (1985). On climatic analogs in paleoclimatic reconstruction. *Quaternary Research* **23**, 275–86.

Clark, N., Blasing, T. J. and Fritts, H. C. (1975). Influence of interannual climatic fluctuations on biological systems. *Nature* **256**(5515): 302–304.

Cleaveland, M. K. (1986). Climatic response of densitometric properties in semiarid site tree rings. *Tree Ring Bulletin* **48**, 13–29.

Cogbill, C. V. (1977). The effect of acid precipitation on tree growth in eastern North America. *Water, Air, and Soil Pollution* **8**, 89–93.

Conkey, L. E. (1982a). Eastern US tree-ring widths and densities as indicators of past climate. PhD. dissertation, University of Arizona, Tucson, 204 pp.

Conkey, L. E. (1982b). Temperature reconstructions in the north-eastern United States. In *Climate from Tree Rings*, M. K. Hughes, P. M. Kelly, J. R. Pilcher and V. C. LaMarche, Jr., eds., pp. 165–168. Cambridge University Press, Cambridge.

Conkey, L. E. (1986). Red spruce tree-ring widths and densities in eastern North America as indicators of climate. *Quaternary Research* **26**, 232–243.

Cook, E. R. (1982). Eastern North America. In *Climate from Tree Rings*, M. K. Hughes, P. M. Kelly, J. R. Pilcher and V. C. LaMarche, Jr., eds., pp. 126–133. Cambridge University Press, Cambridge.

Cook, E. R. (1985). A time-series analysis approach to tree-ring standardization. PhD dissertation, University of Arizona, Tucson, 175 pp.

Cook, E. R. (1987). The use and limitations of dendrochronology in studying effects of air pollution on forests. In *Effects of Atmospheric Pollutants on Forests, Wetlands and Agricultural Ecosystems*, T. C. Hutchinson and K. M. Meema, eds. *NATO ASI Series*, Vol. G16. Springer-Verlag, Berlin and Heidelberg.

Cook, E. R. Personal communications. Tree-Ring Laboratory, Lamont-Doherty Geological Observatory of Columbia University, Palisades, New York.

Cook, E. R. and Jacoby, G. C. (1977). Tree-ring–drought relationships in the Hudson Valley, New York. *Science* **198**, 399–401.

Cook, E. R. and Jacoby, G. C. (1983). Potomac River streamflow since 1730 as reconstructed by tree rings. *Journal of Climate and Applied Meteorology* **22**, 1659–1674.

Cook, E. R. and Peters, K. (1981). The smoothing spline: a new approach to standardizing forest interior tree-ring width series for dendroclimatic studies. *Tree-Ring Bulletin* **41**, 45–53.

Cook, E. R., Johnson, A. H. and Blasing, T. J. (1987). Forest decline: modeling the effect of climate in tree rings. *Tree Physiology* **3**, 27–40.

Cooper, C. F., Blasing, T. J. and Fritts, H. C. (1974). Simulation models of the effects of climatic change on natural eco-systems. In *Proceedings of the Third Conference on the Climatic Assessment Program (CIAP), February 26–March 1, 1974.* US Department of Transportation, Cambridge, MA.

Craighead, F. C. (1927). Abnormalities in annual rings resulting from forest fires. *Journal of Forestry* **25**, 840–842.

Cropper, J. P. (1982). Climate reconstructions from tree rings: comment. In *Climate from Tree Rings*, M. K. Hughes, P. M. Kelly, J. R. Pilcher and V. C. LaMarche, Jr, eds, pp. 65–67. Cambridge University Press, Cambridge.

Cropper, J. P. (1985). Tree-ring response functions. An evaluation by means of simulations. PhD. dissertation, Department of Geosciences, University of Arizona, Tucson, 132 pp.

Dean, J. S. (1986). Dendrochronology. In *Dating and Age Determination of Biological Materials*, M. R. Zimmerman and J. L. Angel, eds., pp. 126–165. Croom Helm, London.

DeWitt, E. and Ames, A. (eds) (1978). Tree-ring chronologies of eastern North America. *Chronology Series* VI, Vol. 1. Laboratory of Tree-Ring Research, University of Arizona, Tucson.

Dieterich, J. H. (1980). Chimney Spring forest fire history. *Research Paper* RM-220, US Department of Agriculture, Forest Service, Rocky Mountains Forest and Range Experiment Station, Fort Collins, CO.

Dieterich, J. H. and Swetnam T. W. (1984). Dendrochronology of a fire scarred ponderosa pine. *Forest Science* **30(1)**, 238–247.

Doane, C. C. and McManus, M. L. (eds) (1981). The gypsy moth: research toward integrated pest management. USDA Expanded Gypsy Moth Research Development Program, Forest Service, Science and Education Agency *Technical Bulletin* 1584.

Douglass, A. E. (1914). A method of estimating rainfall by the growth of trees. In *The Climatic Factor*, E. Huntington, ed., pp. 101–22. *Carnegie Institution of Washington Publication* No. 192.

Douglass, A. E. (1919). *Climatic Cycles and Tree Growth*, Vol. I. A Study of the Annual Rings of Trees in Relation to Climate and Solar Activity. *Carnegie Institution of Washington Publication* No. 289.

Douglass, A. E. (1928). *Climatic Cycles and Tree Growth*, Vol. II. A Study of the Annual Rings of Trees in Relation to Climate and Solar Activity. *Carnegie Institution of Washington, Publication* No. 289.

Douglass, A. E. (1935). Dating Pueblo Bonito and other ruins of the Southwest. *National Geographic Society Contribution Technical Paper, Pueblo Bonito Series* 1.

Douglass, A. E. (1936). *Climatic Cycles and Tree Growth*, Vol. III. A Study of Cycles. *Carnegie Institution of Washington Publication* No. 289.

Douglass, A. E. (1937). Tree rings and chronology. *University of Arizona Bulletin* **8**, Physical Science Series 1.

Douglass, A. E. (1941). Crossdating in dendrochronology. *Journal of Forestry* **39**, 825–831.

Douglass, A. E. (1946). Precision of ring dating in tree-ring chronologies. *University of Arizona Bulletin* **XVII**, No. 3.

Duff, G. H. and Nolan, N. J. (1953). Growth and morphogenesis in the Canadian forest species, I. The controls of cambial and apical activity in *Pinus resinosa* Ait. *Canadian Journal of Botany* **31**, 471–513.

Eckstein, D. (1985). On the application of dendrochronology for the evaluation of forest damage. In *Inventorying and Monitoring Endangered Forests. IUFRO Conference Zurich 1985*, pp. 287–290, Birmensdorf, Eidg. Anstalt fur das forstliche Versuchswesen.

Eckstein, D. and Frisse, E. (1982). The influence of temperature and precipitation on vessel area and ring width of oak and beech. In *Climate from Tree Rings*, M. K. Hughes, P. M. Kelly, J. R. Pilcher and V. C. LaMarche Jr (eds), p. 12. Cambridge University Press, Cambridge.

Eckstein, D., Ogden, J., Jacoby, G. C. and Ash, J. (1981). Age and growth rate determination in tropical trees: the application of dendrochronological methods. *School of Forestry and Environmental Studies Bulletin* No. 94, Yale University, pp. 83–106.

Eckstein, D., Richter, K., Aniol, R. W. and Quiehl, F. (1984). Dendroclimatological investigations of the beech decline in the southwestern part of the Vogelsberg (West Germany). In German, with English abstract. *Forstwissenschaftliches Centralblatt* **103**, 274–290.

Evenden, J. D. (1940). Effects of defoliation by the pine butterfly upon ponderosa pine. *Journal of Forestry* **38**, 949–955.

Ferguson, C. W. (1968). Bristlecone pine: science and esthetics. *Science* **159**(3817), 839–846.

Ferrel, G. T. (1980). Growth of white firs defoliated by Modoc budworm in northeastern California. *Research Paper* PSW-153, US Department of Agriculture, Forest Service, Pacific Southwest Forest and Range Experiment Station, Berkely, California.

Filion, L., Payette, S., Gauthier, L. and Boutin, Y. (1986). Light rings in subarctic conifers as a dendrochronological tool. *Quaternary Research* **26**, 272–279.

Fox, C. A. (1980). The effect of air pollution on western larch as detected by tree-ring analysis. PhD dissertation, Arizona State University, Tempe, 98 pp.

Fox, C. A., Kincaid, W. B., Nash III, T. H., Young, D. L. and Fritts, H. C. (1986). Tree-ring variation in western larch (*Larix occidentalis*) exposed to sulfur dioxide emissions. *Canadian Journal of Forest Research* **16**, 283–292.

Franklin, J. F., Shugart, H. H. and Harmon, M. E. (1987). Tree death as an ecological process. *Bioscience* **37**, 550–556.

Friedland, A. J., Gregory, R. A., Karenlampi, L., Johnson, A. H. (1984). Winter damage to foliage as a factor in red spruce decline. *Canadian Journal of Forest Research* **14**, 963–965.

Fritts, H. C. (1969). Bristlecone pine in the White Mountains of California: growth and ring-width characteristics. *Papers of the Laboratory of Tree-Ring Research* 4. University of Arizona Press, Tucson.

Fritts, H. C. (1971). Dendroclimatology and dendroecology. *Quarternary Research* **1**, 419–449.

Fritts, H. C. (1974). Relationships of ring widths in arid site conifers to variations in monthly temperature and precipitation. *Ecological Monograph* **44**, 411–440.

Fritts, H. C. (1976). *Tree Rings and Climate*. Academic Press, London, 567 pp. Reprinted in: *Methods of Dendrochronology* Vols. II and III, L. Kairiukstis, Z. Bednarz and E. Feliksik (ed). Proceedings of the Task Force Meeting on Methodology of Dendrochronology: East/West Approaches, 2–6 June, 1986, Krakow, Poland. IIASA, Laxenburg, Austria.

Fritts, H. C. (1982). An overview of dendroclimatic techniques, procedures, and prospects. In *Climate from Tree Rings*, M. K. Hughes, P. M. Kelly, J. R. Pilcher and V. C. LaMarche, Jr (eds), pp. 191–197. Cambridge University Press, Cambridge.

Fritts, H. C. Reconstructing large-scale climatic patterns from tree-ring data: a diagnostic analysis. University of Arizona Press, Tucson. (in press).

Fritts, H. C. and Gordon, G. A. (1982). Reconstructed annual precipitation for California. In *Climate from Tree Rings*, M. K. Hughes, P. M. Kelly, J. R. Pilcher and V. C. LaMarche, Jr., eds., pp. 185–191. Cambridge University Press, Cambridge.

Fritts, H. C., Smith, D. G., Budelsky, C. A. and Cardis, J. W. (1965a). The variability of ring characteristics within trees as shown by a reanalysis of four ponderosa pine. *Tree-Ring Bulletin* **27**, 3–18.

Fritts, H. C., Smith, D. G., Cardis, J. W. and Budelsky, C. A. (1965b). Tree-ring characteristics along a vegetation gradient in northern Arizona. *Ecology* **46**, 393–401.

Fritts, H. C., Mossiman, J. E. and Bottorf, C. P. (1969). A revised computer program for standardizing tree-ring indices. *Tree-Ring Bulletin* **29**, 15–20.

Fritts, H. C., Blasing, T. J., Hayden, B. P. and Kutzbach, J. E. (1971). Multivariate techniques for specifying tree-growth and climate relationships and for reconstructing anomalies in paleoclimate. *Journal of Applied Meteorology* **10**, 845–864.

Fritts, H. C., Lofgren, G. R. and Gordon, G. A. (1979). Variations in climate since 1602 as reconstructed from tree rings. *Quaternary Research* **12**, 18–46.

Fritts, H. C., Guiot, J. and Gordon, G. A. (in press). Methods for calibration, verification and reconstruction. In *Methods of Tree-Ring Analysis: Applications in the Environmental Sciences*, L. Kairiukstis and E. Cook, eds. Reidel Press, The Netherlands.

Gale, J. (1986). Carbon dioxide enhancement of tree growth at high elevations. *Science* **231**, 859.

Garfinkel, H. L. and Brubaker, L. B. (1980). Modern climate–tree-growth relationships and climatic reconstruction in subarctic Alaska. *Nature* **286**, 872–874.

Gates, D. M. (1980). *Biophysical Ecology*. Springer-Verlag, New York, 611 pp.

Gates, W. L. and Mintz, Y. (1975). *Understanding Climatic Change: A Program for Action*. Report of the Panel on Climatic Variation of the US Committee for GARP (Global Atmosphere Research Project), National Research Council, National Academy of Sciences, Washington, D.C.

Gemmill, B., McBride, J. R. and Laven, R. D. (1982). Development of tree-ring chronologies in an ozone air pollution-stressed forest in southern California. *Tree-Ring Bulletin* **42**, 23–31.

Giardino, J. R., Shroder, Jr., J. F. and Lawson, M. P. (1984). Tree-ring analysis of movement of a rock-glacier complex on Mount Mestas, Colorado, U.S.A. *Arctic and Alpine Research* **16**, 299–309.

Glock, W. S. (1951). Cambial frost injuries and multiple growth layers at Lubbock, Texas. *Ecology* **32**, 28–36.

Glock, W. S., Studhalter, R. A. and Agerter, S. R. (1960). Classification and Multiplicity of Growth Layers in the Branches of Trees at the Extreme Lower Forest Border. *Smithsonian Miscellaneous Collection* 140(1). *Smithsonian Institution Publication* 4421, Washington, DC.

Gordon, G. A. (1982). Verification of dendroclimatic reconstructions. In *Climate from Tree Rings*, M. K. Hughes, P. M. Kelly, J. R. Pilcher and V. C. LaMarche, Jr, (eds), pp. 58–61. Cambridge University Press, Cambridge.

Gordon, G. A. and Leduc, S. K. (1981). Verification statistics for regression models. *Seventh Conference on Probability and Statistics in Atmospheric Sciences*, Session 8.

Gore, A. P., Johnson, E. A. and Lo, H. P. (1985). Estimating the time a dead tree has been on the ground. *Ecology* **66**, 1981–1983.

Gould, S. J. (1987). *Time's Arrow, Time's Cycle*. Harvard University Press, Cambridge, Mass.

Graumlich, L. J. and Brubaker, L. B. (1986). Reconstruction of annual temperatures (1590–1979) for Longmire, Washington derived from tree rings. *Quaternary Research* **25**, 223–234.

Graumlich, L. J. and Brubaker, L. B. (1987). Increasing net primary productivity in Washington (USA) forests during the last 1000 years. In *Proceedings of the International Symposium on Ecological Aspects of Tree-Ring Analysis*, August 17–21, 1986, Marymount College, Tarrytown, New York, US Department of Energy CONF-8608144, pp. 59–69.

Graybill, D. A. (1979). Revised computer programs for tree-ring research. *Tree-Ring Bulletin* **39**, 77–82.

Graybill, D. A. (1982). Chronology development and analysis. In *Climate from Tree Rings*, M. K. Hughes, P. M. Kelly, J. R. Pilcher and V. C. LaMarche, Jr, (eds), pp. 21–28. Cambridge University Press, Cambridge.

Graybill, D. A. (1987). A network of high elevation conifers in the western US for detection of tree-ring growth response to increasing atmospheric carbon dioxide. *Proceedings of the International Symposium on Ecological Aspects of Tree-Ring Analysis*, August 17–21, 1986, Marymount College, Tarrytown, New York, US Department of Energy CONF-8608144, pp. 463–474.

Greve, U., Eckstein, D. Aniol, R. W. and Scholz, F. (1986). Dendroclimatological investigations on Norway spruce under different loads of air pollution. Allaemeine Forst- und Jagdzetund **157**, 174–179.

Guiot, J. (1986). ARMA techniques for modelling tree-ring response to climate and for reconstructing variations of paleoclimates. *Ecological Modelling* **33**, 149–171.

Guyette, R. and McGinnes, Jr, E. A. (1987). Potential in using elemental concentrations in radial increments of old growth eastern redcedar to examine the chemical history of the environment. *Proceedings of the International Symposium on Ecological Aspects of Tree-Ring Analysis*, August 17–21, 1986, Marymount College, Tarrytown, New York, US Department of Energy CONF-8608144, pp. 671–680.

Hammer, C. U., Clausen, H. B., Friedrich, W. L. and Tauber, H. (1987). The Minoan eruption of Santorini in Greece dated to 1645 B.C.? *Nature* **328**, 517–519.

Harmon, M. E., Franklin, J. F. Swanson, F. J. *et al.* (1986). Ecology of coarse woody debris in temperate ecosystems. *Advances in Ecological Research* **15**, 133–302.

Hecht, A. D. (ed.) (1985). *Paleoclimate Analysis and Modelling*. Wiley, New York.

Heikkinen, O. and Tikkanen, M. (1981). The effect of air pollution on growth in conifers: an example from the surroundings of the Skoldvik oil refinery. *Terra* **39**, 134–144. In Finnish, with English summary.

Holmes, R. L. (1983). Computer-assisted quality control in tree-ring dating and measuring. *Tree-Ring Bulletin* **43**, 69–78.

Holmes, R. L., Stockton, C. W. and LaMarche, Jr, V. C. (1979). Extension of river flow records in Argentina from long tree-ring chronologies. *Water Resources Bulletin* **15**, 1081–1085.

Holmes, R. L., Adams, R. K. and Fritts, H. C. (1986). Tree-ring chronologies of western North America: California, eastern Oregon and northern Great Basin. *Chronology Series* VI, Laboratory of Tree-Ring Research, University of Arizona, Tucson.

Hornbeck, J. W. and Smith, R. B. (1985). Documentation of red spruce growth decline. *Canadian Journal of Forest Research* **15**, 119–1201.

Horton, J. C. and Bicak, C. J. (1987). Modeling for biologists. *Bioscience* **37**, 808–809.

Huber, F. (1976). Problemes d'interdatation chez le pin sylvestre et influence du climat sur la structure de ses accroissements annuels. *Annales des Sciences Forestieres* **33**, 61–86. In French.

Hughes, M. K. (1987). Requirements for spatial and temporal coverage. Introduction. In *Methods in dendrochronology: East/West Approches*, L. Kairiukstis, (ed.) IIASA. Polish Academy of Sciences.

Hughes, M. K., Kelly, P. M., Pilcher, J. R. and LaMarche, Jr, V. C. (eds). (1982). *Climate from Tree Rings*. Cambridge University Press, Cambridge. 223 pp.

Jacoby, G. C. (ed) (1980). *Proceedings of the International Meeting on Stable Isotopes in Tree-Ring Research*, New Paltz, New York. *Carbon Dioxide Effects Research and Assessment Program, Publication* No. 12, US Department of Energy, CONF-790518, UC-11, Washington, DC.

Jacoby, G. C. and Hornbeck, J. W. (compilers) (1987). *Proceedings of the International Symposium on Ecological Aspects of Tree-Ring Analysis*, August 17–21, 1986, Marymount College, Tarrytown, New York, US Department of Energy CONF-8608144, 726 pp.

Jacoby, G. C., Cook, E. R. and Ulan, L. D. (1985). Reconstructed summer degree days in central Alaska and northwestern Canada since 1524. *Quaternary Research* **23**, 18–26.

Jagels, R. and Telewski, F. W. (in the press.) Video image analysis in tree-ring research. In *Methods of Tree-Ring Analysis: Applications in the Environmental Sciences*, L. Kairuikstis, and E. Cook, eds. Reidel Press, The Netherlands.

Jenkins, G. M. and Watts, D. G. (1968). *Spectral Analysis and Its Applications*. Holden-Day, San Francisco.

Johnson, A. H. and Siccama, T. G. (1983). Acid deposition and forest decline. *Environmental Science and Technology* **17**, 294A–305A.

Johnson, A. H., Siccama, T. G., Wang, D., Turner, R. S. and Barringer, T. H. (1981). Recent changes in patterns of tree growth rate in the New Jersey pinelands: a possible effect of acid rain. *Journal of Environmental Quality* **10**, 427–430.

Jones, R. H. (1985). Time series analysis-time domain. In *Probability, Statistics, and Decision Making in the Atmospheric Sciences*, A. H. Murphy and R. W. Katz (eds), pp. 223–260. Westview Press, Boulder.

Jones, P. D., Wigley, T. M. L. and Wright, P. B. (1986). Global temperature variations, 1861–1984. *Nature* **322**, 430–434.

Jonsson, B. and Sundberg, R. (1972). Has the acidification by atmospheric pollution caused a growth reduction in Swedish forests? *Research Note* No. 20. Department of Forest Field Research, Royal College of Forestry, Stockholm.

Jonsson, B. and Svensson, L. G. (1982). A study of the effects of air pollution on forest yield. A follow-up of the report of Jonsson and Sundberg 1972 and a new study based on forest types. *Avdelningen för Skogsuppskattning och Skogsindelning, Sveriges Lantbruksuniversitet* No. 9, 61 pp.

Kairiukstis, L. and Cook, E. R. (eds) (in press.) *Methods of Tree-Ring Analysis: Applications in the Environmental Sciences*. Reidel Press, The Netherlands.

Keen, F. P. (1937). Climatic cycles in eastern Oregon as indicated by tree rings. *Monthly Weather Review* **65**, 175–181.

Keller, R. (1968). Des caractéristiques nouvelles pour l'étude des propriétés mécaniques des bois: les composantes de la densité. *Annales des Sciences Forestieres* **25**, 237–249. In French.

Keller, T. (1980). The effect of a continuous springtime fumigation with sulfur dioxide and carbon dioxide uptake and structure of the annual ring in spruce. *Canadian Journal of Forest Research* **10**, 1–6.

Kelly, P. M. and Sear, C. B. (1984). Climatic impact of explosive volcanic eruptions. *Nature* **311**, 740–743.

Kienast, F. (1982). Analytical investigations based on annual tree rings in damaged forest areas of the Valais (Rhone Valley) endangered by pollution. *Geographica Helvetica* **3**, 143–148. In German, with English summary.

Kienast, F. (1985). Tree-ring analysis, forest damage and air pollution in the Swiss Rhone valley. *Land Use Policy* **2**, 71–77.

Kienast, F. and Schweingruber, F. H. (1986). Dendroecological studies in the Front Range, Colorado, USA. *Arctic and Alpine Research* **18**, 277–288.

Kienast, F., Flühler, H. and Schweingruber, F. H. (1981). Jahrringanalysen an Föhren (*Pinus silvestris* L.) aus immissionsgefährdeten Waldbeständen des Millelwallis (Sacon, Schweiz). *Eidgenoessische Anstalt fuer das Forstliche Versuchswesen Mitteilungen* **57, 4**, 415–432.

Knight, H. A. (1987). The pine decline. *Journal of Forestry* **85**, 25–28.

Koerber, T. W. and Wickman, B. E. (1970). Use of tree-ring measurements to evaluate impact of insect defoliation. In *Tree-Ring Analysis with Special Reference to Northwest America*, J. H. G. Smith and J. Worrall, eds., *University of British Columbia Faculty of Forest Research Bulletin* 7, Vancouver, pp. 101–106.

Kontic, R., Niederer, M., Nippel, C. A., Winkler-Seifert, A. and A. (1986). Dendrochronological analysis on conifers for presentation and interpretation of forest damages (Vallis, Switzerland). Swiss Federal Institute of Forestry Research, Report No. 283, Birmensdorf.

Kuhn, T. S. (1970). *The Structure of Scientific Revolution. International Encyclopedia of Unified Science*, Vol. II, No. 2, University of Chicago Press, Chicago, IL., 210 pp.

Kutzbach, J. E. (1985). Modeling of paleoclimates. *Reviews of Geophysics* **28A**, 159–196.

LaMarche, V. C., Jr, (1966). An 800-year history of stream erosion as indicated by botanical evidence. In *US Geological Survey Professional Paper* 550-D, pp. 83–86, Washington, DC.

LaMarche, V. C., Jr (1968). Rates of slope degradation as determined from botanical evidence, White Mountains, California. Erosion and sedimentation in a semiarid environment. In *US Geological Survey Professional Paper* 325-I, pp. 341–377.

LaMarche, V. C., Jr (1970). Frost-damage rings in subalpine conifers and their application to tree-ring dating problems. In *Tree-Ring Analysis with Special Reference to Northwest America*, J. H. G. Smith and J. Worrall, eds., *University of British Columbia Faculty Forestry Bulletin* 7, Vancouver, pp. 99–100.

LaMarche, V. C., Jr, (1974). Frequency-dependent relationships between tree-ring series along an ecological gradient and some dendroclimatic implications. *Tree-Ring Bulletin* **34**, 1–20.

LaMarche, V. C., Jr (1978). Tree-ring evidence of past climatic variability. *Nature* **276**, 334–338.

LaMarche, V. C., Jr (1982). Sampling strategies. In *Climate from Tree Rings*, M. K. Hughes, P. M. Kelly, J. R. Pilcher and V. C. LaMarche, Jr., eds., pp. 2–6. Cambridge University Press, Cambridge.

LaMarche, V. C., Jr and Harlan, T. P. (1973). Accuracy of tree-ring dating of bristlecone pine for calibration of the radiocarbon time scale. *Journal of Geophysical Research* **78**, 8849–8858.

LaMarche, V. C., Jr and Hirschboeck, K. K. (1984). Frost rings in trees as records of major volcanic eruptions. *Nature* **307**, 121–126.

LaMarche, V. C., Jr, Graybill, D. A., Fritts, H. C. and Rose, M. R. (1984). Increasing atmospheric carbon dioxide: tree-ring evidence for growth enhancement in natural vegetation. *Science* **225**, 1019–1021.

LaMarche, V. C., Jr, Graybill, D. A., Fritts, H. C. and Rose, M. R. (1986). Carbon dioxide enhancement of tree growth at high elevations. In Technical Comments, *Science* **231**, 859–860.

Lamb, H. H. (1977). *Climate: Present, Past and Future, II. Climatic History and the Future*. Methuen, London. 835 pp.

LeBlanc, D. C., Raynal, D. J., White, E. H. and Ketchledge, E. H. (1987). Characterization of historical growth patterns in declining red spruce trees. In *Proceedings of the International Symposium on Ecological Aspects of Tree-Ring Analysis*, August 17–21, 1986, Marymount College, Tarrytown, New York, US Department of Energy CONF-8608144, pp. 360–371.

Leggett, P., Hughes, M. K. and Hibbert, F. A. (1978). A modern oak chronology from North Wales and its interpretation. In *Dendrochronology in Europe: Principles, Interpretation and Applications to Archaeology and History*, J. Fletcher, ed., pp. 187–194, *BAR International Series* 51, Oxford, England.

Lemon, E. R. (ed) (1983). *CO_2 and Plants*. American Association for the Advancement of Science Selected Symposium 84. Westview Press, Boulder, Colorado, 280 pp.

Lenz, O., Schar, E. and Schweingruber, F. H. (1976). Methodische probleme bei der radiographisch-densitometrischen bestimmung der dichte und der jahrringbreitenvon Holz. *Holzforschung* **30**, 114–123. In German.

Lepp, N. W. (1975). The potential of tree-ring analysis for monitoring heavy metal pollution patterns. *Environmental Pollution* **9**, 49–61.

Lofgren, G. R. and Hunt, J. H. (1982). Transfer functions. In *Climate from Tree Rings*, M. K. Hughes, P. M. Kelly, J. R. Pilcher and V. C. LaMarche, Jr., eds., pp. 50–56. Cambridge University Press, Cambridge.

Long, A. (1982). Stable isotopes in tree rings. In *Climate from Tree Rings*, M. K. Hughes, P. M. Kelly, J. R. Pilcher and V. C. LaMarche, Jr., eds., pp. 12–18. Cambridge University Press, Cambridge.

Lorimer, C. G. (1985). Methodological considerations in the analysis of forest disturbance history. *Canadian Journal of Forest Research* **15**, 200–213.

Lough, J. M. and Fritts, H. C. (1987). An assessment of the possible effects of volcanic eruptions on North American climate using tree-ring data, 1602 to 1900 A.D. *Climatic Change* **10**, 219–239.

Madany, M. H., Swetnam, T. W. and West, N. E. (1982). Comparison of two approaches for determining fire dates from tree scars. *Forest Science* **28**, 856–861.

Marchand, P. J. (1984). Dendrochronology of a fir wave. *Canadian Journal of Forest Research* **14**, 51–56.

Marquardt, D. W. and Snee, R. D. (1985). Developing empirical models with multiple regression biased estimation techniques. In *Probability, Statistics, and Decision Making in the Atmospheric Sciences*, A. H. Murphy and R. W. Katz, eds., pp. 45–100, Westview Press, Boulder.

Mass, C. and Schneider, S. H. (1977). Statistical evidence on the influence of sunspots and volcanic dust on long-term temperature records. *Journal of Atmospheric Science* **34**, 1995–2004.

McBride, J. R., Semion, V. and Miller, P. R. (1975). Impact of air pollution on growth of ponderosa pine. *California Agriculture* **29**, 8–10.

McLaughlin, S. B. (1984). FORAST: a regional scale study of forest responses to air pollutants. In *Proceedings of the Conference, Air Pollution and the Productivity of the Forest*. Izaak Walton League, D. Davis (ed.), pp. 241–253.

McLaughlin, S. B. (1985). Effects of air pollution on forests: a critical review. *Journal of the Air Pollution Control Association* **35**, 512–534.

McLaughlin, S. B., Blasing, T. J., Mann, L. K. and Duvick, D. N. (1983). Effects of acid rain and gaseous pollutants on forest productivity: a regional scale approach. *Journal of the Air Pollution Control Association* **33**, 1042–1049.

Meko, D. M. (1981). Applications of Box-Jenkins methods of time series analysis to the reconstruction of drought from tree rings. PhD dissertation, University of Arizona, Tucson, 149 pp.

Miller, P. R. (1973). Oxidant-induced community change in a mixed conifer forest. In *Air Pollution Damage to Vegetation. Advances in Chemistry Series* **122**, 101–117.

Miller, P. R. (1985). Ozone effects in the San Bernardino National Forest. In *Air Pollutant Effects on Forest Ecosystems*, May 8–9, 1985, St Paul, MN, pp. 161–196.

Monserud, R. A. (1986). Time-series analyses of tree-ring chronologies. *Forest Science* **32**, 349–372.

Morrison, I. K. (1984). Acid rain: a review of literature on acid deposition effects in forest ecosystems. *Forestry Abstracts* **45**, 483–506.

Morrow, P. A. and LaMarche, Jr, V. C. (1978). Tree-ring evidence for chronic insect suppression of productivity in subalpine *Eucalyptus*. *Science* **201**, 1244–1245.

Nash, T. H., Fritts, H. C. and Stokes, M. A. (1975). A technique for examining nonclimatic variation in widths of tree rings with special reference to air pollution. *Tree-Ring Bulletin* **35**, 15–24.

Norton, D. A. (1979). Phenological growth characteristics of *Nothofagus solandri* at three altitudes in the Craigieburn Range, New Zealand. *New Zealand Journal of Botany* **22**, 413–424.

Norton, D. A. (1983). A dendroclimatic analysis of three indigenous tree species, South Island, New Zealand. PhD dissertation, University of Canterbury, New Zealand, 439 pp.

Ogden, J. (1982). Australasia. In *Climate from Tree Rings*, M. K. Hughes, P. M. Kelly, J. R. Pilcher and V. C. LaMarche, Jr, (eds), pp. 90–103. Cambridge University Press, Cambridge.

O'Neil, L. C. (1963). The suppression of growth rings in jack pine in relation to defoliation by the Swaine jack-pine sawfly. *Canadian Journal of Botany* **41**, 227–235.

Parker, M. L. and Henoch W. E. S. (1971). The use of Engelmann spruce latewood density for dendrochronological purposes. *Canadian Journal of Forest Research* **1**, 90–98.

Payette, S., Filion, L., Gauthier, L. and Boutin, Y. (1985). Secular climate change in old-growth tree-line vegetation of northern Quebec. *Nature* **315**, 135–138.

Peet, R. K. (1981). Forest vegetation of the Colorado Front Range. *Vegetatio* **45**, 3–75.

Peterson, D. L. (1985). Evaluating the effects of air pollution and fire on tree growth by tree-ring analysis. In *Proceedings of the Eighth Conference on Fire and Forest Meteorology*, Detroit, Michigan, April 29 to May 2, 1985. Society of American Foresters, Bethesda, Maryland, pp. 124–131.

Peterson, D. L., Arbaugh, M. J., Wakefield, V. A. and Miller, P. R. (1987). Evidence of growth reduction in ozone-injured Jeffrey pine (*Pinus jeffreyi* [Grev. and Balf.]) in Sequoia and Kings Canyon National Parks. *Journal of the Air Pollution Control Association* **37**, 906–912.

Phipps, R. L. (1972). *Tree Rings, Stream Runoff and Precipitation in Central New York. US Geological Survey Professional Paper* 800–B, B259–264, Washington, DC.

Phipps, R. L. (1982). Comments on interpretation of climatic information from tree rings, eastern North America. *Tree Ring Bulletin* **42**, 11–22.

Phipps, R. L., Ierley, D. L. and Baker, C. P. (1979). Tree rings as indicators of hydrologic change in the Great Dismal Swamp, Virginia and North Carolina. *US Geological Survey Water Resources Investigations* 78–136, US Geological Survey, Washington, DC.

Pickett, S. T. A. and White, P. S. (eds). (1985). *The Ecology of Natural Disturbance and Patch Dynamics*, Academic Press, New York, 472 pp.

Polge, H. (1963). Une nouvelle methode de determination de la texture du bois: l'analyse densitometrique de cliches radiographiques. *Annales de l'ecole nat. eaux et forets et de la station de recherche et exper.* **20**, 531–581. In French.

Polge, H. (1966). Etablissement des courbes de variation de la densite du bois par l'exploration densitometrique de radiographies d'echantillons preleves a la tariere sur des arbres vivants. *Annales des sciences forestieres* **23**, 1–206. In French.

Polge, H. (1970). The use of x-ray densitometric methods in dendrochronology. *Tree-Ring Bulletin* **30**, 4–10.

Pollanschütz, J. (1971). Die ertragskundlichen Messmethoden zur Erkennung und Beurteilung von forstlichen Rauchschäden. In *Mitt. d. forstl. Bundes-Versuchanst (Wien)*. **92**, 153–206.

Richards, K. S. (1981). Stochastic processes in one-dimensional series: an introduction. In *Concepts and Techniques in Modern Geography*, Institute of British Geographers, London, 57 pp.

Sanders, C. J., Stark, R. W., Mullins, E. J. and Murphy, J. (1985). *Recent Advances in Spruce Budworms Research.* Proceedings of the CANUSA Spruce Budworms Research Symposium. Canadian Forestry Service, Ottawa.

Schmitt, D. M., Grimble, D. G. and Searcy, J. L. (1984). Managing the spruce budworm in eastern North America. *Agriculture Handbook* 620. US Department of Agriculture, Forest Service, Washington, DC, 192 pp.

Schulman, E. (1947). Tree-ring hydrology in southern Califormia. *Laboratory of Tree-Ring Research Series* No. 4, *University of Arizona Bulletin* **18(3)**.

Schulman, E. (1951). Tree-ring indices of rainfall, temperature, and river flow. In *Compendium of Meteorology*, American Meteorological Society, (eds), pp. 1024–1029. Boston, MA.

Schulman, E. (1956). *Dendroclimatic Changes in Semiarid America.* University of Arizona Press, Tucson, 142 pp.

Schweingruber, F. H. (1980). Density fluctuations in annual rings of conifers in relation to climatic-ecological factors, or the problem of the false annual ring. Swiss Federal Institute of Forestry Research, Report No. 213, Birmensdorf, 35 pp. In German, with English summary.

Schweingruber, F. H. (1983). *Der Jahrring: Standort, Methodik, Zeit und Klima in der Dendrochronologie.* Verlag Paul Haupt, Bern, 234 pp. In German.

Schweingruber, F. H. (1986). Abrupt growth changes in conifers. *IUFRO Report*, Lijubliana, Sept. 1986.

Schweingruber, F. H. (personal communications). Swiss Federal Institute of Forestry research (Eidgenössiche Anstalt fur das Forstliche Versuuchswessen), Birmensdorf, ZH Switzerland.

Schweingruber, F. H., Fritts, H. C., Braker, O. U., Drew, L. G. and Schar, E. (1978a).

The x-ray technique as applied to dendroclimatology. *Tree-Ring Bulletin* **38**, 61–91.

Schweingruber, F. H., Braker, O. U. and Schar, E. (1978b). Dendroclimatic studies in Great Britain and in the Alps. In *Evolution of Planetary Atmospheres and the Climatology of the Earth*, pp. 369–72. CERN, Paris.

Schweingruber, F. H., Braker, O. U. and Schar, E. (1979). Dendroclimatic studies on conifers from central Europe and Great Britain. *Boreas* **8**, 427–452.

Schweingruber, F. H., Kontic, R. and Winkler-Seifert, A. (1983). Application of annual ring analysis in investigations of conifer die-back in Switzerland. *Swiss Institute of Forestry Research, Report* No. 253, Aug. 1983, Birmensdorf, pp. 6–29. In German, with English abstract.

Scott, J. T., Siccama, T. G., Johnson, A. H. and Briesch, A. R. (1984). Decline of red spruce in the Adirondacks, New York. *Bulletin of the Torrey Botanical Club* **111**, 438–444.

Self, S., Rampino, M. R. and Barbera, J. J. (1981). The possible effects of large 19th and 20th century volcanic eruptions on zonal and hemispheric surface temperatures. *Journal of Volcanology and Geothermal Research* **11**, 41–60.

Sheffield, R. M. and Cost, N. D. (1987). Behind the decline. *Journal of Forestry* **85**, 29–33.

Sheppard, P. R. and Jacoby, G. C. (1987). Tree ring data. *Journal of Forestry* **85**, 50–51.

Shroder, J. F., Jr, (1978). Dendrogeomorphological analysis of mass movement on Table Cliffs Plateu, Utah. *Quaternary Research* **9**, 168–185.

Shroder, J. F. Jr (1980). Dendrogeomorphology: review and new techniques of tree-ring dating. *Progress in Physical Geography* **4**, 161–188.

Shugart, H. H. (1984). *A Theory of Forest Dynamics: The Ecological Implications of Forest Succession Models*. Springer-Verlag, New York, 278 pp.

Shugart, H. H. (1987). Dynamic ecosystem consequences of tree birth and death patterns. *Bioscience* **37**, 596–602.

Sigafoos, R. S. (1964). Botanical evidence of floods and flood-plain deposition. *US Geological Survey Professional Paper* 485-A, Washington, DC.

Smiley, T. L. (1958). The geology and dating of Sunset Crater, Flagstaff, Arizona. In *Guidebook of the Black Mesa Basin, Northeastern Arizona*, R. Y. Anderson and J. W. Harshberger, eds., New Mexico Geological Society, Socorro, pp. 186–190.

Smith, W. H. (1985). Forest quality and air quality. *Journal of Forestry* **83**, 82–92.

Spencer, D. A. (1964). Porcupine population fluctuations in past centuries revealed by dendrochronology. *Journal of Applied Ecology* **1**, 127–49.

Stahle, D. W., Cleaveland, M. K. and Hehr, J. G. (1985a). A 450-year drought reconstruction for Arkansas, United States. *Nature* **316**, 530–532.

Stahle, D. W., Cook, E. R. and White, J. W. C. (1985b). Tree-ring dating of baldcypress and the potential for millennia-long chronologies in the southeast. *American Antiquity* **50**, 796–802.

Stevens, D. (1975). A computer program for simulating cambium activity and ring growth. *Tree-Ring Bulletin* **35**, 49–56.

Stockton, C. W. (1971). Feasibility of augmenting hydrologic records using tree-ring data. PhD dissertation, University of Arizona, Tucson, 172 pp.

Stockton, C. W. and Fritts, H. C. (1973). Long-term reconstruction of water level changes for Lake Athabasca by analysis of tree rings. *Water Resources Bulletin* **9**, 1006–1027.

Stockton, C. W. and Jacoby, Jr, G. C. (1976). Long-term surfacewater supply and streamflow trends in the Upper Colorado River Basin based on tree-ring analysis.

Lake Powell Research Project Bulletin 18, University of California, Los Angeles Institute of Geophysics and Planetary Physics, 70 pp.

Stockton, C. W. and Meko, D. M. (1975). A long-term history of drought occurrence in western United States as inferred from tree rings. *Weatherwise* **28**, 244–249.

Stockton, C. W. and Meko, D. M. (1983). Drought recurrence in the Great Plains as reconstructed from long-term tree-ring records. *Journal of Climate and Applied Meteorology* **22**, 17–29.

Stockton, C. W., Boggess, W. R. and Meko, D. M. (1985). Climate and tree rings. *Paleoclimate Analysis and Modeling*, Alan D. Hecht (ed.), pp. 71–161., Wiley, New York.

Stokes, M. A. and Smiley, T. L. (1968). *An Introduction to Tree-Ring Dating*. University of Chicago Press, 73 pp.

Swetnam, T. W. (1987). A dendrochronological assessment of western spruce budworm, *Choristoneura occidentalis*, Freeman, in the Southern Rocky Mountains. PhD dissertation, University of Arizona, Tucson, 213 pp.

Swetnam, T. W. and Dieterich, J. H. (1985). Fire history of ponderosa pine forests in the Gila Wilderness, New Mexico. In *Proceedings-Symposium and Workshop on Wilderness Fire*; Nov. 15–18, 1983, Missoula, MT. *General Technical Report* INT-182, U. S. Department of Agriculture, Forest Service, Ogden, UT, pp. 390–397.

Swetnam, T. W., Thompson, M. A. and Sutherland, E. K. (1985). Using dendrochronology to measure radial growth of defoliated trees. *Agriculture Handbook* 639, US Department of Agriculture, Forest Service, Washington, DC, 39 pp.

Taylor, R. M. (1981). Tree-ring analysis in forest productivity studies: an investigation of growth-climate relationships in the New Forest, Hampshire. *Journal of Biogeography* **8**, 293–312.

Taylor, B. L., Tzin Gal-Chen and Schneider, S. H. (1980). Volcanic eruptions and long-term temperature records: an empirical search for cause and effect. *Quarterly Journal of the Royal Meteorological Society* **106**, 175–199.

Telewski, F. W., Wakefield, A. H. and Jaffe, M. J. (1983). Computer assisted image analysis of tissues of ethrel treated *Pinus taeda* seedlings. *Plant Physiology* **72**, 177–181.

Thammincha, S. (1981). Climatic variation in radial growth of Scots pine and Norway spruce and its importance in growth estimation. *Acta Forestalia Fennica* **171**, 57 pp.

Thompson, M. A. (1981). Tree rings and air pollution: a case study of *Pinus monophylla* growing in east-central Nevada. *Environmental Pollution* (Series A) **26**, 251–266.

Tomlinson, G. H. (1983). Air pollutants and forest decline. *Environmental Science and Technology* **17**, 246A–256A.

Trefil, J. S. (1985). Concentric clues from growth rings unlock the past. *Smithsonian* **16**, 47–55.

Veblen, T. T. and Lorenz, D. C. (1986). Anthropogenic disturbance and recovery patterns in montane forests, Colorado Front Range. *Physical Geography* **7**, 1–24.

Villalba, R., Boninsegna, J. A. and Holmes, R. L. (1985). *Cedrela angustifolia* and *Juglans australis*, two new tropical species useful in dendrochronology. *Tree-Ring Bulletin* **45**, 25–35.

Wagener, W. W. (1961). Past fire incidence in Sierra Nevada Forests. *Journal of Forestry* **59**, 739–748.

Webb, G. E. (1983). *Tree Rings and Telescopes: The Scientific Career of A. E. Douglass*. The University of Arizona Press, Tucson, 242 pp.

Webb, T., Kutzback, J. and Street-Perrott, F. A. (1985). 20,000 years of global climatic change: paleoclimatic research plan. In *Global Change*, T. F. Malone and J. G. Roederer, eds. ICSU Press, New York, pp. 182–219.

White, P. S. (1979). Pattern, process, and natural disturbance in vegetation. *Botanical Review* **45**, 229–299.

Wigley, T. M. L. (1982). Oxygen-18, carbon-13 and carbon-14 in tree rings. In *Climate from Tree Rings*, M. K. Hughes, P. M. Kelly, J. R. Pilcher and V. C. LaMarche, Jr., eds., pp. 18–21. Cambridge University Press.

Wigley, T. M. L., Briffa, K. R. and Jones, P. D. (1984). On the average value of correlated time series, with applications in dendroclimatology and hydrometeorology. *Journal of Climate and Applied Meteorology* **23**, 201–213.

Wilson, B. F. (1964). A model for cell production by the cambium of conifers. In *The Formation of Wood in Forest Trees*, M. H. Zimmermann (ed), pp. 19–36. Academic Press, New York.

Wilson, B. F. and Howard, R. A. (1968). A computer model for cambial activity. *Forest Science* **14**, 77–90.

Wood, F. B. Jn. (1988) The need for systems Research on Global Climate Change *Systems Research* **5**, 225–240.

Woodwell, G. M., Hobbie, J. E., Houghton, R. A. *et al.* (1983). Global deforestation: contribution to atmospheric carbon dioxide. *Science* **222**, 1081–1085.

Wycoff, W. R., Crookston, N. L. and Stage, A. R. (1982). User's guide to the stand prognosis model. US Department of Agriculture, Forest Service *General Technical Report* INT-133, Intermountain Forest and Range Experiment Station, Ogden, UT, 112 pp.

Yamaguchi, D. K. (1983). New tree-ring dates for recent eruptions of Mount St. Helens. *Quarternary Research* **20**, 246–250.

Yamaguchi, D. K. (1985). Tree-ring evidence for a two-year interval between recent prehistoric explosive eruptions of Mount St. Helens. *Geology* **13**, 554–557.

Yokobori, M. and Ohta A. (1983). Combined air pollution and pine ring structure observed xylochronologically. *European Journal of Forest Pathology* **13**, 30–45.

Young, C. E. (1979). Tree rings and Kaibab North deer hunting success, 1925–1975. *Journal of Arizona Nevada Academy of Sciences* **14**, 61–65.

Zackrisson, O. (1980). Forest fire history: ecological significance and dating problems in the North Swedish Boreal Forest. In *Proceedings of the Fire History Workshop*, Oct. 20–24, Tucson, Arizona. *General Technical Report* RM-81. US Department of Agriculture, Forest Service. Rocky Mountain Forest and Range Experiment Station, Fort Collins, CO, pp. 120–125.

Zahner, R. and Myers, R. K. (1986a). A computer simulation for estimating drought impacts on southern pine growth rates. *Proceedings, Tree Rings and Forest Mensuration*, ed. by NCASI, Laboratory of Tree-Ring Research, University of Arizona, Tucson. April 9–11, 1986.

Zahner, R. and Myers, R. K. (1986b). Assessing the impact of drought on forest health. In *Forests, the World, & the Profession.* Proceedings of the 1986 Society of American Foresters National Convention, Birmingham, AL, Oct. 5–8. Society of American Foresters, 5400 Grosvenor Lane, Bethesda, MD 29814.

Zahner, R. and Myers, R. K. (1987). Dendroecological analysis of loblolly pine tree-ring chronologies for the near-term survey. Progress Report to Southeastern Forest Experiment Station, USDA Forest Service, Ashville, NC. 20 pp.

Zedaker, S. M., Hyink, D. M. and Smith, D. W. (1987). Growth declines in red spruce. *Journal of Forestry* **85**, 34–36.

Community Structure and Interaction Webs in Shallow Marine Hard-Bottom Communities: Tests of an Environmental Stress Model

B. A. MENGE and T. M. FARRELL

I. INTRODUCTION

A major goal of community ecology is determining the causes of spatial and temporal variation in community structure. Species diversity, species composition, relative abundance, trophic complexity, size structure, and spatial structure are all components of community structure. Differences in

ADVANCES IN ECOLOGICAL RESEARCH Vol. 19
ISBN 0-12-013919-7

community structure occur at several scales in space and time (Sutherland, 1981; DeAngelis and Waterhouse, 1987). Community patterns can vary on spatial scales ranging from centimetres to hundreds of kilometers. On rocky shores, these spatial scales correspond to microspatial variation (biota in holes and open surfaces often differ; Menge *et al.*, 1983), within site variation (changes in zonation, patchiness), and regional and global variation (e.g. gradients in species diversity and food web complexity). Relevant temporal scales of community variation are similarly broad, ranging from hours to centuries.

Community structure can vary due to physical and physiological stress (wave exposure, nutrient availability, tidal height) and biological factors (recruitment, predation, competition, mutualism; and indirect effects of these factors). Combining even a few factors and scales in a study of community variation can yield an investigation of great complexity and difficulty. Nevertheless, the number of sophisticated investigations of community regulation is growing (Brown *et al.*, 1986; Carpenter *et al.*, 1985; Dayton, 1971; Dayton *et al.*, 1984; Dungan, 1986; Hall *et al.*, 1970; Inouye *et al.*, 1980; Lubchenco and Menge, 1978; Lubchenco, 1986; Lynch, 1979; McAuliffe, 1984; McNaughton, 1983; Menge, 1976; Morin, 1983; Robles, 1987; Sebens, 1985, 1986a,b; Sousa, 1979a; Peterson, 1982; Underwood *et al.*, 1983; Wilbur, 1987; Williams, 1981). If we expect to gain an understanding of the forces which regulate communities, we must continue such multifactorial investigations (Connell, 1983; Diamond and Case, 1986; Strong *et al.*, 1984; Schoener 1986; Lubchenco, 1986; Menge and Sutherland, 1987).

Although the quest may initially be local in its focus, the hope is that a global understanding of how communities are structured will emerge from a collection of studies performed at different sites (Connell, 1975; Menge and Sutherland, 1976; Lubchenco and Gaines, 1981). In our view, we have made substantial progress towards this goal, with an increasing number and quality of studies at the level of the community (defined as all species at all trophic levels at a particular location in a given habitat (see Menge, 1982; Jaksic, 1981). Although there are increasing numbers of such studies in terrestrial (McNaughton, 1983; Bormann and Likens, 1979; Brown *et al.*, 1986; Louda, 1982) and freshwater habitats (e.g. Lynch, 1979; McAuliffe, 1984; Morin, 1983; Peckarsky, 1983; Hall *et al.*, 1970; Wilbur, 1987; Zaret, 1980), the majority of these investigations are from marine habitats. The ultimate goal of this review is to use these marine studies of community regulation to evaluate relevant community theory.

II. THEORETICAL FRAMEWORK

The conceptual framework of this review is the body of community theory which deals with community regulation. That is, what determines the relative importance of several ecological processes (disturbance, predation, competition, recruitment) and environmental factors (environmental stress, productivity, habitat complexity) in influencing community structure. The theory of community regulation has expanded greatly during the past three decades (Hutchinson, 1959; Hairston *et al.*, 1960; Paine, 1966; Rosenzweig, 1973; Menge and Sutherland, 1976, 1987; Connell, 1975, 1978; Fretwell, 1977; Glasser, 1979; Huston, 1979; Buss and Jackson, 1979; Oksanen *et al.*, 1981; Lubchenco and Gaines, 1981; Roughgarden, 1986).

Here we focus on the predictions offered by a recent culmination of these efforts (Menge and Sutherland 1987; Fig. 1). This model proposes that variation in community structure depends directly on variation in the effects of abiotic disturbance, competition, and predation; and indirectly on variation in recruitment density (or for clonal organisms, individual growth rates) and environmental stress. The strengths of competition and predation are predicted to be a function of the complexity of interaction webs (the subset of linkages in food webs that have strong effects on the interacting species). Mobile organisms are assumed to be more sensitive to environmental stress than are sessile organisms, leading to the prediction that a given level of stress will have stronger effects on mobile consumers than on their sessile prey. Interaction web complexity is thus assumed to increase with a decrease in environmental stress, and therefore, community structure is indirectly dependent on environmental stress.

The relative importance of physical factors, competition, and predation is also predicted to vary with trophic level. At higher trophic levels, competition and physical factors are predominant under conditions of low and high stress, respectively. Predation has no influence at high trophic levels but increases in importance at lower trophic levels under low but not high environmental stress.

Additional assumptions are (1) important consumers tend to be generalized in diet, (2) consumers are mobile while species in the lowest trophic level (= basal species) are sessile, (3) recruitment densities of major species are positively correlated, and (4) basal species may include both autotrophs and heterotrophs.

Specific predictions include: Under the severest conditions with high recruitment, food web structure was postulated to be under the direct control of the physical environment via physical or physiological disturbance. Consumer presence and/or activity should be prevented while sessile organisms are kept scarce or even eliminated. Under less harsh conditions,

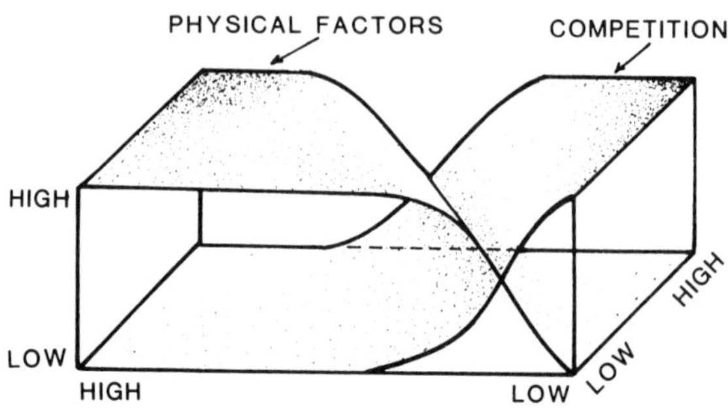

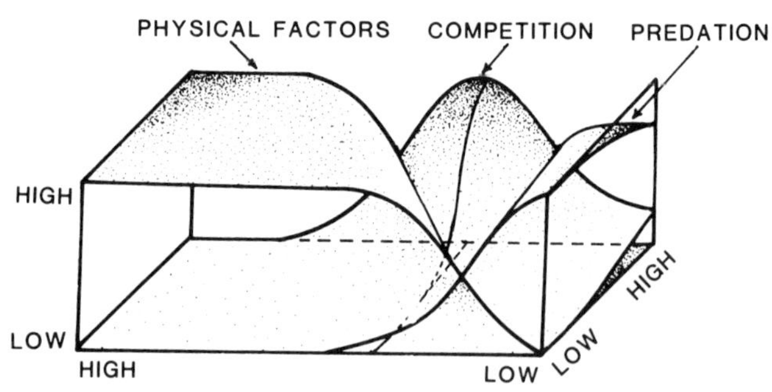

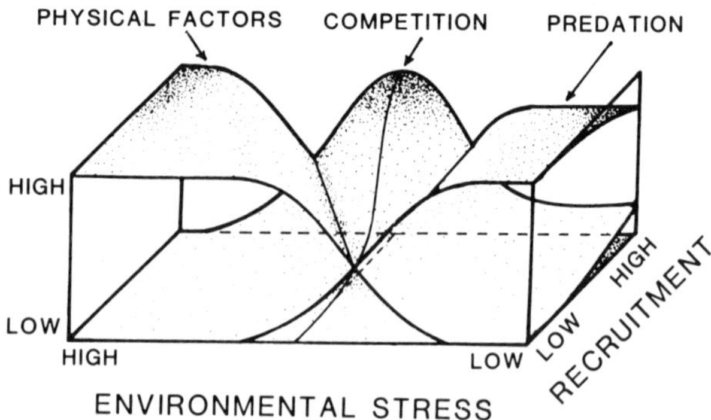

Fig. 1. Model of community regulation (see Figure 3 in Menge and Sutherland 1987). The model predicts the relative importance of physical factors, competition, and predation in controlling community structure as a function of environmental stress and recruitment density (or growth rates of recruits) by trophic level.

consumers were predicted to still be inhibited while sessile organisms, more tolerant of physical or physiological stress, could become dense, leading to competition. Thus, under severe physical conditions, the highest trophic level would be the sessile organisms.

At a more moderate position on the environmental gradient, consumers would no longer be completely suppressed, so that their effects on prey would increase. Increased consumer effectiveness should reduce the intensity of competition among the sessile organisms, enhancing coexistence of prey (Paine, 1966). Finally, near the favorable end of the environmental gradient, consumer effectiveness would peak. This would be a consequence of both relaxed constraints on consumer activity and increases in abundance and diversity of consumers. Strong predation, would, with the exception of organisms capable of escaping consumers, greatly reduce competition among species at lower trophic levels. Escaping prey species would compete as adults, but only after passing through a predation bottleneck as juveniles. Escapes would be mostly gained by growing to large size (see Connell, 1975; Lubchenco and Gaines, 1981; Menge, 1982; Wilbur, 1980; Neill, 1975).

With low recruitment (or slow growth in clonal organisms), some of these predictions are modified (Menge and Sutherland, 1987). The primary effect of low recruitment was predicted to be the reduction of the importance of competition at all levels of environmental stress. The primary controllers of food web structure were thus predicted to be physical and/or physiological disturbance and predation. These models thus suggest (1) that as environmental harshness lessens, physical conditions shift from exerting direct control on food web structure, to exerting indirect control through influences on biotic interactions, to exerting little effect; (2) that at any point along the environmental gradient, several factors will contribute to control of food web structure; and (3) the relative contribution of each factor varies in a predictable manner.

The model reconciles seemingly contradictory predictions that the plants are controlled by competition (Hairston *et al.*, 1960) and that sessile organisms are controlled by predation (Menge and Sutherland, 1976). The model argues that each prediction is correct, depending on environmental conditions. Plants are predicted to compete when omnivory by the top consumers is weak or absent, or when moderate environmental stress allows a less complex three-trophic-level interaction web (i.e. with no omnivory). Plants will not compete when omnivory by top consumers is strong, or when intermediate environmental stress allows a simple two-trophic-level interaction web. Several examples of empirical results from marine systems which support these predictions were presented in Menge and Sutherland (1987).

III. A MODEL SYSTEM: ROCKY INTERTIDAL COMMUNITIES

From the standpoint of community dynamics, rocky intertidal communities are equivalent to the model systems so valuable in other biological sciences (*Drosophila* in genetics, *E. coli* in cell biology, giant axons of squid and large-celled ganglia of opisthobranch molluscs in neurobiology etc.). Few other biological communities have fostered the breadth or depth of knowledge at the individual organism, population, community, ecosystem, and biogeographic levels comparable to that generated by assemblages on rocky shores.

The quality and productivity of research in rocky intertidal communities depend on several factors (Connell, 1972; Paine, 1977). First, the time scale of both organism longevity and major processes tends to be on the order of months and years, rather than decades and centuries as in many terrestrial habitats, particularly forests. This time scale also increases the likelihood of observing and manipulating short-term processes, which might be missed in faster systems with time scales of minutes and hours. For example, one can expect to observe the complete successional sequence in marine hard substratum systems at least once, and probably several times, within one's scientific lifespan.

Second, spatial scales tend to be small relative to the investigator, facilitating investigation at spatial scales ranging from individual organisms to the landscape. Incorporation of variation along environmental gradients is readily done, since intertidal habitats have steep physical and biological gradients. Experimental manipulations affect relatively small portions of the community while realistically mimicking natural processes such as disturbance or grazing. Large sample sizes are therefore easily obtained. Unlike habitats in which organisms interact within the substratum (forest canopy, soil, mud, sand), the primarily two-dimensional landscape of rocky habitats enhances direct observation and manipulation of the community. Finally, the trophic structure of marine communities is generally less affected by humans than the trophic structure of most terrestrial habitats.

Studying rocky intertidal habitats has some disadvantages (sometimes hazardous working conditions, periodic inaccessibility; many species spend part of their life cycles in a separate habitat as plankton—making studies of dispersal, population genetics, and recruitment difficult). These disadvantages limit the scope but not the quality of intertidal community research.

This review consists of four parts. First, we present the patterns of community structure on rocky intertidal shores. Second, we review the evidence for the importance of different processes as causal agents of the patterns. Particular aims include: consideration of each process in relation to the trophic complexity of the community; and determination of the importance of recruitment as a structuring process. In the latter case, it is particularly important to identify the conditions under which the structure of the developed community is independent of, directly dependent on, and indirectly dependent on recruitment. Third, we use the studies surveyed to evaluate the predictions of the community regulation model. Fourth, we identify some limitations of the model and recommend directions for future research.

IV. GLOBAL PATTERNS OF COMMUNITY STRUCTURE ON ROCKY SHORES

To many, the most interesting aspect of community ecology is determining the interacting processes and mechanisms that produce observed patterns in natural communities. Yet, it is striking how infrequently patterns of space utilization, size structure, and trophic relationships have been quantified. Indeed, even the qualitative descriptions of community structure provided in studies are often insufficient to generate a rough mental image of the physical appearance of the system. Both satisfactory investigation of community dynamics and developing general principles depends on quantitative descriptions of natural patterns. The underlying reasons for the dearth of quantitative description are complex and probably include a lack of appreciation of the importance of quantitative description for future comparison and synthesis, and inadequate development of, or failure to adopt, standardized procedures for quantitative description.

Despite inadequacies in the available information, we attempt a qualitative analysis of global patterns of rocky intertidal community structure. We also present a quantitative comparison of space utilization in three intertidal habitats for which comparable information is available. We restricted our survey to studies in which patterns of space use along a vertical (tidal fluctuation) and horizontal (wave exposure) gradients were reasonably clear. We determined species that were dominant (the dominant covers > 50% of the rock surface and hides other species from view unless moved aside) in the three layers of an intertidal community: (1) primary space—on the rock surface, (2) secondary space—the plant understory (plants < 15 cm tall), and (3) the canopy (plants > 15 cm tall). These dominants were then categorized by the general organismal group to which they belong (barnacle, fucoid algae, algal crust, mussel, red algal turf). Different species of acorn barnacle, (or mussel, fucoid, etc.) are functionally similar in general ecological requirements and to some extent, interactions. These ecologically similar (but sometimes taxonomically different) groups of species serve as a focus for our efforts to determine if similar patterns in different regions (but with physically similar environments) result from common ecological processes.

Marine biologists have long recognized that the two most important local physical gradients in intertidal habitats are clines in period of emersion (the vertical or tidal gradient), and in magnitude of wave forces (the horizontal or wave exposure gradient). Although both can be, and have been quantified (Menge, 1976; Denny, 1982; Denny *et al.*, 1985), most workers categorize their study sites subjectively. This is primarily because wave shock has proven difficult to quantify, and the vertical limits of many species are extended well above their "normal" level on the shore by wave splash (Lewis, 1964). As a result, our assignment of a species to a zone and wave shock category is also subjective. We recognize three wave exposure categories (sites with high, intermediate, and low wave forces) and three intertidal zones (high, mid, and low). We do not here consider the "supralittoral fringe", or that universally observed band between land and lower tidal levels which is dominated by blue-green algae and littorine snails (Stephenson and Stephenson, 1972). Finally, we categorized the studies surveyed by climate: cold-temperate sites have typical annual water temperature ranges between 0 and 20°C; warm-temperate sites have temperature ranges between 10 and 20°C; and tropical sites have average temperatures between 20 and 30°C.

Qualitative analysis of global patterns of visually dominant space occupiers indicates the existence of trends along both vertical and horizontal environmental gradients (Fig. 2). In high, wave-exposed zones, barnacles are dominant in 100% of the cold- and warm-temperate sites but only 20% of the tropical sites. Bare- and crust-dominated surfaces increase in frequency in all

Fig. 2. Qualitative patterns of community structure at cold and warm temperate, and tropical sites. Data are arranged by tidal height (high, mid, and low intertidal), and exposure to waves (exposed, intermediate, and protected). Bars are percentage values of the sample of studies ($= N$, listed under each histogram) which are visually dominated by the organism listed to the left of the exposed column. Visual dominants are those organisms covering $\geqslant 50\%$ of the surface when viewed without disturbing the outer layer of organisms. Dominants obscured by overlying covers of visual dominants were frequently not reported in the reviewed studies, and are thus not incorporated in this figure. If all space use categories covered $\leqslant 50\%$ of the surface, the study was assigned to the "none" category. C = canopy species, represented as solid bars; 2° = understory species, represented as cross-hatched bars; 1° = species of short stature occupying the rock surface, represented as open bars. See text for definition of cold, warm temperate, and tropical. Data were taken from the descriptions of Stephenson and Stephenson (1972), Lewis (1964), Menge (1976), Lubchenco and Menge (1978), and Dayton (1971).

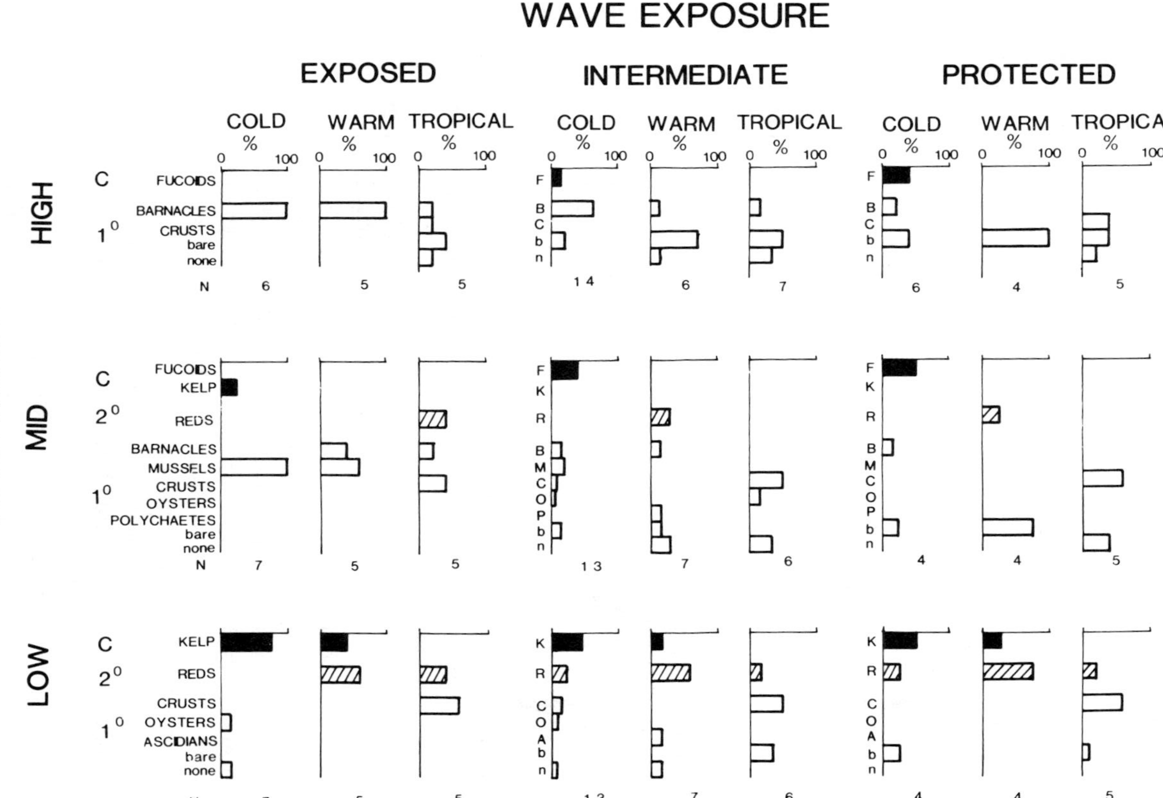
WAVE EXPOSURE
EXPOSED
INTERMEDIATE
PROTECTED
COLD
WARM
TROPICAL
%
0
100
TIDAL HEIGHT
HIGH
MID
LOW
C
1⁰
2⁰
FUCOIDS
KELP
REDS
BARNACLES
MUSSELS
CRUSTS
OYSTERS
POLYCHAETES
ASCIDIANS
bare
none
N
F
K
R
B
M
C
O
P
A
b
n
6 5 5 14 6 7 6 4 5
7 5 5 13 7 6 4 4 5
7 5 5 13 7 6 4 4 5

climatic zones with reduced wave-exposure in the high zone. Fucoid algae increase in dominance frequency with reduced wave-exposure, but only in cold-temperature sites; canopy species are scarce in warm-temperate and tropical high zones.

In mid zones, mussels are dominant in 100% of cold-temperate, wave-exposed sites. Although dominant in 60% of warm-temperate sites, mussels are infrequently or never dominant in other zones or less wave-exposed sites in any climatic region. Fucoid canopies (cold-temperate) and red-algal turfs (warm-temperate, and tropical wave-exposed) dominate or co-dominate (with algal crusts or mosaics) in less exposed mid zone sites (Fig. 2).

In low zones, kelps, red-algal turfs, and crusts most often dominate cold-temperate, warm-temperate, and tropical sites, respectively, regardless of wave-exposure. As noted before (Gaines and Lubchenco, 1982; Vermeij, 1978 and included references), kelps do not occur in tropical climatic regions. Red-algal turfs sometimes visually dominate low zones in all climatic regions and wave-exposures except cold-temperate, wave-exposed sites. However, their absence in the latter is an artifact, since some wave-exposed sites in Oregon (and probably elsewhere) are visually dominated by red-algal turfs (personal observations).

Several points bear emphasis. First, the classical high, mid, and low zonation pattern of barnacles, mussels, and kelp is exclusively a temperate, wave-exposed phenomenon, and near-universal only in cold-temperate sites (Fig. 2). Second, canopy species (fucoids, kelps) are dominants only in temperate zones and extend into mid and high zones only in cold-temperate climatic regions. (Fucoids such as *Sargassum* spp. are sometimes visual dominants on reef crests of coral reef habitats, but do not here qualify as true "low-intertidal" zones, since tidal amplitudes are usually < 1 m, and temporally irregular.) Third, a greater variety of types of organism are visual dominants at intermediate wave-exposures in all climatic regions than at wave-exposed or wave-protected sites, with a few exceptions (high, cold and tropical intermediate *vs* protected; mid tropical intermediate *vs* exposed; low, tropical intermediate *vs* protected). Fourth, there is a general cold-temperate to tropical trend from dominance by organisms of higher stature (tree-like algae or sessile invertebrates) to dominance either by organisms of low, flattened or sheet-like stature (e.g. algal crusts) or by bare rock.

Certain of these trends are illustrated more precisely by quantitative data from two cold-temperate and one tropical site from North and Central America (Fig. 3). In general, the pattern of zonation at temperate wave-exposed sites is the classic arrangement of barnacles (high), mussels (mid), and kelps (low). In contrast, zonation patterns at temperate wave-protected sites are bare space and crusts (high), fucoid algal canopy/crustose algae primary space (mid), and red-algal turf or "free" primary space (low). The striking feature of comparisons between zonation at wave-exposed sites *vs*

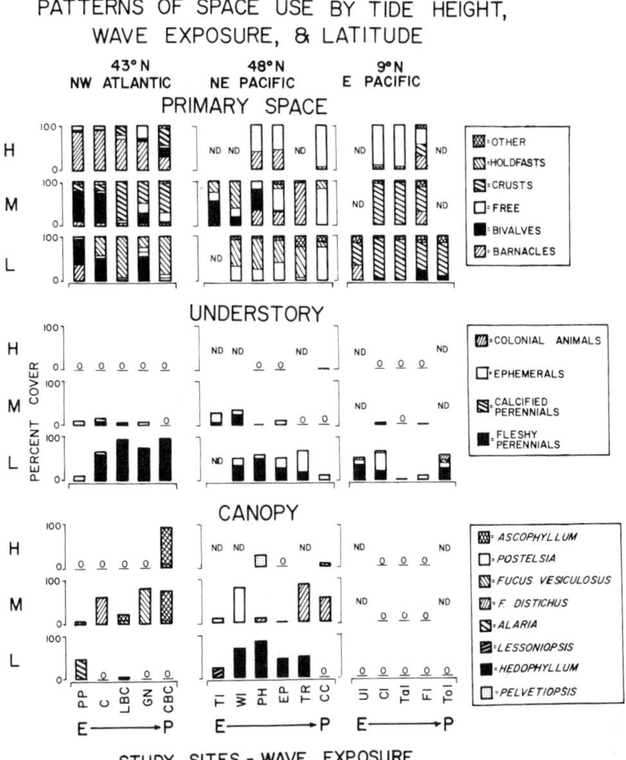

Fig. 3. Quantitative patterns of community structure at two temperate (43°N, NW Atlantic = Menge 1976 and Lubchenco and Menge 1978; 48°N, NE Pacific = Dayton 1971, 1975) and one tropical rocky intertidal region(s) (9°N, E Pacific = Menge and Lubchenco 1981, Lubchenco *et al.* 1984). Data are arranged by tidal height (H = high, M = mid, and L = low), exposure to waves (E = exposed, P = protected), and vertical layer above the surface. ND = no data available, 0 = no organisms in that category occurred at the site. NW Atlantic study site code: PP = Pemaquid Point, Maine; C = Chamberlain, Maine; LBC = Little Brewster Cove, Massachusetts; GN = Grindstone Neck, Maine; and CBC = Canoe Beach Cove, Massachusetts. NE Pacific study site code (all in Washington state): TI = Tatoosh Island, WI = Waadah Island, PH = Portage Head, EP = Eagle Point, TR = Turn Rock, CC = Colin's Cove. E Pacific study sites (all in Panama): UI = Uva Island, CI = Chitre Island, TaI = Taboguilla Island, FI = Flamenco Island, ToI = Tortola Island. Organism codes, starting with PRIMARY SPACE: other = anemones, polychaetes, tunicates, and sponges; holdfasts = holdfasts of understory and canopy algae; crusts = encrusting algae; free = bare rock. UNDERSTORY: colonial animals = arborescent bryozoans and hydrozoans; ephemerals = algae with rapid invasion, growth, and reproduction such as ulvoids, *Porphyra* spp., and *Giffordia* spp.; calcified perennials = foliaceous calcareous algae with slower life cycles such as *Jania* spp., *Corallina* spp., and *Calliarthron* spp.; fleshy perennials = foliaceous noncalcareous algae with slower life cycles such as *Chondrus*, *Mastocarpus*, *Endocladia*, *Dictyota*, *Hymenena*, *Dilsea*, *Constantinea*, and *Gelidium*. CANOPY genera are listed in the code box.

wave-protected sites is the variety in observed patterns at the cold-temperate sites with decreased wave exposure. The tropical site presents a strong contrast. Bare space (high) or algal crusts (mid and low) dominate primary space at all tidal levels and canopy species are totally lacking (Fig. 3). Finally, covers of foliose algae and sessile invertebrates tend to be highest at the most wave-protected sites. Such organisms are nearly absent at sites of intermediate wave exposure.

What are the processes and mechanisms which underlie these variations in abundance of sessile organisms along both local and global environmental gradients? In the following section, we first examine how ecological processes affect community structure with respect to different environmental gradients. We then attempt to identify general features in each case.

V. COMMUNITY REGULATION IN ROCKY INTERTIDAL HABITATS

Community dynamics are usually revealed using two types of experimental manipulation. First, densities or cover of a single species are altered (to levels below or above normal densities). Second, all organisms are removed from a site or plot and allowed to return to normal. The first method is generally used to study maintenance of the mature community while the second method is used to study development of the community (= succession). The second approach can be made more powerful by combining it with the first method; one then studies community development under normal conditions (colonization proceeds without further alteration) and under experimentally altered conditions (colonization proceeds with manipulation of the colonization pattern (exclusion of a predator, herbivore, or sessile species). A thorough study should determine the dynamics underlying both maintenance and development of the community, ideally simultaneously. Unfortunately, this ideal is infrequently achieved.

Although we adhere to the view that experimentation is the most powerful way to reveal the causes of community structure, this approach is not without pitfalls. The many problems and errors which afflict the design (Hurlbert, 1984), execution (Dayton and Oliver, 1980; Virnstein, 1978; Underwood and Denley, 1984), and interpretation (Bender *et al.*, 1984; Schoener, 1983; Roughgarden, 1983; Quinn and Dunham, 1983; Toft and Shea, 1983; Simberloff, 1983; Strong, 1983; Connell, 1983; Wiens, 1977) of field experimentation are well chronicled. These criticisms emphasize that experimentation is most powerful when accompanied by attention to details of natural history, experimental design, execution of perturbations, statistical inference, and interpretation of results. Below, we consider the potential and actual effects of different ecological processes on the structure of the sessile organisms.

VI. ECOLOGICAL PROCESSES

Space on the rock surface ("primary" space) is an important resource for the sessile organisms whose distribution and abundance taken together produce observed patterns of spatial community structure (Connell, 1961a,b, 1970, 1972; Paine, 1966, 1971, 1974, 1976, 1977, 1980, 1984; Dayton, 1971, 1975; Menge, 1976; Lubchenco and Menge, 1978). Processes producing increases in space utilized by sessile species include recruitment and individual growth. Processes producing decreases in space utilized by sessile organisms include competition, predation, and disturbance. The first task in a study of community regulation is to determine which combination of these factors produces observed patterns.

The best means of obtaining the information necessary to test models of community organization is the performance of studies which simultaneously investigate the impact of each process on structure. Ideally, this should include studies done along environmental gradients (tidal height = percentage of time exposed to air, wave-exposure). For many reasons, few such studies are available as yet. Most investigations evaluate the influence of one or a few processes on a restricted portion of the environmental gradients. Indeed, in many published studies it was often unclear just what the important environmental gradients were and where the study was done along these clines. Below, we summarize the influences of recruitment and disturbance on community structure. We then examine the small collection of studies that have rather thoroughly investigated the influence of the major structuring processes, particularly disturbance, competition, and predation.

A. Recruitment

The main process producing increases in sessile species abundance is recruitment of new individuals. Recruitment has been a longstanding concern of marine ecologists (Thorson, 1950; Connell, 1961a; Barnes, 1956; Paine, 1974; Denley and Underwood, 1979; Grosberg, 1982; Connell, 1985; Menge, 1988). It is thus surprising how rarely the relationship between recruitment and adult abundance is determined.

Table 1 summarizes 39 studies in which the relationship between the final adult abundance and initial recruitment density of a species was investigated in a specific habitat. The data are heavily weighted towards barnacles (7 species in 22 studies), most likely because recruitment is relatively easily observed and quantified in cirripedes. Most (34 of 39) studies were done in rocky intertidal regions in cold and warm temperate climates.

We evaluated the contribution of initial recruitment density (defined after Connell, 1985, as the number of settlers surviving for about a week after metamorphosis) in three ways (Table 1, "within-site correlation recruit/adult

Table 1

Relationship between recruitment patterns and variation of adult populations in marine communities. Blank in any column means the information is unavailable; ? indicates that the estimate is uncertain.

Organism (reference)	Latitude	Habitat[a] (level)	Surface (Adult Abundance)	Recruitment Density (#/m^2/yr unless stated otherwise)	Longevity	Within-site Correlation Recruit/Adult Abund.		
						Yes/No	Statistic[b]	Cause[c]
SESSILE ORGANISMS								
BARNACLES ($n = 22$ studies, 7 species)								
Semibalanus balanoides (Connell 1961a,b)	54°N	Int (H,M,L)	Rock (> 80%)	High ($13–320 \times 10^3$)	5 y	N	b = 0·04(H) 0·04(M) 0·01(L)	P,D
S. balanoides (Menge 1976)	42°–44°N	Int (High)	Rock (90–100%)	High ($0–244 \times 10^3$)	5 y(?)	Y	b = 0·07***	R,Ca
S. balanoides (Menge 1976)	42°–44°n	Int (Mid)	Rock (0–10%)	High ($0–154 \times 10^3$)		Y	b = 0·04***	P,Ce
S. balanoides (Lubchenco and Menge 1978)	42°–44°N	Int (Low)	Rock (0–10%)	High ($0–76 \times 10^3$)		N	b = 0·01	P,Ce
S. balanoides (Grant 1977 in Connell 1985)	44°N	Int (High)	Rock (20–100%)	High (47×10^4)	5 y	N	b = 0·03	Ca
S. balanoides (Hatton 1938 in Connell 1985)	49°N	Int (H,M,L)	Rock ($1·9–11 \times 10^4/m^2$)	High ($4–24 \times 10^4$)	5 y	Y	b = 0·39*	R,?
Chthamalus stellatus (Hatton 1938 in Connell 1985)	49°N	Int (H,M,L)	Rock ($9–76 \times 10^2/m^2$)	Interm. ($3·7–24 \times 10^3$)		N	b = 0·09	
C. fissus (Hines 1978, 1979)	35°N	Int (High)	Rock (?)	Interm. ($2·6–10·3 \times 10^3$)	5·6 y	Y		R,Ca
C. fissus (Hines 1978, 1979)	35°N	Int (Low)	Rock (?)	Interm. ($1·7–11·7 \times 10^3$)	5·6 y	N		P,Ce

Balanus glandula (Hines 1978, 1979)	35°N	Int (High)	Rock (?)	Interm. ($1–9 \times 10^3$)	8 y	Y		R,Ca
B. glandula (Hines 1978, 1979)	35°N	Int (Low)	Rock (?)	Interm. ($4–34{\cdot}6 \times 10^3$)	8 y	N		P,Ce
Tetraclita squamosa (Hines 1978, 1979)	35°N	Int (Low)	Rock (?)	Interm. (750–6150)	14 y	N		P,Ce
B. glandula (Gaines and Roughgarden 1985)	35°N	Int (High)	Rock (55%)	High ($170–700 \times 10^3$)		N		Ca,P
B. glandula (Gaines and Roughgarden 1985)	35°N	Int (High)	Rock (10–40%)	Interm. ($7–45 \times 10^3$)		Y		R
B. glandula (Connell 1970, 1985)	48°N	Int (Mid)	Rock Concrete ($0/m^2$)	Interm. ($3–32 \times 10^3$)	5–8 y	N		P
B. glandula (Connell 1985)	48°N	Int (High)	Rock ($9–76 \times 10^4/m^2$)	Interm. ($3{\cdot}7–24 \times 10^3/m^2$)	5–8 y	Y	b = 0·28**	
B. glandula (Grosberg 1982)	36°N	Int (High)	Pilings (?)	High ($\leqslant 75 \times 10^3$)	8 y	Y	p = ***	L
B. crenatus (Grosberg 1982)	36°N	Int (High)	Pilings (?)	High ($\leqslant 90 \times 10^3$)		Y	p = ***	L
B. inexpectatus (Menge 1988)	9°N	Int (Low)	Rock (1–2%)	Low ($\leqslant 1{\cdot}6 \times 10^3$)		Y	p = ***	R,P
C. fissus (Sutherland and Ortega 1986)	9°N	Int (H,M)	Rock (0–70%)	Low		Y ($\leqslant$ 60 days) N ($\geqslant$60 days)		R P

Table 1 *continued*

Organism (reference)	Latitude	Habitat[a] (level)	Surface (Adult Abundance)	Recruitment Density (#/m^2/yr unless stated otherwise)	Longevity	Within-site Correlation Recruit/Adult Abund.		
						Yes/No	Statistic[b]	Cause[c]
BIVALVES (n = 3 studies, 2 species)								
Mytilus edulis (Menge 1976, 1978b, Lubchenco and Menge 1978)	42–44°N	Int (Mid-Low, Wave-exposed	Rock ($\leqslant$ 100%)	High (34–102 × 10^3)	1–5 y	Y	b = 1·72***	R,Ce
M. edulis (Menge 1976, 1978b, Lubchenco and Menge 1978)	42–44°N	Int (Mid-Low, Wave-protected	Rock (0–10%)	High (34–102 × 10^3)	1–5 y	Y	b = 0·98***	P,R
Chama echinata (Menge, unpubl.)	9°N	Int (Low)	Rock (1–2%)	Low ($\leqslant$0.008 × 10^3)		Y	p = ***	R,P
OTHER SOLITARY SESSILE (n = 1 study, 1 species)								
brachiopods (Witman and Cooper 1983)	42°N	Sub (33 m)	Rock ?	High (20–108)	2–3 y	N	r^2 = 0·08 (n = 30)	P,D
COLONIAL SESSILE (n = 7 studies)								
bryozoans, sponges, ascidians (Russ 1980)	38°S	Sub (1–2·2 m)	Plates (100%)	High (59–78%/4 mo, 90–100%/7 mo)	> 1 y	N		P,Ca
bryozoans (Keough 1983, 1984a,b)	35°S	Sub (7 m)	*Pinna* shells, pilings (< 40%)	Low (0·1–1/shell/y)	12–15 mo	Y?		R,P

tunicates (Keough 1983, 1984a,b)	35°C	Sub (7 m)	*Pinna* shells, pilings (< 40%)	Low (< 0·1/ shell/y)	12–48 mo	Y		P,R
Membranipora membranacea (Yoshioka 1982)	33°N	Sub (0–60 m)	*Macrocystis* blades (⩽ 70%)	High (⩽ 1000/blade	6 wk	Y		P,R
(fouling) (Sutherland and Karlson 1977, Karlson 1978)	34°N	Sub (0·3 m)	tile plates	High (44–93% in 90 d)	< 12 mo	N		Ce
(fouling) (Winston and Jackson 1984)	18°N	sub (12–13 m)	settling panels, coral (100%)	Low (20% in 1 y)	> 1 y	N		Ce,P
(fouling) (Sutherland 1980)	10°N	Sub (0–1 m)	settling plates, mangrove roots (?)	Interm. (0–28% in 2 mo)	> 1 y?	N		Ce
MOBILE ORGANISMS (n = 8 studies, 4 species)								
Patella vulgata (Bowman and Lewis 1977)	53°N	Int (High, barn.)	Rock	High (10–1950)	15–17 y	N	$r = 0{\cdot}02^{ns}$ ($n = 8$)	Ca
Patella vulgata Bowman and Lewis 1977)	53°N	Int (High, barn., mus.)	Rock	High (10–1950)	15–17 y	N	$r = 0{\cdot}1^{ns}$ ($n = 8$)	Ca
Patella vulgata (Bowman and Lewis 1977)	53°N	Int (Mid, barn.)	Rock	High (10–1950)	15–17 y	N	$r = 0{\cdot}04^{ns}$ ($n = 5$)	Ca
Patella vulgata (Bowman and Lewis 1977)	53°N	Int (Mid, mus.)	Rock	High (10–1950)	8 y	N	$r = 0{\cdot}47^{ns}$ ($n = 8$)	Ca

Table 1 *continued*

Organism (reference)	Latitude	Habitat[a] (level)	Surface (Adult Abundance)	Recruitment Density (#/m^2/yr unless stated otherwise)	Longevity	Within-site Correlation Recruit/Adult Abund.		
						Yes/No	Statistic[b]	Cause[c]
Patella vulgata (Bowman and (Lewis 1977)	53°N	Int (Mid, rock)	Rock	High (10–1950)	4–5 y	N	$r = 0{\cdot}75^{ns}$ ($n=6$)	Ca
Tegula brunnea (Watanabe 1984)	35°N	Sub (1–13 m)	algal turf	High ($\leqslant$ 200/mo.)		Y		R,P
T. pulligo (Watanabe 1984)	35°N	Sub (1–13 m)	algal turf	High ($\leqslant$226/mo.)		Y		R,P
T. montereyi (Watanabe 1984)	35°N	Sub (1–13 m)	algal turf	Interm. ($\leqslant$ 33/mo.)		Y		R,?

[a]Int = intertidal; Sub = subtidal. Levels: H = High; M = Mid; L = Low.

[b]b = slope of regression between adult and recruitment density; * = $p<0{\cdot}05$, ** = $p<0{\cdot}01$, *** = $p<0{\cdot}001$; ns = not significant; p = *** means there is a significant correlation between adult and recruitment densities; r = correlation coefficient; r^2 = coefficient of determination.

[c]Ca = intraspecific competition, Ce = interspecific competition, D = disturbance, L = larval behavior, P = predation, R = recruitment.

abund."). First, in some cases, we determined the significance of the slope of the regression between final adult abundance and recruitment density using data presented in the study. Second, in other cases, we used the significance of the regression or correlation coefficient as presented in the original paper. Third, in the remaining cases, the authors did not present quantitative estimates but stated a relationship between recruitment and final adult abundance.

A summary of these data suggests that there is no consistent relationship between relative recruitment density and the presence of a significant correlation between final adult abundance and recruitment (Table 2: χ^2, $p = 0{\cdot}214$; Fisher Exact Test, $p = 0{\cdot}33$). Further analysis revealed no relationship between the significance of a recruit/adult abundance regression and either latitude (χ^2, $p = 0{\cdot}926$) or level on the shore (χ^2, $p = 0{\cdot}31$).

This conclusion must be qualified in at least two ways. First, most analyses performed by us or the original authors considered only a single source of variation in final adult abundance, recruitment density. As suggested by the far right column in Table 1, final adult abundance may also be influenced by factors other than recruitment, including predation, disturbance, and competition (intraspecific and interspecific). Thus, the correlation between recruitment and adult abundance may be statistically significant, but of little biological importance.

For example, we analyzed the causes of variation in final abundance of the mussel *Mytilus edulis* in experiments done in New England (Menge, 1976) using multiple regression. The analysis indicates that at exposed sites, recruitment explains 25% of the variance in final mussel abundance while 75% of the variance remains unexplained. At sheltered sites, predation explains 52%, recruitment explains 11%, and substratum inclination explains 5·5% of the variance in final adult abundance while 31·5% of the variance remains unexplained. In both cases, the regression between recruitment density and final adult abundance was highly significant ($p < 0{\cdot}006$ or $\ll$), yet relatively little variation was actually explained by this factor (25% and 11% at exposed and sheltered sites, respectively). Even this analysis is limited, however, since other factors (disturbance, competition) may also influence mussel cover (and presumably explain some of the unexplained variance in mussel abundance).

A second qualification is that the influence of recruitment density is dependent on the time scale. Many of the studies in Table 1 would probably provide results similar to the analysis of Sutherland and Ortega (1986, personal communication). These workers found that recruitment density was strongly correlated with postrecruitment density up to 60 days, but thereafter, the correlation became increasingly weak (Table 1). This presumably reflects the increasing influence of postrecruitment sources of mortality such as predation, competition, and desiccation stress in studies of longer duration (Sutherland and Ortega, 1986). Similarly, in studies in New

England, the community occurring on the mid or low shore after natural or experimental clearance of space becomes increasingly dissimilar to that initially established. Barnacles typically settle heavily initially, but are soon replaced by mussels, algae or free space depending on site and shore level and the set of processes which occur in each circumstance (Menge, 1976; Lubchenco and Menge, 1978).

Analysis of relative recruitment densities by latitude suggests a relationship between broad climatic regime and recruitment (Table 2B). All 26 cases with high recruitment and 8 of 9 cases with intermediate recruitment densities were in cold or warm temperate communities. Five of 7 cases with low recruitment occurred in tropical regions. Although the numbers are still too low (especially in tropical areas) to be completely convincing, the relationship between latitude and recruitment rate deserves further investigation.

Table 2
Recruitment density and adult abundance.

A. Relation between final adult abundance and recruitment density (sessile species).

Within site correlation between final adult Abundance and recruitment?	Relative recruitment density (number of studies)			
	Low	Intermediate	High	*n*
Yes	4	2	9	15
No	1	5	9	15
n	5	7	18	30

Pearson $\chi^2 = 3{\cdot}1$, 2 df, $p = 0{\cdot}214$
Fisher Exact Test (intermediate and high combined): $p = 0{\cdot}33$

B. Relative recruitment density by latitude

	Latitude (°N or S)[a]												
	Cold temperate					Warm temperate					Tropical		
	54	53	49	48	42–44	38	36	35	34	33	18	10	9
Recruitment level													
High	3-b	5-l	1-b	2-b	4-b 2-c 1-a	*1-f*	2-b	2-g 1-b	1-f	1-e	0	0	0
Int.	0	0	1-b	0	0	0	0	6-b 1-g	0	0	0	1-f	0
Low	0	0	0	0	0	0	0	*2-f*	0	0	1-f	0	3-b[b] 1-c

[a]Entries are number of studies and category. Code to categories: a = brachiopod, b = barnacle, c = bivalve, e = bryozoan, f = fouling community, g = gastropod, l = limpet.
[b]Number includes *Tetraclita panamensis* study by Sutherland (1987).

Under what conditions is recruitment a controlling factor? First, recruitment can influence adult community structure if the recruiting organism is generally free of interspecific sources of mortality (predation, interspecific competition). This is seen in studies of barnacles occurring in high intertidal zones which are free of competitors and predators (New England, Menge, 1976; central California, Hines, 1978, 1979; Grosberg, 1982; Gaines and Roughgarden, 1985), and mussels or surfgrasses occurring in mid or low intertidal zones which are either naturally or artificially free of predators and are dominant competitors for space (Menge, 1976; Lubchenco and Menge, 1978; Dayton, 1971; Turner 1983a,b).

Second, recruitment can influence adult community structure if the recruiting organism either swamps its predators with high recruitment or keeps pace with the rate of predation. Yoshioka (1982) provided an example of keeping pace in a study which shows a high correlation between recruitment rate and adult abundance of the bryozoan *Membranipora membranacea*. Related to this are situations where a species recruits densely to a site and thereafter resists invasion (Dayton *et al.*, 1984; Turner, 1984; Sutherland and Karlson, 1977; Sutherland, 1978, 1981; Lubchenco and Menge, 1978).

Third, recruitment can influence adult community structure if recruitment rates of all species in the community are low and those organisms that do settle are prevented from exerting competitive dominance. Examples include studies of subtidal assemblages of fouling invertebrates on small, isolated bivalve shells (Keough, 1983, 1984a,b), where apparently both low settlement and slow growth prevent interactions between sessile organisms, and intertidal studies (Gaines and Roughgarden, 1985; Connell, 1985; Menge *et al.*, 1985; Menge *et al.*, 1986a,b; Menge, 1988), where settlement was low and competition between invertebrates did not occur.

There are few examples for mobile species. Watanabe (1984) indicates that recruitment strongly affects guild structure of subtidal *Tegula* spp. in central California. Bowman and Lewis (1975) present information which indicates that recruitment does not affect population density of *Patella vulgata* (Table 1). Victor (1986) suggests that abundance of bluehead wrasses on coral reefs off the Caribbean coast of Panama reflects settlement patterns and not local biotic interactions.

We conclude that recruitment can influence community structure in shallow-water marine communities, but that in many cases this affect is overridden by various post-recruitment mortality sources. We further emphasize two points needing further investigation.

(1) Rates of recruitment and their potential community consequences may, in certain circumstances, be important. In particular, we suggest that the structure of depauperate communities, and ones in which interspecific competition for space is lacking, are likely to be more strongly affected by variable recruitment rates than are more speciose communities, or ones with

frequent and intense competition for space. In speciose communities, it is likely that initial patterns established by recruitment will be obliterated by interactions with other species, while the likelihood of such effects seems less in depauperate communities. Similarly, intense competition is more likely to alter initially established patterns than is weak competition.

(2) Spatial scales must be specified when considering the community role of recruitment. For example, variation in recruitment over space may be greater when comparing small areas on the scale of tens to hundreds of cm^2 than when comparing larger areas on the scale tens to hundreds of m^2. Comparison between studies focused on large scale variation and those focused on small scale variation are inappropriate and bound to lead to confusion and disagreement. We recommend that future studies explicitly consider the relationship between recruitment and adult community structure on the same size of study plot, and how the spatial scale of the investigation affects conclusions.

B. Growth

Growth subsequent to recruitment is another source of increase in cover of sessile organisms. This can be especially important in those organisms which possess clonal growth forms and can acquire large areas through growth (colonial invertebrates and many marine plants; see Jackson, 1977, 1978). Since colonial invertebrates are scarce in most intertidal regions (Jackson, 1977), and others have reviewed the community role of these organisms (Keough and Connell, 1984), we do not discuss these further. Further, Sousa (1984) has reviewed the asexual abilities of plants and plant-like sessile invertebrates (mussel beds) to acquire space. Paine (1974) found that competitive dominance of the mussel *Mytilus californianus* was expressed (in the absence of the seastar *Pisaster ochraceus*) by slow, growth-caused overgrowth ("cascading") of low intertidal organisms by mussels higher on the shore. We note, with Sousa (1984), that this is a slow process, and as a possible result, quite susceptible to disturbance prior to full coverage of the surface.

Finally, mutualisms (+ + interactions) and commensalisms (+ 0 interactions) can foster increases in abundance of sessile organisms. Few examples of mutualisms are available (Vance, 1978; Osman and Haugsness, 1981; Bloom, 1975), and we know of no studies demonstrating a community effect of such interactions. As others have noted, this should be a fertile area for future research. In the following sections, we consider those processes which reduce the abundance of sessile organisms.

C. Disturbance

Much evidence suggests that community structure can be strongly affected by physical disturbance (defined as sources of mortality imposed by the abiotic environment, e.g. wave-action, sand scour or burial, cobble scour; excludes direct and indirect biotic sources of mortality, e.g. predation and limpet bulldozing, respectively). Recent reviews by Sousa (1984) and Keough and Connell (1984) have thoroughly covered this subject. They note that disturbance regimes can be quantified with respect to the size of the clearance, intensity or degree of damage, and frequency and seasonality of the disturbance. Patterns of colonization are affected by patch characteristics (patch type [type 1 = those surrounded by occupied areas, type 2 = those isolated from occupied areas]; location; surface characteristics; size and shape; and time of creation), life history characteristics of potential colonists, and mobile consumers. As noted by Sousa (1984), few studies have quantified regimes of physical disturbance and recovery in rocky intertidal habitats, and fewer still have incorporated experimental investigation of these processes. Among the more thorough investigations of disturbance and its community effects are Sousa (1979a,b), Paine and Levin (1981), and Dethier (1984).

Although all communities are subject to physical disturbance, the relative importance of this process in structuring food webs, compared to other processes, is unknown on either a local or geographic scale. The importance of disturbance is best understood in wave-exposed sites in the Pacific Northwest region of North America. Here much of the structure, including species diversity, depends on regular removal of mussels and barnacles, and probably surfgrass and macroalgae, by waves (Paine and Levin, 1981; Dayton, 1971, 1975; Dethier, 1984; Paine, 1979; Turner, 1984). However, predation (Paine, 1966, 1974, 1976, 1980; Dayton, 1971), and grazing (Dayton, 1971; Gaines, 1985; Dethier and Duggins, 1984; Paine, 1980, 1984) are also important. Predation evidently maintains the structure of at least the upper half of the low intertidal algal zone by eliminating competitively superior mussels (Paine, 1974, 1980, 1984). Furthermore, the experiments of Paine and those of Dayton (1975) indicate that without physical disturbance, wave-exposed sites in this region would be dominated by mussels in the mid zone, and by one or two species of large brown algae (*Hedophyllum*, *Lessoniopsis*, *Laminaria*) or surfgrass (*Phyllospadix scouleri*) in the low zone.

With increasing protection from waves in Washington state (and Oregon: Turner, 1983a,b, 1984; Gaines, 1985, B. Menge and T. Farrell, pers. obs.), the situation becomes more complicated. The importance of wave-induced disturbance decreases while the importance of predation and grazing evidently increases, at least at some sites (Dayton, 1971). However, at present we can do little but guess at the relative effects of these and other processes on community structure at wave-protected sites, even in this reasonably well-studied region.

In New England, the importance of disturbance similarly increases with increased wave-exposure (Menge, 1976, 1978a, 1983; Lubchenco and Menge, 1978; B. Menge and J. Lubchenco, unpublished data). Here too, wave-exposed sites are dominated by barnacles (*Semibalanus* [= *Balanus*] *balanoides*; high zone), mussels (*Mytilus edulis*; mid and low zones), and kelp (an overstory of *Alaria esculenta* lies over the mussels). Wave-caused disturbance was generally severe during winter from 1971 to 1976, but its impact was more variable than at Tatoosh Island in Washington state (Paine and Levin, 1981; B. Menge, unpublished data). During one mild winter (1973–1974), the mussel bed persisted largely intact; few bare patches were produced (Fig. 4a,b). The following winter was much more severe, however, and the mussel bed was not merely disturbed—it was nearly totally destroyed. Rather than creating holes in a uniform landscape of mussels, waves removed nearly all mussels, leaving only small (< 1 m diameter) patches of mussels surrounded by bare rock. Although opportunistic algae invaded the cleared space (as usually occurs during winter; Lubchenco and Menge, 1978), and barnacles settled densely shortly thereafter (Fig. 4c), mussels slowly increased in cover through the following summer and autumn. Recovery to mussel dominance did not occur until the following summer (1976; Fig. 4d).

Physical disturbance occurs irregularly at more protected sites in New England, and can denude large areas of fucoid algae (*Fucus vesiculosus*, *F. distichus*) in the mid and low intertidal. For example, winter storms shifted large boulders at Grindstone Neck in 1974–1975, scouring the surface clear of the dominant stand of *F. vesiculosus* (B. Menge and J. Lubchenco, unpublished data). Similar effects of boulder/cobble scour have been observed in the low zone, which is dominated by *Chondrus crispus* at all but wave-exposed sites (Lubchenco and Menge, 1978). In general, recovery is rapid, generally taking about 1 year. In fact, all denuded surfaces, whether experimental or natural, generally recovered in periods < 1 year in New England, with the exception of wave-protected sites dominated by *Ascophyllum nodosum* (Menge, 1977; B. Menge and J. Lubchenco, unpublished data). As has been shown by others, recovery of this long-lived (12–19 years) alga is slow, taking (e.g.) a minimum of 6 years (Keser *et al.*, 1981).

Comparison between New England and Washington state indicates that zones in New England tend to be quantitatively dominated by single species, while those in Washington (and other west coast sites; B. Menge, unpublished data) tend to be quantitatively dominated by species mosaics. If New England experiences a similar gradient of wave exposure, why do a few species come to dominate most space in this system? That is, why don't other species invade disturbed patches and persist for several years as they do on the west coast (Paine and Levin, 1981)? A partial answer is that disturbance

Fig. 4. Photographs of the Pemaquid Point, Maine study site. For orientation, arrows point to two topographic features common to all photos. a. September 1973: the mid and low intertidal mussel bed is virtually continuous. b. May 1984: the mussel bed remains intact after a mild winter. c. August 1975: small patches of mussels (black) remain after catastrophic loss in winter 1974–75. Barnacles (white) settled densely in spring 1975, filling up the bare rock surface exposed after mussels were torn loose. d. May 1976: mussel abundance has increased considerably, although recovery was not complete until late summer 1976.

regime (variation in wave shock) in New England is more variable (Menge, 1976; Wethey, 1985). In particular, extreme, catastrophic disturbance from ice scour is a regular event (occurring 30–40% of the winters) in New England (Wethey, 1985). In contrast, sea ice never forms along the northeastern Pacific coast. Wethey argues that this disturbance regime prevents the evolution of long-lived, slow-growing, late-maturing sessile invertebrates which occupy mid and low intertidal habitats along the Pacific west coast. Thus, the anemones (Sebens, 1983), long-lived large brown algae (Dayton, 1975), mussels (Suchanek, 1981; Paine and Levin, 1981), and barnacles (Wethey, 1985) which are so conspicuous in rocky intertidal communities on the west coast are lacking from New England rocky shore habitats (Menge, 1976; Wethey, 1985). The only exception is *Ascophyllum*, which is evidently sufficiently resistant to ice scour to persist. However, this persistence is probably restricted to wave-protected sites, where ice movement is more gentle due to calmer conditions.

To test the hypothesis that catastrophic disturbance causes extinction of invading long-lived sessile invertebrates which are restricted to mid and low intertidal zones, Wethey (1985) performed computer simulations. These showed that with frequent ice scour (35–40% of the winters), long-lived, late-maturing acorn barnacles rapidly became extinct; barnacles which mature within their first year and which can inhabit a broad range of habitats, like *Semibalanus balanoides*, persist. Wethey's simulations thus support the idea that intertidal zones in New England tend to be dominated by fast-growing, early-maturing species because longer-lived, later-maturing species, even if they gained a foothold, would be eliminated by regular catastrophic disturbances.

This interpretation cannot fully explain why diversity in New England is so low, however. It is conceivable, for instance, that multispecific mosaics of opportunistic species could occur in New England. In fact, few other sessile species are co-dominants; discrete mosaics are lacking. A disturbance-based interpretation of biogeographic differences in community structure between New England and the west coast is therefore only part of the story. Different physical disturbance regimes explain some, but not all, of the high diversity on the west coast and the low diversity in New England. As noted above for the west coast, other factors evidently must be considered, particularly in more wave-protected habitats.

Although wave disturbance declines with increasing protection from waves along local gradients, desiccation and heat stress are potentially more severe with reduced wave action (Lewis, 1964; Connell, 1972, 1974; Sousa 1984; Underwood and Denley, 1984). These factors can directly cause massive mortality (Glynn, 1968; Loya, 1976; B. Menge pers. obs.) and indirectly influence community structure by affecting populations of mobile consumers (Dayton, 1971; Menge, 1978b, 1983, pers. obs.; Garrity, 1984;

Leigh *et al.*, 1987). Unfortunately, the relative importance of this factor compared to others is unknown; further study is needed.

For physical disturbance to have a large direct influence on community structure, the system must include as dominant space occupants species that are susceptible to the disturbance. This point has not generally been discussed, probably partly because it seems obvious, and partly because few studies have been conducted in rocky intertidal habitats which were not spatially dominated by large, sessile invertebrates and macroalgae. However, studies on the Pacific coast of Panama have revealed that the community is dominated by algal crusts and solitary sessile invertebrates of low relief (Lubchenco *et al.*, 1984; Menge and Luchenco, 1981; Menge *et al.*, 1986a). Although a typical wave-exposure gradient exists in this region, physical disturbance from waves is unimportant because the dominant space occupants (algal crusts) are not susceptible to being dislodged by even the most severe wave activity (Menge and Lubchenco, 1981). However, desiccation and heat stress probably have important direct effects at sites of intermediate and low wave-exposure, particularly in the high and mid zones (Garrity, 1984; Lubchenco *et al.*, 1984; Menge *et al.*, unpublished data). Further, strong wave action, desiccation, and heat stress all probably have important indirect effects, primarily by influencing the activity patterns of the many consumer species in this system (Lubchenco *et al.*, 1984; Menge and Lubchenco, 1981; Menge *et al.*, 1983, 1986a,b).

We conclude that, although physical disturbance is known to be important in many situations, too little is known to permit evaluation of its importance relative to other processes. In particular, the importance of physical stress in wave-protected habitats needs much work. We suggest that a major focus in any such study should be the relationship between the most prominent source of physically caused mortality and the dominant space occupants.

VII. BIOTIC INTERACTIONS: PREDATION AND COMPETITION

Recent reviews offer syntheses of the effects of consumers and/or competition on community structure in nearshore rocky marine habitats (Branch, 1981; Connell, 1975, 1983; Dayton, 1984; Gaines and Lubchenco, 1982; Hawkins and Hartnoll, 1983; Lawrence, 1975; Lubchenco and Gaines, 1981; Menge, 1982; Menge and Sutherland, 1976, 1987; Paine 1977, 1980; Sih *et al.*, 1985; Schoener, 1983; Underwood, 1979). In general, the evidence indicates that consumers can have exceedingly important controlling effects on prey in hard-bottom habitats (Sih *et al.*, 1985). Moreover, competition appears frequent in nature (Schoener, 1983; Connell, 1983), although rarely having the influence predicted by competition theory (Wiens, 1977, 1984). As some of these reviewers note, however, the relative importance of these agents in structuring communities is unclear. This is because there exist few studies which consider both factors within the same study. Sih *et al.* (1985) found only 17 studies in which both predation and competition among the sessile prey were manipulated. Of these, 9 were in 5 geographically distinct marine rocky intertidal habitats (Dayton, 1971, 1975; Jara and Moreno, 1984; Lubchenco, 1980, 1983; Lubchenco and Menge, 1978; Menge, 1976; Paine, 1971; Sousa, 1979b). Given the limited data base, an evaluation of how ecological processes interact with environmental constraints to structure rocky intertidal (and other) communities must be advanced with caution. Below, we attempt such a synthesis.

VIII. TESTS OF A MODEL OF COMMUNITY ORGANIZATION

Evaluation of the predictions listed in the "THEORETICAL FRAMEWORK" section involves (1) determining whether or not the model adequately describes the organization of a subset of communities whose dynamics are already known, and (2) determining the predictive value of the model—can it predict with any accuracy the dynamics of unstudied communities. Ideally, we would like a model which generated predictions on the basis of easily obtained information, such as diversity, abundance, number of consumer species, etc.

To assess the descriptive ability of the model, four phenomena need to be examined in relation to trophic complexity and environmental stress: (1) the direct impact of physical and/or physiological disturbance, (2) the effectiveness of consumers as controlling agents, (3) the strength of intraspecific and interspecific competition, and (4) the role of structuring agents at both juvenile and adult stages of basal species. From this information, measures of the relative importances of the different structuring agents (physical factors, competition, and predation) and trophic complexity must be derived. If, for example, the importance of competition decreased, and predation increased, with increasing trophic complexity from intermediate to low environmental stress, then we would conclude that the model incorporates a certain level of explanatory or descriptive capability.

Even this first step in evaluation of the model is no simple task, and unfortunately the majority of experimental multi-species studies done so far cannot be used. Despite our claim that rocky intertidal communities are the best understood systems in the world, there is much in these systems that is not known. For example, the nature and strength of interactions among consumers in most intertidal communities are poorly known. Clearly, determining the descriptive value of this community model (and, we suggest, any community model) requires a formidable amount of information. Useful shortcuts are few, but important ones include selecting potentially dominant consumers and resources as the first foci of investigation on the basis of preliminary studies of food web structure, abundances, and previous studies in similar communities.

A. Interaction Webs: Sources

To test the model of community regulation summarized earlier, we analyzed 25 experimental investigations of marine rocky intertidal food webs. We also examined the extent that 11 marine hard-bottom subtidal food webs supported the predictions.

Criteria for inclusion were: a study must have presented reasonably complete information on (1) food web structure, (2) experiments examining the effects of predation and competition for space, (3) patterns of space occupancy with respect to gradients of wave exposure and level on the shore. Some studies also included information on interactions among consumers and the effects of physical stress and disturbance on the consumer and/or basal species, and variation in the effects of consumers at different stages of the life history of sessile prey species ("bottlenecks"). This information was included when available, but as will be seen, its lack is a potential source of variation in calculations of the relative importance of each factor. An investigation may actually include several separate food webs, particularly in those systems which have been the focus of concerted study. For example, different intertidal zones frequently have distinct food webs because the vertical ranges of many species are restricted.

These criteria were deemed the absolute minimum necessary to even crudely evaluate the Menge–Sutherland model. Unfortunately, it was often difficult or impossible to extract interaction webs from some reasonably well-studied communities. For example, the work done in warm-temperate eastern Australia (Underwood, 1978, 1980, 1981; Denley and Underwood, 1979; Jernakoff, 1983; Underwood and Jernakoff, 1981, 1984; Underwood *et al.*, 1983; Moran *et al.*, 1984; Fairweather *et al.*, 1984; Caffey, 1982, 1985) is under-represented due in part to our unfamiliarity with the system. None the less, with the assistance of Dr A. J. Underwood, we were able to extract an interaction web (# 25) from these studies. Such problems also plagued our efforts to incorporate studies done in Japan, Chile, South Africa, New Zealand, and Tasmania.

Table 3 summarizes the location, physical conditions, and literature source for each web. Table 4 provides the coded scientific name, type of organism, trophic status, and web membership of each dominant sessile and mobile organism in each web. The data set is dominated by northern-hemisphere cold temperate investigations, but examples are available from warm temperate (northern- and southern-hemisphere) and tropical regions.

Table 3
Characteristics of food webs used in derivation of interaction webs.

Web number	Location	Habitat		References
		Wave conditions	Level on shore	
A ROCKY INTERTIDAL FOOD WEBS				
1	New England	Exposed	High	Menge 1976
2	New England	Intermediate	High	Menge 1976
3	New England	Sheltered	High	Menge 1976, unpublished
4	New England	Exposed	Mid	Menge 1976, 1978a,b
5	New England	Intermediate	Mid	Menge 1976, 1977; Lubchenco 1986
6	New England	Sheltered	Mid	Menge 1976, 1978a,b; Lubchenco 1980, 1983, 1986
7	New England	Exposed	Low	Lubchenco and Menge 1978; Menge 1983
8	New England	Intermediate	Low	Lubchenco and Menge 1978; Menge 1983
9	New England	Sheltered	Low	Lubchenco and Menge 1978; Menge 1982, 1983
10	Scotland	Intermediate	High	Connell 1961a,b
11	Scotland	Intermediate	Mid	Connell 1961a,b
12	Pacific Northwest (USA)	Exposed	Upper Mid	Paine and Levin 1981, Suchanek 1979
13	Pacific Northwest (USA)	Exposed	Lower Mid	Dayton 1971; Paine 1966, 1974, 1976, 1980, 1984
14	Pacific Northwest (USA)	Intermediate	Lower Mid	Dayton 1971; Paine 1966, 1974, 1976, 1980, 1984, Frank 1982; Marsh 1986
15	Pacific Northwest (USA)	Sheltered	Lower Mid	Dayton 1971; Louda 1979; Menge 1972a,b; Menge and Menge 1974
16	Pacific Northwest (USA)	Exposed	Low	Dayton 1973, 1975; Paine 1984
17	Pacific Northwest (USA)	Intermediate	Low	Dayton 1973, 1975; Louda 1979; Paine 1969; Dethier and Duggins 1984; Duggins and Dethier 1985

18	Pacific Northwest (USA)	Sheltered	Low	Dayton 1975; Menge 1972a,b; Menge and Menge 1974; Louda 1979
19	Southern California	Intermediate	Low	Sousa 1979a,b; Fawcett 1984
20	Gulf of California	Sheltered	Lower Mid	Dungan 1986, 1987
21	Gulf of Panama	Intermediate	High	Menge and Lubchenco 1981; Lubchenco *et al.* 1984; Garrity and Levings 1981
22	Gulf of Panama	Intermediate	Mid	Menge and Lubchenco 1981; Lubchenco *et al.* 1984; Garrity and Levings 1981, 1983; Levings and Garrity 1983
23	Gulf of Panama	Intermediate	Low	Menge and Lubchenco 1981; Lubchenco *et al.* 1984; Menge *et al.* 1985, 1986a,b
24	Southern California	Sheltered	High	Robles 1987
25	Eastern Australia	Intermediate	Mid	Underwood *et al.* 1983; Fairweather *et al.* 1984; Creese 1982

		Habitat			References
		Turbulence	Depth	Substratum inclination	
B ROCKY SUBTIDAL FOOD WEBS					
1	New England (Inner East Point)	Low	6–9 m	Vertical	Sebens 1985, 1986a,b, unpub. MS
2	New England (Shag Rocks)	Moderate–High	6–9 m	Vertical	Sebens 1985, 1986a,b unpub. MS
3	New England (Halfway Rock)	High	10–13 m	Vertical	Sebens 1985, 1986a,b unpub. MS
4	New England (Inner East Point)	Low	6–9 m	Horizontal	Sebens 1985, unpub. MS
5	New England (Shag Rocks)	Moderate–High	6–9 m	Horizontal	Sebens 1985, unpub. MS
6	New England (Murray Rock, Star Island)	Moderate	11–18 m	Inclined	Witman 1985, 1987
7	New England (Murray Rock, Star Island)	High	4–8 m	Inclined	Witman 1985, 1987
8	Gulf of Panama (Contadora I.)	Moderate	0–6 m	Horizontal	Wellington 1982
9	Gulf of Panama (Contadora I.)	Low	6–10 m	Horizontal	Wellington 1982

Table 4

List of species in food webs classified by type of organism, trophic status, scientific name and web number.

Taxon code	Name	Type of Organism	Trophic Status	Web Numbers
A Intertidal food webs		*Sessile animals*		
Ai	*Abietinaria* sp.	Colonial hydrozoan	Basal	23
Bg	*Balanus glandula*	Barnacle (acorn)	Basal	12–15, 17, 18
Bi	*Balanus inexpectatus*	Barnacle (acorn)	Basal	23
Ca	*Chthamalus anisopoma*	Barnacle (acorn)	Basal	20
Cc	*Chamaesipho columna*	Barnacle (acorn)	Basal	25
Cd	*Chthamalus dalli*	Barnacle (acorn)	Basal	12–15, 17, 18
Ce	*Chama echinata*	Oyster	Basal	23
Cf	*Chthamalus fissus*	Barnacle (acorn)	Basal	19, 21, 22
Cp	*Catophragmus pilsbryi*	Barnacle (acorn)	Basal	23
Cs	*Chthamalus stellatus*	Barnacle (acorn)	Basal	10, 11
Ei	*Euraphia imperatrix*	Barnacle (acorn)	Basal	21
gv	gray vermetid	Vermetid gastropod	Basal	23
Mc	*Mytilus californianus*	Mussel	Basal	12–14, 16, 17, 24
Me	*Mytilus edulis*	Mussel	Basal	4–9, 12, 24
Mo	*Modiolus capax*	Mussel	Basal	23
Oi	*Ostrea iridescens*	Oyster	Basal	23
Op	*Ostrea palmula*	Oyster	Basal	22
os	orange sponge	Sponge	Basal	23
Pp	*Pollicipes polymerus*	Barnacle (gooseneck)	Basal	12
Sb	*Semibalanus balanoides*	Barnacle (acorn)	Basal	1–11
Sc	*Semibalanus cariosus*	Barnacle (acorn)	Basal	12, 13, 15
Tr	*Tesseropora rosea*	Barnacle (acorn)	Basal	25
Tp	*Tetraclita panamensis*	Barnacle (acorn)	Basal	22

		Algae		
ac	algal crusts	Mixed	Basal	16
Ae	*Alaria esculenta*	Brown kelp	Basal	7
An	*Ascophyllum nodosum*	Brown fucoid	Basal	3, 6
br	blade reds	Red	Basal	16, 17
ca	canopy annuals	Brown kelp	Basal	16–18
Cc	*Chondrus crispus*	Red foliose	Basal	8, 9
cc	coralline crustose algae	Red crust	Basal	23
ep	ephemerals	Mixed foliose	Basal	4–9, 12–18, 22
fa	foliose algae	Mixed	Basal	24
Fd	*Fucus distichus*	Brown fucoid	Basal	5, 11, 14, 15, 18
fr	foliose reds	Red	Basal	16, 17
Fv	*Fucus vesiculosus*	Brown fucoid	Basal	6
Gc	*Gigartina canaliculata*	Red foliose	Basal	19
Ge	*Gelidium coulteri*	Red foliose	Basal	19
Gp	*Gelidium pusillum*	Red foliose	Basal	23
Gl	*Gigartina leptorhyncos*	Red foliose	Basal	19
Hi	*Hildenbrandia* spp.	Red crust	Basal	22, 23
Hs	*Hedophyllum sessile*	Brown kelp	Basal	16, 17
Ja	*Jania* spp.	Red foliose	Basal	23
Le	*Lessoniopsis littoralis*	Brown kelp	Basal	16
Lp	*Laurencia pacifica*	Red foliose	Basal	19
ma	macroalgae	Mixed	Basal	25
mi	microalgae (sporelings)	Mixed	Basal	11–23, 25
pr	polysiphonous reds	Red foliose	Basal	23
Ra	*Ralfsia* sp.	Brown crust	Basal	20–23
Sh	*Schizothrix calcicola*	Blue-green algae	Basal	22, 23
Ul	*Ulva* spp.	Green algae	Basal	19
ul	ulvoids	Green algae	Basal	23

Table 4 *continued*

Taxon code	Name	Type of Organism	Trophic Status	Web Numbers
		Vascular plants		
Ps	*Phyllospadix scouleri*	Surfgrass	Basal	16, 17
		Mobile animals		
Ab	*Acanthina brevidentata*	Gastropod	Predator	21–23
Af	*Asterias forbesi*	Seastar	Predator	8, 9
Ah	*Acanthochitona hirudiniformis*	Chiton	Herbivore	23
At	*Acmaea mitra*	Limpet	Herbivore	16
Al	*Arenaria melanocephala*	Black turnstone	Predator	14
As	*Acanthina spirata*	Gastropod	Predator	19, 20
Av	*Asterias vulgaris*	Seastar	Predator	8, 9
Az	*Aphriza virgata*	Surfbird	Predator	14
Bd	*Bodianus diplotaenia*	Wrasse	Omnivore	22, 23
Bp	*Balistes polylepis*	Triggerfish	Omnivore	23
Cb	*Cancer borealis*	Crab	Omnivore	9
Cf	*Ceratostoma foliatum*	Gastropod	Predator	13
Cl	*Cellana tramoserica*	Limpet	Herbivore	25
Cm	*Carcinus maenas*	Crab	Herbivore	9
Ct	*Chiton stokesii*	Chiton	Herbivore	22
Co	*"Collisella" scabra*	Limpet	Herbivore	24
Dh	*Diodon hystrix*	Porcupinefish	Omnivore	21–23
Ev	*Echinometra vanbrunti*	Echinoid	Omnivore	23
Fi	*Fissurella virescens*	Limpet	Herbivore	22, 23
Fo	*Fissurella longifissa*	Limpet	Herbivore	22, 23
Gg	*Grapsus grapsus*	Crab	Herbivore	22, 23
Hb	*Haematopus bachmani*	Oystercatcher	Predator	14
Ke	*Kyphosus elegans*	Chub (fish)	Herbivore	22
Kt	*Katharina tunicata*	Chiton	Herbivore	17, 18

La	*Littorina acutispira*	Gastropod	Herbivore	25
Ld	*Lottia digitalis*	Limpet	Herbivore	13–15
Lh	*Leptasterias hexactis*	Seastar	Predator	12, 15, 18
Li	*Lottia limatula*	Limpet	Herbivore	24
Ll	*Littorina littorea*	Gastropod	Herbivore	6, 9
Lo	*Littorina obtusata*	Gastropod	Herbivore	6
Lp	*Lottia pelta*	Limpet	Herbivore	12–18
Ls	*Lottia strigatella*	Limpet	Herbivore	12–15
Lt	*Lottia strongiana*	Limpet	Herbivore	21
Mm	*Mopalia muscosa*	Chiton	Herbivore	24
Mo	*Morula marginalba*	Gastropod	Predator	25
Nc	*Nucella canaliculata*	Gastropod	Herbivore	12, 13
Ne	*Nucella emarginata*	Gastropod	Herbivore	12–15
Nm	*Nucella lamellosa*	Gastropod	Herbivore	15, 18
Nn	*Nerita funicalata*	Gastropod	Herbivore	22
Np	*Nucella lapillus*	Gastropod	Predator	3, 5, 6, 8, 9, 11
Ns	*Nerita scabricosta*	Gastropod	Herbivore	21, 22
Nu	*Nuttalina californica*	Chiton	Herbivore	24
Ob	*Octopus bimaculoides*	Cephalopod	Predator	19
Ov	*Ozius verreauxii*	Crab	Predator	23
Pc	*Pachygrapsus crassipes*	Crab	Herbivore	19
Ph	*Pycnopodia helianthodes*	Seastar	Predator	16, 17
Pi	*Panulirus interruptus*	Lobster	Predator	24
Pl	*Patelloida latistrigata*	Limpet	Herbivore	25
Po	*Pisaster ochraceus*	Seastar	Predator	13–18
Pp	*Purpura pansa*	Gastropod	Predator	21, 22
Pt	*Pachygrapsus transversus*	Crab	Herbivore	23
Pv	*Patella vulgata*	Limpet	Herbivore	11
Sa	*Stegastes acapulcoensis*	Damselfish	Herbivore	22
	Stegastes acapulcoensis	Damselfish	Omnivore	23
Sd	*Searlesia dira*	Gastropod	Predator	15, 17, 18
Sg	*Siphonaria gigas*	Pulmonate limpet	Herbivore	22

Table 4 *continued*

Taxon code	Name	Type of Organism	Trophic Status	Web Numbers
Sm	*Siphonaria maura*	Pulmonate limpet	Herbivore	23
Sp	*Strongylocentrotus purpuratus*	Echinoid	Herbivore	16, 17
Sr	*Scarus perrico*	Parrotfish	Omnivore	23
Tf	*Tegula funebralis*	Gastropod	Herbivore	17
Te	*Tectura fenestrata*	Limpet	Herbivore	19
Tl	*Tonicella lineata*	Chiton	Herbivore	16
Tm	*Thais melones*	Gastropod	Predator	21–23
Ts	*Tectura scutum*	Limpet	Herbivore	13, 15–18
Tt	*Thais triangularis*	Gastropod	Predator	23
B Subtidal food webs		*Sessile organisms*		
As	*Alcyonium siderium*	Octocoral	Basal	2, 3
Ap	*Aplidium pallidum*	Colonial tunicate	Basal	1–3
cx	"complex"	Amphipod tubes/detritus	Basal	1–5
Hp	*Halichondria panicea*	Sponge	Basal	2, 3
Mm	*Modiolus modiolus*	Mussel	Basal	6, 7
Ms	*Metridium senile*	Anemone	Basal	2
Me	*Mytilus edulis*	Mussel	Basal	4, 5
Pd	*Pocillopora damicornis*	Coral	Basal	8, 9
Pg	*Pavona gigantea*	Coral	Basal	8, 9
		Algae		
cc	coralline crusts	Red crust	Basal	3–7
Ch	*Chondrus crispus*	Red foliose	Basal	4, 5
fr	foliose reds	Red foliose	Basal	6, 7
Ld	*Laminaria digitata*	Brown kelp	Basal	6, 7
Li	*Lithothamnion glaciale*	Coralline crust	Basal	5
Ls	*Laminaria saccharina*	Brown kelp	Basal	6, 7

mt	mixed algal turf	Mixed foliose algae	Basal	6–9
Ph	*Phyllophora* spp.	Red foliose	Basal	4, 5
Py	*Phymatolithon rugulosum*	Coralline crust	Basal	5
Wm	*Waernia mirabilis*	Red fleshy crust	Basal	1–4
		Mobile animals		
Ae	*Aeolidia papillosa*	Nudibranch	Predator	2, 3
Af	*Asterias forbesi*	Seastar	Predator	1–5
Ar	*Arothron* spp.	Pufferfish	Predator	8, 9
Av	*Asterias vulgaris*	Seastar	Predator	1–5, 6, 7
Cb	*Cancer borealis*	Crab	Predator	4–6
Ci	*Cancer irroratus*	Crab	Predator	4, 6
Cs	*Coryphella salmonacea*	Nudibranch	Predator	2, 3
gr	Serranidae	Groupers	Predator	8, 9
Ha	*Homarus americanus*	Lobster	Predator	4–6
ja	Carangidae	Jacks	Predator	8, 9
La	*Lacuna vincta*	Gastropod	Herbivore	7
lu	Lutjanidae	Snappers	Predator	8, 9
Ma	*Macrozoarces americanus*	Eelpout (fish)	Predator	1–5
Pl	*Prionurus laticlavus*	Surgeonfish	Omnivore	8, 9
Sa	*Stegastes acapulcoensis*	Damselfish	Omnivore	8, 9
Sd	*Strongylocentrotus droebachiensis*	Sea Urchin	Omnivore	2, 3, 5, 6
Sg	*Scarus ghobban*	Parrotfish	Omnivore	8, 9
Sp	*Scarus perrico*	Parrotfish	Omnivore	8, 9

B. Interaction Webs: Methods

The interaction webs extracted from these studies are presented in Table 5. Web membership is by definition restricted to those species involved in "strong" interactions (*sensu* MacArthur, 1972a; Paine, 1980). That is, a web includes those sessile organisms that actually or potentially dominate space, and those consumers whose foraging activities have strong effects on community structure. Potentially dominant space occupiers and consumers with strong effects on prey were usually revealed by field experiments, although in some cases, our experience with a system or particularly convincing observational data were used to ascertain the importance of a consumer or prey species.

The interaction webs were used to calculate an index of web complexity. We reasoned that trophic complexity would be influenced by (1) the number of trophic levels ("# LVL" in Table 5), (2) the number of types of effective consumers ("# TYP"), (3) the number of species per type of effective consumer ("SPP/TYPE"), and (4) the number of trophic levels that each type feeds upon ("# LVLS FED ON"). A consumer "type" is defined as a consumer that differs from others at higher taxonomic levels (crabs, fishes, chitons, limpets, predaceous gastropods, and sea urchins are all different types). In calculating the number of levels fed upon, we counted herbivorous and predaceous versions of a given taxonomic type separately. For example, predaceous fishes and herbivorous fishes were counted as a single (taxonomic) type in the "# TYP" column of Table 5, while the predaceous and herbivorous representatives of these types were counted separately in the "# LVLS FED ON" column. SPP/TYPE was calculated as follows (e.g. web 6): (2[species of herbivorous snail] + 1[species of predaceous whelk]) ÷ 2(types of consumers) = 1·5. Similary, # LVLS FED ON was calculated as (e.g. Web 8): (2[# levels of seastar prey] + 1[# levels of whelk prey]) ÷ 2(# of consumer types) = 1·5. We made no effort to weight the four categories used in calculating the index of web complexity. Thus, the index (SCORE) = # LEVELS + # TYPES + SPP./TYPE + # LEVELS FED ON.

Estimates of the relative importance of physical factors (physiological disturbance, or "PS" in Table 6, and physical disturbance, or "PD"), competition ("− −"), and predation ("+ −") were obtained by determining whether (= 1) or not (= 0) each actually or potentially dominant basal species was strongly affected by each factor. Here dominance refers to abundances (usually percentage of the surface covered) > 10% in experiments or under natural conditions. Species were strongly affected if they exhibited statistically significant changes in experiments or, in those papers not subjected to statistical analysis, exhibited striking differences from abundances in control treatments. Some effects listed in Table 6 were based on unpublished data (e.g., in New England); others were based on our experience with a system. We also list for each species whether (= 1) or not

Table 5

Interaction web matrices taken from studies of community regulation in rocky intertidal habitats. See Table 3 for names of species codes. # LVL = number of trophic levels in web; # TYP = number of types of consumer in web; $\bar{x}$SPP/TYPE = mean number of species per type of consumer; $\bar{x}$# LVLS FED ON = mean number of trophic levels fed on by each type of consumer at each level. Trophic complexity SCORE is the sum of (# LVL) + (# TYP) + (mean # SPP/TYPE) + (mean # LVLS FED ON). "NONE" = no effective consumers occur in the web.

PREY SPP. (CODE)[a]	CONSUMER SPECIES	WEB COMPLEXITY				SCORE
		# LVL	# TYP	$\bar{x}$SPP/ TYPE	$\bar{x}$# LVLS FED ON	
INTERTIDAL INTERACTION WEBS						
WEB 1.	NEW ENGLAND (= NE) EXPOSED HIGH					
	NONE					
Sb		1	0	0	0	1
WEB 2.	NE INTERMEDIATE HIGH					
	NONE					
Sb		1	0	0	0	1
WEB 3.	NE PROTECTED HIGH					
	Np					
Sb	X	2	1	1	1	5
An						
WEB 4.	NE EXPOSED MID					
	NONE					
Me		1	0	0	0	1
Sb						
ep						
WEB 5.	NE INTERMEDIATE MID					
	Np					
Me	X	2	1	1	1	5
Sb	X					
Fd						
ep						

Table 5 *continued*

PREY SPP. (CODED)[a]	CONSUMER SPECIES						WEB COMPLEXITY				SCORE
							# LVL	# TYP	$\bar{x}$SPP/ TYPE	$\bar{x}$# LVLS FED ON	
WEB 6.	NE PROTECTED MID										
	Np	Ll	Lo								
Me	X						2	2	1·5 [=(2+1)÷2]	1·0 [=(1+1)÷2]	6·5
Sb	X										
An		X	X								
Fv		X	X								
ep		X	X								
WEB 7.	NE EXPOSED LOW										
	NONE										
Me							1	0	0	0	1
Sb											
Ae											
ep											
WEB 8.	NE INTERMEDIATE LOW										
	Np	Av	Af								
Me	X	X	X				3	2	1·5 [=(2+1)÷2]	1·5 [=(2+1)÷2]	8
Sb	X	X	X								
Cc											
ep											
Np		X	X								
WEB 9.	NE PROTECTED LOW										
	Np	Av	Af	Cb	Cm	Ll					
Me	X	X	X	X	X		3	4	1·5 [=(2+2+1+1)÷4]	1·5 [=(2+2+1+1)÷4]	10
Sb	X	X	X	X	X						
Cc						X					
ep						X					
Np		X	X	X	X						

WEB 10. SCOTLAND INTERMEDIATE HIGH

	NONE					
Cs		1	0	0	0	1
Sb						

WEB 11. SCOTLAND INTERMEDIATE MID

	Np	Pv					
Cs	X		2	2	1·0	1·0	6
Sb	X				[=(1+1)÷2]	[=(1+1)÷2]	
Fd		X					
mi		X					

WEB 12. PACIFIC NORTHWEST (= PNW) EXPOSED UPPER MID

	Ne	Nc	Lh	Lp	Ls					
Mc	X	X	X			3	3	1·7	1·3	9
Me	X	X	X					[=(1+2+2)÷3]	[=(2+1+1)÷3]	
Pp	X	X	X							
Sc	X	X	X							
Bg	X	X	X							
Cd	X	X	X							
ep				X	X					
mi				X	X					
Ne			X							
Nc			X							
Lp			X							
Ls			X							

WEB 13. PNW EXPOSED LOWER MID

	Ne	Nc	Cf	Po	Lp	Ld	Ls	Ts					
Mc	X	X	X	X					3	3	2·7	1·3	10
Sc	X	X	X	X							[=(1+3+4)÷3]	[=(2+1+1)÷3]	
Bg	X	X	X	X									
Cd	X	X	X	X									
ep					X	X	X	X					
mi					X	X	X	X					
Ne				X									
Nc				X									
Cf				X									
Lp				X									
Ld				X									
Ls				X									
Ts				X									

Table 5 *continued*

PREY SPP. (CODED)[a]	CONSUMER SPECIES									WEB COMPLEXITY # LVL	# TYP	$\bar{x}$SPP/ TYPE	$\bar{x}$# LVLS FED ON	SCORE
WEB 14. PNW INTERMEDIATE LOWER MID														
	Ne	Po	Hb	Al	Az	Lp	Ld	Ls						
Mc	X	X	X	X	X					3	4	2·3	1·5	10·8
Bg	X	X	X	X	X							[=(1+3+1+4) ÷4]	[=(2+2+1+4) ÷4]	
Cd	X	X												
Fd						X	X	X						
ep						X	X	X						
mi						X	X	X						
Ne		X												
Lp		X	X	X	X									
Ld		X	X	X	X									
Ls		X	X	X	X									
WEB 15. PNW PROTECTED LOWER MID														
	Ne	Nm	Sd	Po	Lh	Lp	Ld	Ls	Ts					
Sc	X	X	X	X	X					4	3	3·0	2·0	12
Bg	X	X	X	X	X							[=(2+3+4)÷3]	[=(2+2+1)÷3]	
Cd	X	X	X	X	X									
Fd						X	X	X	X					
mi						X	X	X	X					
ep						X	X	X	X					
Ne				X	X									
Nm				X	X									
Sd				X	X									
Lp			X	X	X									
Ld			X	X	X									
Ls			X	X	X									
Ts			X	X	X									

WEB 16. PNW EXPOSED LOW

	Po	Ph	Lp	Ts	Tl	At	Sp					
Ms	X							3	4	1·8	1·3	10·1
ep			X	X	X	X	X			[=(2+1+1+3) ÷4]	[=(2+1+1)÷3]	
mi			X	X	X	X	X					
ac			X	X	X	X	X					
Le							X					
Hs							X					
Ps							X					
ca							X					
fr							X					
br							X					
Lp	X											
Ts	X											
Tl	X											
At	X											
Sp		X										

WEB 17. PNW INTERMEDIATE LOW

	Sd	Po	Ph	Tf	Lp	Ts	Kt	Sp					
Mc	X	X							4	6	1·3	1·4	12·7
Bg	X	X									[=(2+1+1 +1+2+1)÷6]	[=(3+1+2+1 +1+1+1)÷7]	
Cd	X	X											
ep				X	X	X	X	X					
mi				X	X	X	X	X					
Ps							X	X					
Hs							X	X					
ca							X	X					
fr							X	X					
br							X	X					
Tf		X											
Lp	X	X											
Ts	X	X											
Kt		X											
Sp			X										
Sd		X											

Table 5 *continued*

PREY SPP. (CODED)[a]	CONSUMER SPECIES							WEB COMPLEXITY # LVL	# TYP	x̄SPP/ TYPE	x̄# LVLS FED ON	SCORE
WEB 18. PNW PROTECTED LOW												
	Nm	Sd	Po	Lh	Kt	Lp	Ts					
Bg	X	X	X	X				4	4	1·8	1·6	11·4
Cd	X	X	X	X						[=(2+2+1 +2)÷4]	[=(3+2+1 +1+1)÷5]	
Fd					X	X	X					
ca					X							
ep					X	X	X					
mi					X	X	X					
Kt			X	X								
Lp		X	X	X								
Ts		X	X	X								
Nm			X	X								
Sd			X	X								
WEB 19. SOUTHERN CALIFORNIA INTERMEDIATE LOW												
	Ob	Pc	As	Te								
Cf			X					3	4	1·0	1·3	9·3
Ul		X		X						[=(1+1+1+1) ÷4]	[=(1+2+1+1) ÷4]	
mi		X		X								
Gc												
Gl												
Ge												
La												
WEB 20. GULF OF CALIFORNIA PROTECTED LOWER MID												
	As	Lt										
Ca	X							2	2	1·0	1·0	6
Ra		X								[=(1+1)÷2]	[=(1+1)÷2]	
mi		X										

WEB 21. GULF OF PANAMA (GOP) INTERMEDIATE HIGH

	Ab	Tm	Pp	Dh	Ns					
Cf	X	X		X		4	3	1·7 [=(1+3+1)÷3]	1·7 [=(3+1+1)÷3]	10·4
Ei	X	X		X						
mi					X					
Ns			X	X						
Ab				X						
Tm				X						
Pp				X						

WEB 22. GOP INTERMEDIATE MID

	Ab	Tm	Pp	Ns	Nn	Sg	Fi	Ct	Gg	Ov	Ke	Sa	Bd	Dh					
Cf	X	X									X				4	6	2·3 [=(4+2+3+2+2+1)÷6]	1·5 [=(3+2+2+1+1+1+1+1)÷8]	13·8
Tp	X	X											X	X					
Op	X	X											X	X					
Ra				X	X		X	X											
Sh				X	X		X	X											
Hi				X	X		X	X											
ep				X	X	X	X	X	X		X	X							
Ns			X							X			X	X					
Nn			X							X			X	X					
Sg														X					
Fi		X								X			X	X					
Ct										X			X	X					
Gg													X						
Ab										X				X					
Tm										X				X					
Pp														X					

Table 5 *continued*

PREY SPP. (CODED)[a]	CONSUMER SPECIES																	WEB COMPLEXITY				SCORE
																		# LVL	# TYP	x̄SPP/ TYPE	x̄# LVLS FED ON	
WEB 23. GOP INTERMEDIATE LOW																						
	Ab	Tm	Tt	Fi	Fo	Sm	Ah	Ev	Ov	Pt	Gg	Ke	Sa	Sr	Bd	Bp	Dh	4	6	2·8 [=(6+3+3+1+3+1)÷6]	1·4 [=(3+1+2+1+1+1+1)÷7]	14·3
Bi	X	X	X									X		X	X	X	X					
Cp	X	X														X						
Mo	X	X	X												X	X	X					
Oi	X	X	X												X	X	X					
Ce	X	X	X												X	X	X					
os								X				X	X		X							
gv	X	X	X												X	X	X					
Ai								X				X	X		X							
Ra				X	X		X	X														
Hi								X														
Ja								X		X	X	X	X	X								
Gp								X		X	X	X	X	X								
pr								X		X	X	X	X									
cc								X														
ul								X				X	X									
mi				X	X	X	X	X		X	X			X								
Sh				X	X		X	X														
Ab		X															X					
Tm																	X					
Tt																	X					
Fi		X													X		X					
Fo		X													X		X					
Sm	X	X																				
Ah		X															X					
Pt																X	X					
Gg															X							
[Ev]															X	X	X					

WEB 24. SOUTHERN CALIFORNIA INTERMEDIATE HIGH

	Pi	Li	Lc	Mm	Nu					
Me	X					3	2	2·5 [=(1+4)÷2]	1·5 [=(2+1)÷2]	9
Mc	X									
fa										
Li	X									
Co	X									
Mm	X									
Nu	X									

WEB 25. EASTERN AUSTRALIA INTERMEDIATE MID

	Mo	Cl	Pl	La					
Tr	X	X			3	3	1·3 [=(1+2+1)÷3]	1·3 [=(2+1+1)÷3]	8·6
Cc	X								
mi		X	X	X					
ma		X	X	X					
Cl									
Pl	X								
La	X								

SUBTIDAL INTERACTION WEBS

WEB 1. NEW ENGLAND CALM VERTICAL

	Av					
Ap	X	2	1	1	1	5
cx	X					
Wm						

WEB 2. NEW ENGLAND TURBULENT VERTICAL (FEW URCHINS)

	Av	Sd	Ae	Cs	Ma					
Ap	X	X				3	4	1·25 [=(1+1+1+2)÷4]	1·0 [=(1+1+1+1)÷4]	9·25
As				X						
Ms			X							
cx		X								
Hp		X								
Wm										
[Sd]					X					

Table 5 *continued*

PREY SPP. (CODED)[a]	CONSUMER SPECIES					WEB COMPLEXITY # LVL	# TYP	x̄SPP/ TYPE	x̄# LVLS FED ON	SCORE
WEB 3.	NEW ENGLAND TURBULENT VERTICAL (DENSE URCHINS)									
	Av	Sd	Ae	Cs						
Ap	X	X				2	3	1·3 [=(1+1+2)÷3]	1·0 [=(1+1+1)÷3]	7·3
As				X						
Ms			X							
cx		X								
Hp		X								
Wm		X								
cc		X								
WEB 4.	NEW ENGLAND CALM HORIZONTAL (FEW URCHINS)									
	Av	Cb	Ci	Ha	Ma					
Ch						4	4	1·25 [=(1+2+1+1)÷4]	2·0 [=(3+2+2+1)÷4]	11·25
Ph										
fr										
cc										
Me	X	X	X	X						
[Sd]	X	X	X	X	X					
Cb				X						
Ci				X						
Av				X						
WEB 5.	NEW ENGLAND TURBULENT HORIZONTAL (DENSE URCHINS)									
	Av	Sd	Cb	Ha						
Li		X				4	4	1·0 [=(1+1+1+1)÷4]	1·75 [=(3+2+1+1)÷4]	10·75
Py		X								
Ch		X								
Ph		X								
Me	X	X	X	X						
[Sd]			X	X						

Cb				X								
Av			X	X								
WEB 6.	NEW ENGLAND SHALLOW											
	NONE											
Mm								1	0	0	0	1
Ld												
Ls												
mt												
WEB 7.	NEW ENGLAND DEEP											
	Av	Sd	La	Cb	Ci	Ta	Ha					
Mm	X			X	X	X	X	4	6	1·2 [=(1+1+2 +1+1+1)÷6]	1·8 [=(3+2+2 +2+1+1)÷6]	13
Ld		X	X									
Ls		X	X									
mt		X	X									
fr		X										
cc		X										
La	X											
[Sd]				X	X	X	X					
Cb							X					
Ci							X					
WEB 8.	GULF OF PANAMA SHALLOW											
	Sa	Ar	Sp	Sg								
Pd		X	X	X				2	1	4	1	8
Pg	X	X	X	X								
mt	X		X	X								
WEB 9.	GULF OF PANAMA DEEP											
	Sa	Ar	Sp	Sg	gr	sn	ja					
Pd		X	X	X				3	2	3·5 [=(4+3)÷2]	1·0 [=(1+1)÷2]	9·5
Pg		X	X	X								
mt			X	X								

[a]Key to trophic position of prey species: normal font = sessile animal; italic = plant; boldface = herbivore; italicized boldface = carnivore; brackets = omnivore

Table 6

Frequency of strong (controlling) interactions in relation to trophic level. See Table 3 for key to web numbers.

Web No.	Basal Dominants (actual or potential)[a]	Agents: Basal Dominants[b] P D	P S	− −	+ −	− − due to coex. esc.	Consumer Dominants	Agents: Consumers[c] P D or S	− −	+ −	% Importance[d] Phys.	Comp.	Pred
Intertidal interaction webs													
1	*Semibalanus balanoides*	1	1	1	0	0		0	0	0	67	33 (3)	0
2	*S. balanoides*	1	1	1	0	0		0	0	0	67	33 (3)	0
3	*S. balanoides*	0	1	0	1	0	*Nucella lapillus*	?	?	?			
	Ascophyllum nodosum	1	1	1	0	0					67	17 (5)	17
4	*Mytilus edulis*	1	0	1	0	0	*N. lapillus*	1	0	0			
	S. balanoides	0	0	1	0	0							
	ephemerals	1	1	1	0	0					57	43 (7)	0
5	*F. distichus*	1	0	1	0	0	*N. lapillus*	?	?	?			
	M. edulis	1	0	1	1	1							
	S. balanoides	1	0	1	1	1							
	ephemerals	1	0	1	0	0					36	36 (10)	18
6	*A. nodosum*	0	0	1	0	1	*N. lapillus*	0	?	?			
	F. vesiculosus	0	0	1	1	1	*Littorina littorea*						
	M. edulis	0	0	0	1	0		0	?	?			
	S. balanoides	0	0	0	1	0							
	ephemerals	0	0	0	1	0					0	33 (6)	67

7	*M. edulis*	1	0	1	0	0	*N. lapillus*	1	0	0			
	S. balanoides	0	0	1	0	0							
	Alaria esculenta	1	0	1	0	0							
	ephemerals	1	0	1	0	0					50	50	0
												(8)	
8	*Chondrus crispus*	1	1	1	0	1	*N. lapillus*	?	?	?			
	M. edulis	1	0	1	1	1							
	S. balanoides	0	0	1	1	0	*Asterias vulgaris*	1	?	?			
	ephemerals	1	0	1	1	1					42	33	25
												(12)	
9	*C. crispus*	0	1	1	0	1	*N. lapillus*	?	?	?			
	M. edulis	0	0	0	1	0	*A. forbesi*	1	?	?			
	S. balanoides	0	0	0	1	0							
	ephemerals	0	0	1	1	0	*Carcinus maenas*	?	?	?			
							Cancer borealis	?	?	?	14	29	43
												(7)	
10	*Chthamalus stellatus*	0	1	1	0	0		0	0	0			
	S. balanoides	0	1	1	0	0					50	50	0
												(4)	
11	*C. stellatus*	0	0	1	1	1	*N. lapillus*	?	?	?			
	S. balanoides	0	0	1	1	0							
	F. vesiculosus	0	0	0	1	0	*Patella vulgata*	?	?	?			
	mircoalgae	0	0	0	1	0					0	33	67
												(6)	
12	*Mytilus californianus*	1	0	1	0	1							
	M. edulis	1	0	1	1	0	*Leptasterias hexactis*	?	?	?			
	Pollicipes polymerus	0	0	1	0	1	*N. canaliculata*	?	?	?			
	S. cariosus	0	0	1	0	1							

Table 6 *continued*

Web No.	Basal Dominants (actual or potential)[a]	Agents: Basal Dominants[b] P D	P S	− −	+ −	− − due to coex. esc.	Consumer Dominants	Agents: Consumers[c] P D or S	− −	+ −	% Importance[d] Phys.	Comp.	Pred
12	*B. glandula*	0	0	1	1	0	*N. emarginata*	?	?	?			
(cont)	*C. dalli*	0	0	1	0	1							
	ephemerals	1	1	1	1	0	*Lottia pelta*	?	?	?			
	microalgae	0	0	0	1	0	*L. strigatella*	?	?	?	20	47 (15)	27
13	*M. californianus*	0	0	0	1	0	*Pisaster*						
	S. cariosus	0	0	0	1	0	*ochraceus*	?	?	?			
	B. glandula	0	0	0	1	0	*N. emarginata*	?	?	?			
							N. canaliculata	?	?	?			
	C. dalli	0	0	0	1	0	*Ceratostoma*						
							foliatum	?	?	?			
							Lottia pelta	?	?	?			
							L. digitalis	?	?	?			
							L. strigatella	?	?	?			
							Tectura scutum	?	?	?	0	0 (4)	100
14	*M. californianus*	0	0	0	1	0	*P. ochraceus*	0	1	0			
	B. glandula	0	0	0	1	0	*Haematopus*						
	C. dalli	0	0	0	1	0	*bachmani*	0	?	?			
							Arenaria						
							melanocephala	0	?	?			
							Aphriza virgata	0	?	?			

							N. emarginata	1	?	?			
							L. pelta	0	?	1			
							L. digitalis	0	?	1			
							L. strigatella	0	?	0			
							T. scutum	0	?	1	13	13	75
												(8)	
15	*S. cariosus*	0	0	0	1	0	*P. ochraceus*	0	1	0			
	B. glandula	0	0	0	1	0	*L. hexactis*	0	1	0			
	C. dalli	0	0	0	1	0	*Searlesia dira*	0	?	?			
	Fucus distichus	0	0	0	1	0	*N. emarginata*	0	?	?			
							N. lamellosa	0	?	?			
							L. pelta	0	?	?			
							L. digitalis	0	?	?			
							L. strigatella	0	?	?			
							T. scutum	0	?	?	0	33	67
												(6)	
16	*M. californianus*	0	0	0	1	0							
	Lessoniopsis												
	littoralis	1	0	1	0	1	*P. ochraceus*	1*	?	?			
	Hedophyllum						*Pycnopodia*						
	sessile	1	0	1	0	1	*helianthodes*	1*	?	?			
	B. glandula	0	0	0	1	0	*Strongylocentrotus*						
	C. dalli	0	0	0	1	0	*purpuratus*	0	?	1			
	canopy ephemerals	1	0	1	0	1	*L. pelta*	0	?	?			
	foliose red algae	1	0	1	0	1	*T. scutum*	0	?	?			
	blade red algae	1	0	1	0	1	*Acmaea mitra*	0	?	?			
	coralline crusts	1	0	1	1	1	*Tonicella lineata*	0	?	?	42	32	26
												(19)	
17	*M. californianus*	0	0	0	1	0	*P. ochraceus*	0	?	0			
	B. glandula	0	0	0	1	0	*P. helianthodes*	1*	?	?			
	C. dalli	0	0	0	1	0	*Searlesia dira*	0	?	?			
	H. sessile	1	0	1	1	1	*S. purpuratus*	0	?	1			

Table 6 *continued*

Web No.	Basal Dominants (actual or potential)[a]	Agents: Basal Dominants[b] P D	P S	− −	+ −	− − due to coex. esc.	Consumer Dominants	Agents: Consumers[c] P D or S	− −	+ −	% Importance[d] Phys.	Comp.	Pred.
17	canopy ephemerals	0	0	1	1	1	*Katharina*						
(cont)	foliose reds	0	0	1	1	1	*tunicata*	0	?	1*			
	blade reds	1	1	1	1	1	*L. pelta*	0	?	1			
	P. scouleri	1	1	1	0	0	*T. scutum*	0	?	1			
							Tegula						
							funebralis	0	?	1	26	22	52
18	*B. glandula*	0	0	0	1	0	*P. ochraceus*	0	1	0		(23)	
	C. dalli	0	0	0	1	0	*L. hexactis*	0	1	0			
	F. distichus	0	1	0	1	0	*S. dira*	0	?	?			
	canopy annuals	0	1	0	1	0	*N. lamellosa*	0	?	?			
	ephemerals	0	1	0	1	0	*K. tunicata*	0	?	?			
	microalgae	0	1	0	1	0	*L. pelta*	0	?	1			
							T. scutum	0	?	1	29	14	57
19	*Gigartina*											(14)	
	canaliculata	1	0	1	0	1	*Octopus*						
	Gigartina						*bimaculoides*	0	?	?			
	leptorhyncos	1	0	1	0	1							
	Gelidium coulteri	1	0	1	0	1	*Pachygrapsus*						
	Laurencia						*crassipes*	0	?	?			
	pacifica	1	0	1	0	1	*Acanthina*						
	Ulva spp.	1	1	1	1	1	*spirata*	0	?	0			
	microalgae	0	1*	1	1	0	*Tectura*						
	Chthamalus fissus	1	0	1	1	1	*fenestrata*	0	?	1	42	37	21
20	*Ralfsia* sp.	0	0	1	1	1	*Acanthina*					(19)	
	Chthamalus						*angelica*	?	?	?			
	anisopoma	0	0	1	1	0	*Lottia*				0	50	50
							strongiana	?	?	?		(4)	

21	*Chthamalus fissus*	0	1	0	1	0	*Diodon hystrix*	0	?	0			
	Euraphia imperatrix	0	1	0	1	0	*Purpura pansa*	1	?	1			
	microalgae	0	1	0	1	0	*Thais melones*	1	0*	1			
							Acanthina brevidentata	1	0*	1			
							Nerita scabricosta	1	?	1	50	0 (14)	50
22	*Ralfsia* sp.	0	1	1	1	1	*D. hystrix*	0	?	0			
	Schizothrix calcicola	0	0	1	1	1	*Bodianus diplotaenia*	0	?	?			
	Hildenbrandia spp.	0	0	1	1	1	*Kyphosus elegans*	0	?	?			
	Tetraclita panamensis	0	0	0	1	0	*Stegastes acapulcoensis*	0	1*	1*			
	Chthamalus fissus	0	0	0	1	0	*Ozius verreauxii*	0	?	?			
	Ostrea palmula	0	0	0	1	0	*Grapsus grapsus*	0	?	1*			
	ephemerals	0	1	0	1	0	*T. melones*	1	?	1			
	microalgae	0	1	0	1	0	*A. brevidentata*	1	0*	1			
							P. pansa	1	0*	1			
							Siphonaria gigas	1	?	1			
							Fissurella virescens	1	?	1			
							N. scabricosta	0	?	1			
							N. funiculata	0	?	1			
							Chiton stokesii	1	?	1	31	14 (31)	62
23	*Ralfsia* sp.	0	1	1	1	1	*D. hystrix*	0	?	0			
	S. calcicola	0	0	1	1	1	*B. diplotaenia*	0	?	?			
	Hildenbrandia spp.	0	0	1	1	1	*Balistes polylepis*	0	?	0*			
	coralline crusts	0	0	1	1	1	*S. acapulcoensis*	0	1*	0			
	Balanus inexpectatus	0	0	0	1	0	*K. elegans*	0	?	?			
	Catophragmus pilsbryi	0	0	0	1	0	*Scarus perrico*	0	?	0			
	Chama echinata	0	0	0	1	0	*G. grapsus*	0	?	1			
							Pachygrapsus transversus	1	?	1			

Table 6 *continued*

Web No.	Basal Dominants (actual or potential)[a]	Agents: Basal Dominants[b] P D	P S	− −	+ −	− − coex. esc. due to	Consumer Dominants	Agents: Consumers[c] P D or S	− −	+ −	% Importance[d] Phys.	Comp.	Pred
23	*Ostrea iridescens*	0	0	0	1	0	*O. verreauxii*	0	?	1			
(cont)	*Modiolus capax*	0	0	0	1	0	*T. melones*	0	?	1			
	gray vermetid	0	0	0	1	0	*A. brevidentata*	0	?	1			
	red vermetid	0	0	0	1	0	*T. triangularis*	?	?	1			
	orange sponge	0	1	0	1	0	*Echinometra*						
	Abietinaria sp.	0	0	0	1	0	*vanbrunti*	1	?	1			
	Gelidium sp.	0	0	0	1	0	*F. virescens*	0	0*	1			
	polysiphonous red	0	0	0	1	0	*F. longifissa*	0	0*	1			
	Jania spp.	0	1	0	1	0	*S. maura*	0	0*	1			
	ulvoids	0	1	0	1	0	*Acanthochitona*						
	microalgae	0	1	0	1	0	*hirudiniformis*	0	?	0	18	13 (40)	68
24	*M. edulis*	0	0	0	1	0	*Panulirus*	1	?	?	25	25	50
	M. californianus	0	0	0	1	0	*interruptus*					(4)	
	foliose algae	0	1	1	0	1							
25	*Tesseropora rosea*	0	0	1	1	1	*Morula marginalba*	?	?	?	17	33	50
	Chamaesipho columna	0	0	1	1	0						(6)	
	foliose algae	0	1	0	1	0							
Subtidal interaction webs													
1	*Aplidium pallidum*	0	0	1	1	0	*Asterias vulgaris*	1	0	1	0	75	25
	Waernia mirabilis	0	0	1	0	0						(4)	
	complex	0	0	1	0	0							
2	*A. pallidum*	0	0	1	1	0	*A. vulgaris*	1	0	1	0	55	45
	Alcyonium siderium	0	0	1	1	1	*Strongylocentrotus*	0	0	1		(11)	
	Metridium senile	0	0	1	1	1	*droebachiensis*						
	complex	0	0	1	1	0	*Aeolidia*	?	?	?			

	Halichondria	0	0	1	1	0	*papillosa*						
	panicea						*Coryphella*	?	?	?			
	Waernia mirabilis	0	0	1	0	1	*salmonacea*						
							Macrozoarces						
							americanus	?	?	?			
3	*A. pallidum*	0	0	0	1	0	*A. vulgaris*	1	0	1	0	27	73
	A. siderium	0	0	0	1	0	*S. droebachiensis*	0	0	1		(11)	
	M. senile	0	0	0	1	0	*A. papillosa*	?	?	?			
	complex	0	0	0	1	0	*C. salmonacea*	?	?	?			
	H. panicea	0	0	0	1	0							
	W. mirabilis	0	0	1	1	1							
	Lithothamnion	0	0	1	1	1							
	glaciale												
	Phymatolithon	0	0	1	1	1							
	rugulosum												
4	*Chondrus crispus*	0	0	1	0	1	*A. vulgaris*	1	0	1	0	83	17
	Phyllophora spp.	0	0	1	0	1	*Cancer borealis*	0	0	1		(6)	
	foliose reds	0	0	1	0	1	*C. irroratus*	0	0	1			
	L. glaciale	0	0	1	0	0	*Homarus*	0	0	1			
	P. rugulosum	0	0	1	0	0	*americanus*						
	Mytilus edulis	0	0	0	1	0	*M. americanus*	?	?	?			
5	*L. glaciale*	0	0	1	1	1	*A. vulgaris*	1	0	1	0	29	71
	P. rugulosum	0	0	1	1	1	*S. droebachiensis*	0	0	1		(7)	
	C. crispus	0	0	0	1	0	*C. borealis*	0	0	1			
	Phyllophora spp.	0	0	0	1	0	*H. americanus*	0	0	1			
	M. edulis	0	0	0	1	0							
6	*Modiolus modiolus*	1	0	1	0	0					43	57	0
	Laminaria digitata	1	0	1	0	0						(7)	
	L. saccharina	1	0	1	0	0							
	mixed algal turf	0	0	1	0	0							
7	*M. modiolus*	0	0	1	1	1	*A. vulgaris*	1	0	1	0	50	50
	L. digitata	0	0	1	1	0	*S. droebachiensis*	0	0	1		(12)	
	L. saccharina	0	0	1	1	0	*Lacuna vincta*	?	?	?			
	mixed algal turf	0	0	1	1	1	*C. borealis*	?	?	?			
	foliose reds	0	0	1	1	1	*C. irroratus*	?	?	?			

Table 6 *continued*

Web No.	Basal Dominants (actual or potential)[a]	Agents: Basal Dominants[b] PD	PS	− −	+ −	− − due to coex. esc.	Consumer Dominants	Agents: Consumers[c] PD or S	− −	+ −	% Importance[d] Phys.	Comp.	Pred
	coralline crusts	0	0	1	1	1	*Tautogolabrus adspersus*	?	?	?			
							H. americanus	?	?	?			
8	*Pocillopora damicornis*	1	0	1	0	1	*Stegastes acapulcoensis*	0	1	0	25	25 (4)	50
	Pavona gigantea	0	0	0	1	0	*Arothron* spp.	?	?	?			
	mixed algal turf	0	0	0	1	0	*Scarus perrico*	?	?	?			
							S. ghobban	?	?	?			
9	*P. damicornis*	0	0	0	1	0	*S. acapulcoensis*	0	0	1	0	0 (3)	100
	P. gigantea	0	0	0	1	0	*Arothron* spp.	?	?	?			
	mixed algal turf	0	0	0	1	0	*S. perrico*	?	?	?			
							S. ghobban	?	?	?			
							Lutjanidae	?	?	?			
							Carangidae	?	?	?			
							Serranidae	?	?	?			

[a]Basal dominants are species which cover $\geqslant 10\%$ of the space, either under usual conditions or at some point during field experiments.

[b]"Agents: Basal Dominants" code, or factors having a controlling effect on the abundance of Basal Dominants: "PD" = physical disturbance; "PS" = physiological disturbance; "− −" = interspecific competition; "+ −" = predation; "− − due to coex. esc." = competition occurred among adults which had passed through a predator-vulnerable stage as juveniles.

[c]"Agents: Consumers" code, or factors having a controlling effect on the abundance of Consumers: "PDorS" = physical or physiological disturbance. Others as defined in footnote b. "?" in body of table indicates that the effect of a particular agent on the species is unknown. "*" means that the assignment of a 0 or 1 is based on inferential, not experimental evidence.

[d]"% Importance" is estimated as the percent of the nonzero table entries (basal dominants only) which are physical factors ("Phys."), competition ("Comp."), or predation ("Pred.'). The number of parentheses below each set of percentages is the total number of controlling interactions.

(= 0) competition for space occurred among adult organisms after they had achieved coexistence escapes from co-occurring consumers ("– – due to coex. esc." in Table 6). The effects of each of these agents on mobile consumers were similarly determined. The % (relative) importance of physical factors, competition, and predation for each web was calculated as the proportion of the total number of strong effects (listed in parentheses in Table 6).

C. Interaction Webs: Results

The data (Fig. 5a) were subjected to both linear and quadratic regression. Linear regressions between web complexity and the percent importance of physical factors, competition, and predation, were all highly significant ($p < 0{\cdot}005$ or less). Web complexity is strongly correlated to physical factors (correlation coefficient [r] = $-0{\cdot}546$), competition ($r = -0{\cdot}612$), and predation ($r = 0{\cdot}753$).

Quadratic regressions provided a marginally better fit to the data (Fig. 5a). Both physical factors and competition are inversely correlated with web complexity ($r = -0{\cdot}633$ and $-0{\cdot}618$, respectively), while predation is positively correlated with web complexity ($r = 0{\cdot}770$). Encouragingly, the shape of these empirical curves is roughly similar to those predicted by the model (Fig. 1c). The relative importance of physical factors decreases sharply with web complexity. The relative importance of competition, although negatively correlated with web complexity, is slightly convex upward. The relative importance of predation increases with web complexity. Despite these supportive trends, however, much of the variance remains unexplained by these relationships (Fig. 5a). For example, in the linear analyses, only 27% of the variance in physical factors, 35% of the variance in competition, and 55% of the variance in predation is explained by interaction web complexity.

A potential source of variation not included in the above analysis is recruitment density (e.g. Fig. 1). The model predicts that as recruitment density decreases, the relative importance of competition declines (Menge and Sutherland, 1987). Unfortunately, as noted earlier, the evaluation of the role of recruitment in structuring communities is hindered by the lack of information. No studies are available in which recruitment density was manipulated as part of the experimental design. However, the studies summarized in Tables 3–6 can be qualitatively categorized as having high, intermediate, or low recruitment. Since the only studies with low recruitment (webs 21–23) occurred at sites of intermediate wave exposure, I compared the average importance of competition at these three sites to that at the eight sites with intermediate wave exposure and high recruitment. As predicted, average percent of competition with high recruitment was higher (32·1%) than with low recruitment (9·0%). Although the difference is significant (1-way ANOVA, $F = 11{\cdot}04$, 1,9 df, $p = 0{\cdot}009$), and the data are thereby

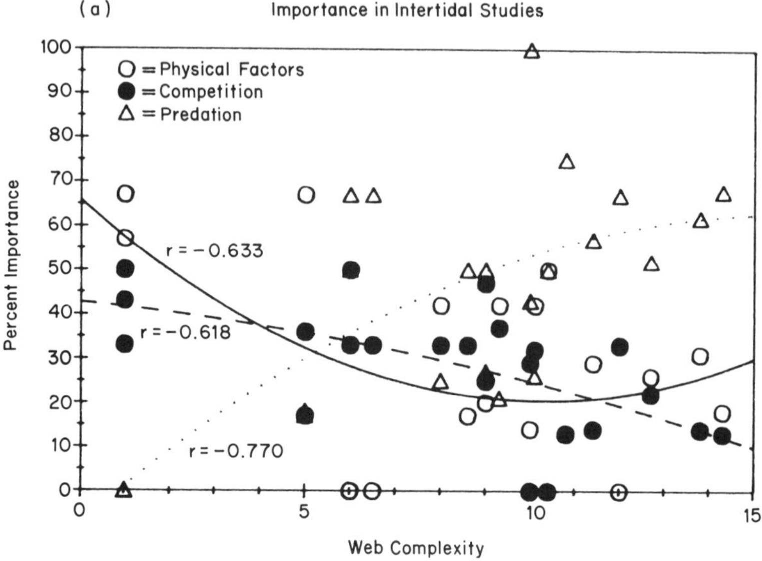

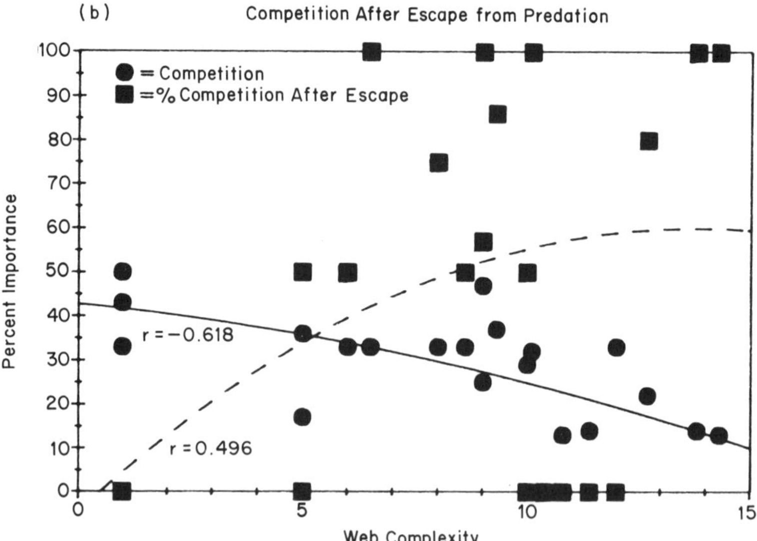

Fig. 5. Test of the Menge and Sutherland (1987) model of community organization. Data were obtained from the webs listed in Table 3 using the method described in Figure 8 in Menge and Sutherland (1987) and the text, and detailed in Tables 4–6. Correlation coefficients (r) are shown next to each quadratic regression line. Solid line = physical factors; dashed line = competition; dotted line = predation. a. Percentage importance of physical factors, competition, and predation (Table 6) in relation to interaction web complexity (Table 5). b. Percent importance of competition and "escape" competition (i.e. competition among adults after passing through a predation "bottleneck" as juveniles). Competition = solid line, escape competition = dashed line.

consistent with the prediction of the model, small sample size and other differences between these studies suggest these results should be regarded with skepticism.

The prediction that the proportion of space occupiers which compete as adults after passing through a predator-vulnerable stage (that is, have achieved a coexistence escape) increases with increasing trophic complexity is also supported by this analysis (Fig. 5b; $F = 6{\cdot}75$, $\mathrm{df} = 1{,}23$, $p = 0{\cdot}016$). However, only 19% of the variance in proportion escaping is explained by the index.

The prediction that trophic complexity increases with a decrease in environmental stress is supported by this analysis. Although sample sizes are again small, interaction web complexity generally increases with decreases in environmental stress. Thus, complexity (mean ± 1 SE, sample size in parentheses) along horizontal and vertical gradients is as follows:

LEVEL ON SHORE	WAVE EXPOSURE		
	Exposed	*Intermediate*	*Sheltered*
High intertidal	1·0 (1)	4·1 ± 3·1 (3)	7·0 ± 2·0 (2)
Mid intertidal	6·7 ± 2·9 (3)	8·8 ± 1·6 (5)	8·2 ± 1·9 (3)
Low intertidal	5·6 ± 4·6 (2)	11·1 ± 1·5 (4)	10·7 ± 0·7 (2)

Although exposed sites are evidently less complex than intermediate sites, protected sites are not still more complex except in the high zone, perhaps because physiological stress is greater at wave-protected than at intermediate sites. Taken together, this information suggests that in rocky intertidal habitats, effective trophic complexity tends to be greatest in low zones at intermediate wave exposures and to decrease with tidal level and with both increased and decreased wave exposure.

Two other predictions of the model—that top consumers should be most strongly influenced by competition and that trophically intermediate consumers should be influenced by both competition and predation—could not be tested owing to lack of information.

These results indicate that the model is capable of describing some major organizational features of rocky intertidal communities, a conclusion previously qualitatively reached for a smaller set of studies (Menge and Sutherland, 1976, 1987). Besides incorporating more recent information, the present analysis further incorporates a promising attempt to quantify several key aspects of community structure and organization in such a way as to permit greater objectivity in efforts to evaluate and ultimately test the predictions of the model. We emphasize, however, that our test is not conclusive, even for rocky intertidal communities. Since the factors regulating many key species in these systems are still unknown, and much of the variance in the importance of these factors remains unexplained, our conclusions are tentative.

D. Subtidal Interaction Webs

In a set of nine subtidal food webs, only the percentage importance of physical factors are significantly correlated with interaction web complexity ($p = 0{\cdot}03$; 44% of the variance is explained). Even this relationship is suspect, however, because the correlation is primarily due to a single point having an influence of 50%. Neither competition nor predation are significantly correlated with increasing web complexity.

The lack of correspondence between the model predictions and the results from these nine webs is probably partly due to small sample size. Another possibility is that different environmental factors become important with depth. The predictions of the model are made explicitly for food webs occurring in environments with large variation in environmental stress (water turbulence, desiccation, heat, unstable substratum). Subtidally, such conditions may be less important in all but the shallowest depths, in which case the assumptions of the model would not be met. In this case, altered or new models should be developed to predict how community structure should vary with other environmental factors such as nutrients/productivity and habitat complexity (Oksanen *et al.*, 1981; Tilman, 1982; Huston, 1979).

IX. APPLICABILITY OF THE MODELS

How broadly does the model of community organization discussed here apply? Connell's (1975) and Menge and Sutherland's (1976, 1987) models grew out of the classic model of Hairston *et al.* (1960). There are, however, some critical differences between the models, which may reflect fundamental differences between terrestrial and marine communities (Menge and Sutherland, 1987). For example, unlike marine (and some freshwater) benthic communities, there are no sessile animals in terrestrial habitats. Another difference is that true omnivory (consumption of both vegetation and animal tissue) is rare in terrestrial habitats but frequently found in major consumers in marine habitats. For instance, omnivorous consumers having strong community effects include seastars (*Stichaster australis*: Paine, 1971; Barker, 1979; *Acanthaster planci*: Birkeland, 1982; Yamaguchi, 1974; Colgan, 1987), fishes (Menge and Lubchenco, 1981; Menge *et al.*, 1985, 1986a,b; Lubchenco *et al.*, 1984); and sea urchins (Lawrence, 1975; Sammarco, 1980; Briscoe and Sebens, 1988). Hence, marine food webs may be more complex since connections between non-adjacent trophic levels are more likely. For example, apparently few terrestrial predators consume or even damage plants; herbivores thus seem to be the major source of mortality or damage for primary producers. In contrast, marine "predators" (= secondary consumers) may consume algae, sessile animals, and mobile animals. Thus, plants in terrestrial habitats may have fewer types of consumers to contend with than do plants in marine habitats. Although this conjecture cannot be evaluated at present, its implication is that terrestrial and aquatic communities may differ markedly in the ways they are organized.

X. FUTURE DIRECTIONS

Identification of ecological generalizations begins with a search for repeated patterns in nature. Once a pattern is detected, further investigation, both theoretical and empirical, focuses on detection of the processes that produce the patterns. The great variety of natural systems demands rather large sample sizes before the generality of a model can be evaluated. Once the descriptive ability of a model is delimited, predictions can be applied to unstudied systems and tested to determine the robustness of the model.

Community ecology is presently in the earliest stages of this process; hence, generalizations are still tentative and their domain uncertain. Simply put, we presently have an insufficient sample of studies of community structure and organization. We view the relative success of our tests of models of community regulation with guarded optimism. The lack of correspondence between subtidal interaction web dynamics and the predictions of the environmental stress model (Menge and Sutherland, 1987) raises two issues. First, we need to determine the domain of the model predictions; in which habitats are the model's predictions applicable? The results in this paper suggest that the predictions are probably restricted to habitats in which variation in interaction web structure responds primarily to environmental stress. Besides shallow marine benthic regions, this set of habitats may include streams, shallows of lakes and ponds, freshwater and terrestrial communities at high elevations and some deserts. The similarities between the environmental stress models (Connell, 1975; Menge and Sutherland, 1976, 1987) and models for stream and lake biota (Lodge *et al.*, 1987, Peckarsky, 1983; Schlosser, 1987) support this conclusion.

Second, further effort is necessary to continue the development and testing of models in situations in which the environmental stress model evidently does not apply (marine subtidal, most terrestrial habitats, planktonic habitats). A good start has been made in the development of models which investigate the role of variation in productivity or some related measure on community structure (Carpenter and Kitchell, 1984, 1987; Carpenter *et al.*, 1985; Huston, 1979; Oksanen *et al.*, 1981; Tilman, 1982). Thus, a future goal should be the integration of such models into a family of models that can make accurate predictions of how communities will be regulated given a particular set of environmental and biotic conditions (Colwell, 1984; MacArthur, 1972b; May, 1986; Schoener, 1986).

Many questions remain unanswered. First, we believe that a major effort must be mounted at the level of food webs. That is, studies are needed which focus on all space-occupying species and the factors that influence the observed patterns of space use. Such studies should specify the range of environmental conditions in which the food web occurs. Detailed quantitative knowledge of patterns of space utilization, size structure, and trophic interrelationships, and how these vary with space and time are necessary. Once such quantitative natural history is known, experiments which simultaneously investigate as many structuring processes as possible (ideally, all) can be designed and carried out. We recognize the immense amount of work hidden by these simple statements, but as noted earlier can offer few shortcuts. Pragmatically, the most profitable approach under the usual limitations of funds, personnel, and time is probably first to perform experimental studies at a rather coarse level of resolution and then later to sharpen the focus of the study at a finer scale (Menge *et al.*, 1986a).

Even in the best studied systems, much remains to be learned. For example, competition and predation are the most-studied interactions in most studies of community organization, yet Tables 1–6 indicate that little is known about these interactions among trophically higher organisms in most rocky intertidal communities. As noted by Connell (1983), Underwood (1978) and others, intraspecific competition is rarely investigated, yet in some respects, may be more important in structuring communities than interspecific competition. Mutualism, commensalism, and other non-traditional interactions have begun to attract more attention (e.g. Boucher *et al.*, 1982), but we still know virtually nothing about the community roles of such processes. Since mutualism, for example, may be of major importance in structuring communities where predation is severe, community-level investigations should begin to include these interactions among the factors included in their designs.

Dayton and Tegner (1984) argue that further progress in understanding community patterns will depend heavily on gaining insight into the roles of environmental constraints. We agree. In particular, our understanding of community variation at a biogeographic level will be poor until such knowledge is obtained. We encourage all efforts to evaluate the effects of such environmental constraints as climate, productivity, spatial and temporal heterogeneity, and history in structuring communities, at both large and small scales.

ACKNOWLEDGEMENTS

The research of B. A. Menge has been supported by the National Science Foundation. The manuscript was written with the support of NSF grants OCE-8415609 and OCE-8811369.

REFERENCES

Barker, M. F. (1979). Breeding and recruitment in a population of the New Zealand starfish *Stichaster australis* (Verrill). *J. exp. mar. Biol. Ecol.* **41**, 195–211.

Barnes, H. (1956). *Balanus balanoides* L. in the Firth of Clyde: the development and annual variation in the larval population and the causative factors. *J. Anim. Ecol.* **25**, 72–84.

Bender, E. A., Case, T. J. and Gilpin, M. E. (1984). Perturbation experiments in community ecology: theory and practice. *Ecology* **65**, 1–13.

Birkeland, C. (1982). Terrestrial runoff as a cause of outbreaks of *Acanthaster planci* (Echinodermata: Asteroidea). *Mar. Biol.* **69**, 175–185.

Bloom, S. A. (1975). The motile escape response of a sessile prey: a sponge-scallop mutualism. *J. exp. mar. Biol. Ecol.* **17**, 311–321.

Briscoe, C. S. and Sebens, K. P. (1988). Omnivory in *Stronglyocentrotus droebachiensis* (Muller) (Echinodermata: Echinoidea): predation on subtidal mussels. *J. exp. mar. Biol. Ecol.* **115**, 1–24.

Bormann, F. H. and Likens, G. P. (1979). *Pattern and Process in a Forested Ecosystem*. Springer-Verlag, New York. 253 pp.

Boucher, D. H., James, S. and Keeler, K. H. (1982). The ecology of mutualism. *Ann. Rev. Ecol. Syst.* **13**, 315–347.

Bowman, R. S. and Lewis, J. R. (1977). Annual fluctuations in the recruitment of *Patella vulgata* L. *J. mar. biol. Ass. U.K.* **57**, 793–815.

Branch, G. M. (1981). The biology of limpets: physical factors, energy flow, and ecological interactions. *Oceanogr. Mar. Biol. Ann. Rev.* **19**, 235–380.

Brown, J. H., Davidson, D. W., Munger, J. C. and Inouye, R. S. (1986). Experimental community ecology: the desert granivore system. In: *Community Ecology* (Ed. by J. Diamond and T. Case), pp. 41–61. Harper and Row, New York.

Buss, L. W. and Jackson, J. B. C. (1979). Competitive networks: nontransitive competitive relationships in cryptic coral reef environments. *Am. Nat.* **113**, 223–234.

Caffey, H. M. (1982). No effect of naturally-occurring rock types on settlement or survival in the intertidal barnacle *Tesseropora rosea* (Krauss). *J. exp. mar. Biol. Ecol.* **63**, 119–132.

Caffey, H. M. (1985). Spatial and temporal variation in settlement and recruitment of intertidal barnacles. *Ecol. Monogr.* **55**, 313–332.

Carpenter, S. R., Kitchell, J. F. and Hodgson, J. R. (1985). Cascading trophic interactions and lake productivity. *BioScience* **35**, 634–639.

Carpenter, S. R. and Kitchell, J. F. (1984). Plankton community structure and limnetic primary production. *Am. Nat.* **124**, 159–172.

Carpenter, S. R. and Kitchell, J. F. (1987). The temporal scale of variance in limnetic primary production. *Am. Nat.* **129**, 417–433.

Colgan, M. W. (1987). Coral reef recovery on Guam (Micronesia) after catastrophic predation by *Acanthaster planci. Ecology* **68**, 1592–1605.

Colwell, R. K. (1984). What's new? Community ecology discovers biology. In: *A New Ecology, Novel Approaches to Interactive Systems* (Ed. by P. W. Price, C. N. Slobodchikoff and W. S. Gaud), pp. 387–396. Wiley, New York.

Connell, J. H. (1961a). Effects of competition, predation by *Thais lapillus*, and other factors on natural populations of the barnacle *Balanus balanoides. Ecol. Monogr.* **31**, 61–104.

Connell, J. H. (1961b). The influence of interspecific competition and other factors on the distribution of the barnacle *Chthamalus stellatus. Ecology* **42**, 710–723.

Connell, J. H. (1970). On the role of natural enemies in preventing competitive exclusion in some marine animals and in rain forest trees. In: *Dynamics of Populations* (Ed. by P. J. den Boer and G. R. Gradwell), pp. 298–312. Proc. Advanced Study Institute on Dynamics of Numbers in Populations, Oosterbeek, The Netherlands, September 7–18, 1970. Pudoc, Wageningen, The Netherlands.

Connell, J. H. (1972). Community interactions on marine rocky intertidal shores. *Ann. Rev. Ecol. Syst.* **3**, 169–192.

Connell, J. H. (1974). Ecology: field experiments in marine ecology. In: *Experimental marine biology* (Ed. by R. Mariscal), pp. 21–54. Academic Press, New York.

Connell, J. H. (1975). Some mechanisms producing structure in natural communities: a model and evidence from field experiments. In: *Ecology and Evolution of Communities*. (Ed. by M. L. Cody, J. M. Diamond), pp. 460–490. Belknap Press of Harvard University Press, Cambridge, MA.

Connell, J. H. (1978). Diversity in tropical rain forests and coral reefs. *Science* **199**, 1302–1310.

Connell, J. H. (1983). On the prevalence and relative importance of interspecific competition: evidence from field experiments. *Am. Nat.* **122**, 661–696.

Connell, J. H. (1985). The consequence of variation in initial settlement vs. post-settlement mortality in rocky intertidal communities. *J. exp. mar. Biol. Ecol.* **93**, 11–45.

Creese, R. G. (1982). Distribution and abundance of the acmaeid limpet *Patelloida latistrigata*, and its interaction with barnacles. *Oecologia (Berl.)* **52**, 85–96.

Dayton, P. K. (1971). Competition, disturbance, and community organization: the provision and subsequent utilization of space in a rocky intertidal community. *Ecol. Monogr.* **41**, 351–389.

Dayton, P. K. (1973). Two cases of resource partitioning in an intertidal community: making the right prediction for the wrong reason. *Am. Nat.* **107**, 662–670.

Dayton, P. K. (1975). Experimental evaluation of ecological dominance in a rocky intertidal algal community. *Ecol. Monogr.* **45**, 137–159.

Dayton, P. K. (1984). Processes structuring some marine communities: are they general? In: *Ecological communities: conceptual issues and the evidence* (Ed. by D. R. Strong Jr., D. Simberloff, L. G. Abele, and A. B. Thistle), pp. 181–197. Princeton Univ. Press, Princeton, New Jersey.

Dayton, P. K. and Oliver, J. S. (1980). Problems in the experimental analyses of population and community patterns in marine benthic environments. In: *Marine benthic dynamics* (eds. K. R. Tenore and B. C. Coull), pp. 93–120. Univ. of South Carolina Press, Columbia, South Carolina.

Dayton, P. K., Currie, V., Gerrodette, T., Keller, B. D., Rosenthal, R. and ven Tresca, D. (1984). Patch dynamics and stability of some California kelp communities. *Ecol. Monogr.* **54**, 253–289.

Dayton, P. K. and Tegner, M. J. (1984). The importance of scale in community ecology: a kelp forest example with terrestrial analogs. In: *A new ecology: novel approaches to interactive systems* (Ed. by P. W. Price, C. N. Slobodchikoff and W. S. Gaud), pp. 457–481. Wiley, New York.

DeAngelis, D. L. and Waterhouse, J. C. (1987). Equilibrium and nonequilibrium concepts in ecological models. *Ecol. Monogr.* **57**, 1–21.

Denley, E. J. and Underwood, A. J. (1979). Experiments on factors influencing settlement, survival and growth of two species of barnacles in New South Wales. *J. exp. mar. Biol. Ecol.* **36**, 269–293.

Denny, M. W. (1982). Forces on intertidal organisms due to breaking ocean waves: design and application of a telemetry system. *Limnol. Oceanogr.* **27**, 178–183.

Denny, M. W., Daniel, T. L. and Koehl, M. A. R. (1985). Mechanical limits to size in wave-swept organisms. *Ecol. Monogr.* **55**, 69–102.

Dethier, M. N. (1984). Disturbance and recovery in intertidal pools: maintenance of mosaic patterns. *Ecol. Monogr.* **54**, 99–118.

Dethier, M. N. and Duggins, D. O. (1984). An indirect "commensalism" between marine herbivores and the importance of competitive hierarchies. *Am. Nat.* **124**, 205–219.

Diamond, J. and Case, T. J. (1986). *Community ecology*. Harper and Row, New York, 665 pp.

Duggins, D. O. and Dethier, M. N. (1985). Experimental studies of herbivory and algal competition in a low intertidal habitat. *Oecologia (Berl.)* **67**, 183–191.

Dungan, M. L. (1986). Three-way interactions: barnacles, limpets, and algae in a Sonoran Desert rocky intertidal zone. *Am. Nat.* **127**, 292–316.

Dungan, M. L. (1987). Indirect mutualism: complementary effects of grazing and predation in a rocky intertidal community. In: *Predation: direct and indirect impacts on aquatic communities* (Ed. by W. C. Kerfoot and A. Sih), pp. 188–200. Univ. Press of New England, Hanover, New Hampshire.

Fairweather, P. G. and Underwood, A. J. (1983). The apparent diet of predators and biases due to different handling times of their prey. *Oecologia (Berl.)* **56**, 169–179.

Fairweather, P. G., Underwood, A. J. and Moran, M. J. (1984). Preliminary investigations of predation by the whelk *Morula marginalba*. *Mar. Ecol. Prog. Ser.* **17**, 143–156.

Fawcett, M. H. (1984). Local and latitudinal variation in predation on an herbivorous marine snail. *Ecology* **65**, 1214–1230.

Frank, P. W. (1975). Latitudinal variation in the life history features of the black turban snail *Tegula funebralis* (Prosobranchia: Trochidae). *Mar. Biol.* **31**, 181–192.

Frank, P. W. (1982). Effects of winter feeding on limpets by Black Oystercatchers, *Haematopus bachmani*. *Ecology* **63**, 1352–1362.

Fretwell, S. D. (1977). The regulation of plant communities by food chains exploiting them. *Perspect. Biol. Med.* **20**, 169–185.

Gaines, S. D. (1985). Herbivory and between-habitat diversity: the differential effectiveness of defenses in a marine plant. *Ecology* **66**, 473–485.

Gaines, S. D. and Lubchenco, J. (1982). A unified approach to marine plant-herbivore interactions. II. Biogeography. *Ann. Rev. Ecol. Syst.* **13**, 111–138.

Gaines, S. D. and Roughgarden, J. (1985). Larval settlement rate: a leading determinant of structure in an ecological community of the marine intertidal zone. *Proc. Nat. Acad. Sci. U.S.A.* **82**, 3707–3711.

Gaines, S. D. and Roughgarden, J. (1987). Fish in offshore kelp forests affect recruitment to intertidal barnacle populations. *Science* **235**, 479–481.

Garrity, S. D. and Levings, S. C. (1981). A predator–prey interaction between two physically and biologically constrained tropical rocky shore gastropods: direct, indirect and community effects. *Ecol. Monogr.* **51**, 267–286.

Garrity, S. D. and Levings, S. C. (1983). Homing to scars as a defense against predators in the pulmonate limpet *Siphonaria gigas* (Gastropoda). *Mar. Biol.* **72**, 319–324.

Garrity, S. D. (1984). Some adaptations of gastropods to physical stress on a tropical rocky shore. *Ecology* **65**, 559–574.

Glasser, J. W. (1979). The role of predation in shaping and maintaining the structure of communities. *Am. Nat.* **113**, 631–641.

Glynn, P. W. (1968). Mass mortalities of echinoids and other reef-flat organisms coincident with midday, low water exposures in Puerto Rico. *Mar. Biol.* **1**, 226–243.

Grosberg, R. K. (1982). Intertidal zonation of barnacles: the influence of planktonic zonation of larvae on vertical distribution of adults. *Ecology* **63**, 894–899.

Hairston, N. G., Smith, F. E. and Slobodkin, L. B. (1960). Community structure, population control, and competition. *Am. Nat.* **94**, 421–425.

Hall, D. J., Cooper, W. E. and Werner, E. E. (1970). An experimental approach to the production dynamics and structure of freshwater animal communities. *Limnol. Oceanogr.* **15**, 839–928.

Hawkins, S. J. and Hartnoll, R. G. (1983). Grazing of intertidal algae by marine invertebrates. *Oceanogr. Mar. Biol. Ann. Rev.* **21**, 195–282.

Highsmith, R. C. (1980). Geographic patterns of coral bioerosion: a productivity hypothesis. *J. exp. mar. Biol. Ecol.* **46**, 177–196.

Hines, A. H. (1978). Reproduction in three species of intertidal barnacles from central California. *Biol. Bull.* **154**, 262–281.

Hines, A. H. (1979). The comparative reproductive ecology of three species of intertidal barnacles. In: *Reproductive ecology of marine invertebrates* (Ed. by S. E. Stancyk) pp. 213–234. Belle W. Baruch Library in Marine Science, vol. 9. Univ. of South Carolina Press, Columbia, SC.

Hurlbert, S. H. (1984). Pseudoreplication and the design of ecological field experiments. *Ecol. Monogr.* **54**, 187–211.

Huston, M. (1979). A general hypothesis of species diversity. *Am. Nat.* **113**, 81–101.

Hutchinson, G. E. (1959). Homage to Santa Rosalia, or why are there so many kinds of animals? *Am. Nat.* **93**, 145–159.

Inouye, R. S., Byers, G. S. and Brown, J. H. (1980). Effects of predation and competition on survivorship, fecundity, and community structure of desert annuals. *Ecology* **61**, 1344–1351.

Jackson, J. B. C. (1977). Competition on marine hard substrata: the adaptive significance of solitary and colonial strategies. *Am. Nat.* **111**, 743–767.

Jackson, J. B. C. (1978). Morphological strategies of sessile animals. In: *Biology and systematics of colonial organisms* (Ed. by G. Larwood and B. R. Rosen), pp. 499–555. Academic Press, London.

Jaksic, F. M. (1981). Abuse and misuse of the term "guild" in ecological studies. *Oikos* **37**, 397–400.

Jara, H. F. and Moreno, C. A. (1984). Herbivory and structure in a midlittoral rocky intertidal community: a case in southern Chile. *Ecology* **65**, 28–38.

Jernakoff, P. (1983). Factors affecting the recruitment of algae in a midshore region dominated by barnacles. *J. exp. mar. Biol. Ecol.* **67**, 17–31.

Keough, M. J. (1983). Patterns of recruitment of sessile invertebrates in two subtidal habitats. *J. exp. mar. Biol. Ecol.* **66**, 213–245.

Keough, M. J. (1984a). Effects of patch size on the abundance of sessile marine invertebrates. *Ecology* **65**, 423–437.

Keough, M. J. (1984b). The dynamics of the epifauna of *Pinna bicolor*: interactions between recruitment, predation, and competition. *Ecology* **65**, 677–688.

Keough, M. J. and Connell, J. H. (1984). Disturbance and patch dynamics of subtidal marine animals on hard substrata. In: *Natural disturbance: an evolutionary perspective* (Ed. by S. T. A. Pickett and P. S. White), pp. 125–151. Academic Press, New York.

Keser, M., Vadas, R. L. and Larson, B. R. (1981). Regrowth of *Ascophyllum nodosum* and *Fucus vesiculosus* under various harvesting regimes in Maine, U.S.A. *Bot. Mar.* **24**, 29–38.

Lawrence, J. (1975). On the relationships between marine plants and sea urchins. *Oceanogr. Mar. Biol. Ann. Rev.* **13**, 213–286.

Leigh, Jr., E. G., Paine, R. T., Quinn, J. F. and Suchanek, T. H. (1987). Wave energy and intertidal productivity. *Proc. Nat. Acad. Sci. USA* **84**, 1314–1318.

Levings, S. C. and Garrity, S. D. (1983). Diel and tidal movement of two co-occurring neritid snails: differences in grazing patterns on a tropical rocky shore. *J. exp. mar. Biol. Ecol.* **67**, 261–278.

Lewis, J. R. (1964). *The Ecology of Rocky Shores*. English University Press, London. 323 pp.

Lodge, D. M., Brown, K. M., Klosiewski, S. P., Stein, R. A., Covich, A. P., Leathers, B. K., and Bronmark, C. (1987). Distribution of freshwater snails: spatial scale and the relative importance of physicochemical and biotic factors. *Am. Malacol. Bull.* **5**, 73–84.

Louda, S. M. (1979). Distribution, movement and diet of the snail *Searlesia dira* in the intertidal community of San Juan Island, Puget Sound, Washington. *Mar. Biol.* **51**, 119–131.

Louda, S. M. (1982). Distribution ecology: variation in plant recruitment over a gradient in relation to insect seed predation. *Ecol. Monogr.* **52**, 25–41.

Loya, Y. (1976). Recolonization of Red Sea corals affected by natural catastrophes and man-made perturbations. *Ecology* **57**, 278–289.

Lubchenco, J. (1980). Algal zonation in a New England rocky intertidal community: an experimental analysis. *Ecology* **61**, 333–344.

Lubchenco, J. (1983). *Littorina* and *Fucus*: effects of herbivores, substratum heterogeneity, and plant escapes during succession. *Ecology* **64**, 1116–1123.

Lubchenco, J. (1986). Relative importance of competition and predation: early colonization by seaweeds in New England. In: *Community Ecology:* (Ed. by J. Diamond and T. Case), pp. 537–555. Harper and Row, New York.

Lubchenco, J. and Gaines, S. D. (1981). A unified approach to marine plant-herbivore interactions. I. Populations and communities. *Ann. Rev. Ecol. Syst.* **12**, 405–437.

Lubchenco, J. and Menge, B. A. (1978). Community development and persistence in a low rocky intertidal zone. *Ecol. Monogr.* **48**, 67–94.

Lubchenco, J., Menge, B. A., Garrity, S. D. *et al.* (1984). Structure, persistence and role of consumers in a tropical rocky intertidal community (Taboguilla Island, Bay of Panama). *J. exp. mar. Biol. Ecol.* **77**, 23–73.

Lynch, M. (1979). Predation, competition, and zooplankton community structure: an experimental study. *Limnol. Oceanogr.* **24**, 253–272.

MacArthur, R. H. (1972a). *Geographical ecology*. Harper and Row, New York.

MacArthur, R. H. (1972b). Coexistence of species. In: *Challenging biological problems* (Ed. by J. Behnke), pp. 253–259. Oxford University Press, Oxford.

Marsh, C. P. (1986). Rocky intertidal community organization: the impact of avian predators on mussel recruitment. *Ecology* **67**, 771–786.

May, R. M. (1986). The search for patterns in the balance of nature: advances and retreats. *Ecology* **67**, 1115–1126.

McAuliffe, J. R. (1984). Competition for space, disturbance, and the structure of a benthic stream community. *Ecology* **65**, 894–908.

McNaughton, S. J. (1983). Serengeti grassland ecology: the role of composite environmental factors and contingency in community organization. *Ecol. Monogr.* **53**, 291–320.

Menge, B. A. (1972a). Foraging strategy of a starfish in relation to actual prey availability and environmental predictability. *Ecol. Monogr.* **42**, 25–50.

Menge, B. A. (1972b). Competition for food between two intertidal starfish species and its effect on body size and feeding. *Ecology* **53**, 635–644.

Menge, B. A. (1976). Organization of the New England rocky intertidal community: role of predation, competition and environmental heterogeneity. *Ecol. Monogr.* **46**, 355–393.

Menge, B. A. (1977). Ecological implications of patterns of rocky intertidal community structure and behavior along an environmental gradient. In: *Ecology of fouling communities* (Ed. by J. D. Costlow), pp. 155–180. US Office of Naval Research; US Government Printing Office, Washington DC (in Russian and English).

Menge, B. A. (1978a). Predation intensity in a rocky intertidal community: relation between predator foraging activity and environmental harshness. *Oecologia (Berl.)* **34**, 1–16.

Menge, B. A. (1978b). Predation intensity in a rocky intertidal community: effect of an algal canopy, wave action and desiccation on predator feeding rates. *Oecologia (Berl.)* **34**, 17–35.

Menge, B. A. (1982). Asteroidea: effects of feeding on the environment. In: *Echinoderm Nutrition* (Ed. by M. Jangoux and J. Lawrence), pp. 521–551. A. A. Balkema, Rotterdam, The Netherlands.

Menge, B. A. (1983). Components of predation intensity in the low zone of the New England rocky intertidal region. *Oecologia (Berl.)* **58**, 141–155.

Menge, B. A. (1988). Relative importances of recruitment and other causes of variation in rocky intertidal community structure: a multivariate evaluation. (Unpublished.)

Menge, B. A. and Lubchenco, J. (1981). Community organization in temperate and tropical rocky intertidal habitats: prey refuges in relation to consumer pressure gradients. *Ecol. Monogr.* **51**, 429–450.

Menge, B. A. and Sutherland, J. P. (1976). Species diversity gradients: synthesis of the roles of predation, competition and temporal heterogeneity. *Am. Nat.* **110**, 351–369.

Menge, B. A. and Sutherland, J. P. (1987). Community regulation: variation in disturbance, competition, and predation in relation to environmental stress and recruitment. *Am. Nat.* **130**, 730–757.

Menge, B. A., Ashkenas, L. R. and Matson, A. (1983). Use of artificial holes in studying community development in cryptic marine habitats in a tropical rocky intertidal region. *Mar. Biol.* **77**, 129–142.

Menge, B. A., Lubchenco, J. and Ashkenas, L. R. (1985). Diversity, heterogeneity, and consumer pressure in a tropical rocky intertidal community. *Oecologia (Berl.)* **65**, 394–405.

Menge, B. A., Lubchenco, J. Ashkenas, L. R. and Ramsey, F. (1986a). Experimental separation of effects of consumers on sessile prey on a rocky shore in the Bay of Panama: direct and indirect consequences of food web complexity. *J. exp. mar. Biol. Ecol.* **100**, 225–269.

Menge, B. A., Lubchenco, J, Gaines, S. D. and Ashkenas, L. R. (1986b). A test of the Menge–Sutherland model of community organization in a tropical rocky intertidal food web. *Oecologia (Berl.)* **71**, 75–89.

Menge, J. L., and Menge, B. A. (1974). Role of resource allocation, aggression and spatial heterogeneity in coexistence of two competing intertidal starfish. *Ecol. Monogr.* **44**, 189–209.

Moran, M. J., Fairweather, P. G. and Underwood, A. J. (1984). Growth and mortality of the predatory intertidal gastropod *Morula marginalba* Blainville (Muricidae): the effects of different species of prey. *J. exp. mar. Biol. Ecol.* **75**, 1–18.

Morin, P. J. (1983). Predation, competition, and the composition of larval anuran guilds. *Ecol. Monogr.* **53**, 119–138.

Neill, W. E. (1975). Experimental studies of microcrustacean competition, community composition and efficiency of resource utilization. *Ecology* **56**, 809–826.

Olson, R. R. and McPherson, R. (1987). Potential vs. realized larval dispersal: fish predation on larvae of the ascidian *Lissoclinum patella* (Gottschaldt). *J. exp. mar. Biol. Ecol.* **110**, 245–256.

Oksanen, L., Fretwell, S. D., Arruda, J. and Niemela, P. (1981). Exploitation ecosystems in gradients of primary productivity. *Am. Nat.* **118**, 240–261.

Osman, R. W. and Haugsness, J. A. (1981). Mutualism among sessile invertebrates: a mediator of competition and predation. *Science* **211**, 846–848.

Paine, R. T. (1966). Food web complexity and species diversity. *Am. Nat.* **100**, 65–75.

Paine, R. T. (1971). A short-term experimental investigation of resource partitioning in a New Zealand rocky intertidal habitat. *Ecology* **52**, 1096–1106.

Paine, R. T. (1974). Intertidal community structure: experimental studies on the relationship between a dominant competitor and its principal predator. *Oecologia (Berl.)* **15**, 93–120.

Paine, R. T. (1976). Size-limited predation: an observational and experimental approach with the *Mytilus-Pisaster* interaction. *Ecology* **57**, 858–873.

Paine, R. T. (1977). Controlled manipulations in the marine intertidal zone, and their contributions to ecological theory. Academy of Natural Sciences, Philadelphia, Spec. Publ. **12**, 245–270.

Paine, R. T. (1979). Disaster, catastrophe, and local persistence of the sea palm *Postelsia palmaeformis. Science* **205**, 685–687.

Paine, R. T. (1980). Food webs: linkage, interaction strength and community infrastructure. *J. Anim. Ecol.* **49**, 667–685.

Paine, R. T. (1984). Ecological determinism in the competition for space. *Ecology* **65**, 1339–1357.

Paine, R. T. and Levin, S. A. (1981). Intertidal landscapes: disturbance and the dynamics of pattern. *Ecol. Monogr.* **51**, 145–178.

Palmer, A. R. and Strathmann, R. R. (1981). Scale of dispersal in varying environments and its implications for life histories of marine invertebrates. *Oecologia (Berl.)* **48**, 308–318.

Pimm, S. (1982). *Food webs.* Chapman and Hall, New York.

Peckarsky, B. L. (1983). Biotic interactions or abiotic limitations? A model of lotic community structure. In: *Dynamics of Lotic Ecosystems* (Ed. by T. D. Fontaine III and S. M. Bartell), pp. 303–323. Ann Arbor Science, The Butterworth Group, Ann Arbor, Michigan.

Peterson, C. H. (1982). The importance of predation and intra- and interspecific competition in the population biology of two infaunal suspension-feeding bivalves, *Protothaca staminea* and *Chione undatella*. *Ecol. Monogr.* **52**, 437–475.

Quinn, J. F. and Dunham, A. E. (1983). On hypothesis testing in ecology and evolution. *Am. Nat.* **122**, 602–617.

Robles, C. (1987). Predator foraging characteristics and prey population structure on a sheltered shore. *Ecology* **68**, 1502–1514.

Rosenthal, R. J., Clarke, W. D. and Dayton, P. K. (1974). Ecology and natural history of a stand of giant kelp, *Macrocystis pyrifera*, off Del Mar, California. *Fish. Bull.* **72**, 670–684.

Rosenzweig, M. L. (1973). Exploitation in three trophic levels. *Am. Nat.* **107**, 275–294.

Root, R. B. (1967). The niche exploitation pattern of the Blue-grey gnatcatcher. *Ecol. Monogr.* **37**, 317–350.

Roughgarden, J. (1983). Competition and theory in community ecology. *Am. Nat.* **122**, 583–601.

Roughgarden, J. (1986). A comparison of food-limited and space-limited animal competition communities. In: *Community Ecology* (Ed. by J. Diamond and T. J. Case), pp. 492–516. Harper and Row, New York.

Sammarco, P. W. (1980). *Diadema* and its relationship to coral spat mortality: grazing, competition, and biological disturbance. *J. exp. mar. Biol. Ecol.* **45**, 245–272.

Schlosser, I. J. (1987). A conceptual framework for fish communities in small warmwater streams. In: *Community and Evolutionary Ecology of North American Stream Fishes* (Ed. by W. J. Matthews and D. C. Heins), pp. 17–26. Oklahoma University Press, Norman, Okalhoma.

Schoener, T. W. (1983). Field experiments on interspecific competition. *Am. Nat.* **122**, 240–285.

Schoener, T. W. (1986). Overview: kinds of ecological communities—ecology becomes pluralistic. In: *Community Ecology* (Ed. by J. Diamond and T. J. Case), pp. 467–479. Harper and Row, New York.

Sebens, K. P. (1983). Population dynamics and habitat suitability of the intertidal sea anemones *Anthopleura elegantissima* and *A. xanthogrammica*. *Ecol. Monogr.* **53**, 405–433.

Sebens, K. P. (1985). The ecology of the rocky subtidal zone. *Amer. Sci.* **73**, 548–557.

Sebens, K. P. (1986a). Spatial relationships among encrusting marine organisms in the New England subtidal zone. *Ecol. Monogr.* **56**, 73–96.

Sebens, K. P. (1986b). Community ecology of vertical rock walls in the Gulf of Maine, U.S.A.: small-scale processes and alternative community states. In: *The Ecology of Rocky Coasts* (Ed. by P. G. Moore and R. Seed), pp. 346–371. Columbia Univ. Press, New York.

Sebens, K. P. (1988). Alternative stable states in the New England subtidal community: predator effects and topography. (Unpublished).

Sih, A., Crowley, P., McPeek, M., Petranka, J. and Strohmeier, K. (1985). Predation, competition, and prey communities: a review of field experiments. *Ann. Rev. Ecol. Syst.* **16**, 269–311.

Simberloff, D. (1983). Competition theory, hypothesis testing, and other community ecological buzzwords. *Am. Nat.* **122**, 626–635.

Sousa, W. P. (1979a). Disturbance in marine intertidal boulder fields: the nonequilibrium maintenance of species diversity. *Ecology* **60**, 1225–1239.

Sousa, W. P. (1979b). Experimental investigations of disturbance and ecological succession in a rocky intertidal algal community. *Ecol. Monogr.* **49**, 227–254.

Sousa, W. P. (1984). The role of disturbance in natural communities. *Ann. Rev. Ecol. Syst.* **15**, 353–391.

Sousa, W. P., Schroeter, S. C. and Gaines, S. D. (1981). Latitudinal variation in intertidal algal community structure: the influence of grazing and vegetative propagation. *Oecologia (Berl.)* **48**, 297–307.

Stephenson, T. A. and Stephenson, A. (1972). *Life Between Tidemarks on Rocky Shores.* W. H. Freeman, San Francisco.

Strong, D. R., Jr. (1983). Natural variability and the manifold mechanisms of ecological communities. *Am. Nat.* **122**, 636–660.

Strong, D. R., Jr., Simberloff, D. Abele, L. G. and Thistle, A. B., eds. (1984). *Ecological Communities: Conceptual Issues and the Evidence.* Princeton University Press, Princeton, New Jersey.

Suchanek, T. H. (1979). The *Mytilus californianus* community: studies on the composition, structure, organization and dynamics of a mussel bed. PhD Dissertation, Univ. of Washington, Seattle. 285 pp.

Suchanek, T. H. (1981). The role of disturbance in the evolution of life history strategies in the intertidal mussels *Mytilus edulis* and *Mytilus californianus. Oecologia (Berl.)* **50**, 143–152.

Sutherland, J. P. (1978). Functional roles of *Schizoporella* and *Styela* in the fouling community at Beaufort, North Carolina. *Ecology* **59**, 257–264.

Sutherland, J. P. (1981). The fouling community at Beaufort, North Carolina: a study in stability. *Am. Nat.* **118**, 499–519.

Sutherland, J. P. (1987). Recruitment limitation in a tropical intertidal barnacle: *Tetraclita panamensis* (Pilsbury) on the Pacific coast of Costa Rica. *J. exp. mar. Biol. Ecol.* **113**, 267–282.

Sutherland, J. P. and Karlson, R. H. (1977). Development and stability of the fouling community at Beaufort, North Carolina. *Ecol. Monogr.* **47**, 425–446.

Sutherland, J. P. and Ortega, S. (1986). Competition conditional on recruitment and temporary escape from predators on a tropical rocky shore. *J. exp. mar. Biol. Ecol.* **95**, 155–166.

Thorson, G. (1950). Reproductive and larval ecology of marine bottom invertebrates. *Biol. Rev.* **25**, 1–45.

Tilman, D. (1982). *Resource Competition and Community Structure.* Monographs in Population Biology 17. Princeton Univ. Press, Princeton, New Jersey.

Toft, C. A. and Shea, P. J. (1983). Detecting community-wide patterns: estimating power strengthens statistical inference. *Am. Nat.* **122**, 618–625.

Turner, T. (1983a). Facilitation as a successional mechanism in a rocky intertidal community. *Am. Nat.* **121**, 729–738.

Turner, T. (1983b). Complexity of early and middle successional stages in a rocky intertidal surfgrass community. *Oecologia (Berl.)* **60**, 56–65.

Turner, T. (1984). Stability of rocky intertidal surfgrass beds: persistence, preemption, and recovery. *Ecology* **66**, 83–92.

Underwood, A. J. (1978). An experimental evaluation of competition between three species of intertidal prosobranch gastropods. *Oecologia (Berl.)* **33**, 185–202.

Underwood, A. J. (1979). The ecology of intertidal gastropods. *Adv. Mar. Biol.* **16**, 111–210.

Underwood, A. J. (1980). The effects of grazing by gastropods and physical factors on the upper limits of distribution of intertidal macroalgae. *Oecologia (Berl.)* **46**, 201–213.

Underwood, A. J. (1981). Structure of a rocky intertidal community in New South Wales: patterns of vertical distribution and seasonal changes. *J. exp. mar. Biol. Ecol.* **51**, 57–85.

Underwood, A. J. and Denley, E. J. (1984). Paradigms, explanations and generalizations in models for the structure of intertidal communities on rocky shores. In: *Ecological Communities: Conceptual Issues and the Evidence* (Ed. by D. R. Strong Jr, D. Simberloff, L. G. Abele and A. B. Thistle), pp. 151–180. Princeton Univ. Press, Princeton, New Jersey.

Underwood, A. J. and Jernakoff, P. (1981). Effects of interactions between algae and grazing gastropods on the structure of a low-shore intertidal algal community. *Oecologia (Berl.)* **48**, 221–233.

Underwood, A. J. and Jernakoff, P. (1984). The effects of tidal height, wave-exposure, seasonality and rock-pools on grazing and the distribution of intertidal macroalgae in New South Wales. *J. exp. mar. Biol. Ecol.* **75**, 71–96.

Underwood, A. J., Denley, E. J. and Moran, M. J. (1983). Experimental analyses of the structure and dynamics of mid-shore rocky intertidal communities in New South Wales. *Oecologia (Berl.)* **56**, 202–219.

VanBlaricom, G. R. (1982). Experimental analyses of structural regulation in a marine sand community exposed to oceanic swell. *Ecol. Monogr.* **52**, 283–305.

Vance, R. R. (1978). A mutualistic interaction between a sessile marine clam and its epibionts. *Ecology* **59**, 679–685.

Vermcij, G. J. (1978). *Biogeography and Adaptation.* Harvard University Press, Cambridge, MA. 332 pp.

Victor, B. C. (1986). Larval settlement and juvenile mortality in a recruitment-limited coral reef fish population. *Ecol. Monogr.* **56**, 145–160.

Virnstein, R. W. (1978). Predator caging experiments in soft sediments: caution advised. In: *Estuarine Interactions* (Ed. by M. L. Wiley), pp. 261–273. Academic Press, New York.

Watanabe, J. M. (1984). The influence of recruitment, competition, and benthic predation on spatial distributions of three species of kelp forest gastropods (Trochidae: *Tegula*). *Ecology* **65**, 920–936.

Wellington, G. M. (1982). Depth zonation of corals in the Gulf of Panama: control and facilitation by resident reef fishes. *Ecol. Monogr.* **52**, 223–241.

Wethey, D. S. (1985). Catastrophe, extinction, and species diversity: a rocky intertidal example. *Ecology* **66**, 445–456.

Wiens, J. A. (1977). On competition and variable environments. *Am. Sci.* **65**, 590–597.

Wiens, J. A. (1984). On understanding a non-equilibrium world: myth and reality in community patterns and processes. In: *Ecological Communities: Conceptual Issues and the Evidence* (Ed. by D. R. Strong Jr., D. Simberloff, L. G. Abele and A. B. Thistle), pp. 439–457. Princeton Univ. Press, Princeton, New Jersey.

Wilbur, H. M. (1980). Complex life cycles. *Ann. Rev. Ecol. Syst.* **11**, 67–83.

Wilbur, H. M. (1987). Regulation of structure in complex systems: experimental temporary pond communities. *Ecology* **68**, 1437–1452.

Williams, A. H. (1981). An analysis of competitive interactions in a patchy back-reef environment. *Ecology* **62**, 1107–1120.

Witman, J. D. (1985). Refuges, biological disturbance, and rocky subtidal community structure in New England. *Ecol. Monogr.* **55**, 421–445.

Witman, J. D. (1987). Subtidal coexistence: storms, grazing, mutualism, and the zonation of kelps and mussels. *Ecol. Monogr.* **57**, 167–187.

Yamaguchi, M. (1974). Growth of juvenile *Acanthaster planci* (L.) in the laboratory. *Pac. Sci.* **28**, 123–138.

Yoshioka, P. M. (1982). Role of planktonic and benthic factors in the population dynamics of the bryozoan *Membranipora membranacea*. *Ecology* **63**, 457–468.

Zaret, T. M. (1980). Predation and freshwater communities. Yale University Press, New Haven, Conn. 187 pp.

Herbivores and Plant Tannins

E. A. BERNAYS, G. COOPER DRIVER and M. BILGENER

I. SUMMARY

This review summarizes the historical background to the interest in tannins, which has increased considerably over the last fifteen years. Distribution of tannins is discussed in taxonomic terms, and correlative evidence shown to be particularly difficult with them because of the co-occurrence of other factors important to animals, both vertebrate and invertebrate. A brief account of the chemistry is followed by some details on the potential modes of action, and the vexed question of quantitative evaluation of tannins in plants is discussed. Some recent papers dealing with the problems and practical guides are referred to so that the ecologist may better make

ADVANCES IN ECOLOGICAL RESEARCH Vol. 19
ISBN 0-12-013919-7

decisions as to what assays should be undertaken and whether assays are worth doing at all.

Biological activity of tannins on herbivores is covered in separate sections on vertebrates and arthropods. In both groups tannins seem to have detrimental effects, but probably these are more common in vertebrates. Some animals are inhibited from feeding, while many suffer poor growth if high concentrations occur in the diet. There is no definitive evidence for the mode of action being antidigestive. Deleterious effects include erosion of gut epithelia and toxicity from the breakdown products of hydrolyzable tannins. The review concludes with an overview and summary of potential roles in plants other than defensive activity against herbivores, and suggestions for research needs.

II. INTRODUCTION

A. Historical Development

There has been knowledge of tanning for the production of leather since prehistoric times, although the nature of the plant extracts and the scientific basis for the change in hide that they cause have only been established in the last hundred years. The word tannin is defined in the *Shorter Oxford Dictionary* as "any member of the astringent vegetable substances which possesses the property of combining with animal hide and converting it into leather". One of the most important sources of tannin is oak galls caused by insects, and the tannin purified from this source is often referred to as gallotannin. Tannins have other commercial uses besides rendering animal skins tough and resistant to decay: they form black complexes with iron salts, making them useful in production of ink, and are used as antidotes for poisoning from alkaloids, glycosides and heavy metal ions, because of their chemical affinity for them. In addition, they provide an important part of the flavor of many foods and drinks.

Although important to humans, and widespread in nature, their biological roles are still controversial. Botanical works in the first half of this century usually considered tannins as excretory products, but sometimes because of their widespread occurrence in bark, as protective agents against invading microorganisms. With respect to herbivores, Haberlandt (1914) and Nierenstein (1934) suggested that tannins could be protective, and there were reports of tannin "toxicity" in domestic animals (Marsh *et al.*, 1918; Smith, 1959). On the other hand early experimental work with insects showed phagostimulatory effects of tannins (Gornitz, 1954; Grevillius, 1905; Lagerheim, 1900). In the 1960s a number of papers appeared on the apparently deleterious effects of certain tannins or tannin-containing plants on mammals and birds (e.g. Booth and Bell, 1968; Chang and Fuller, 1964; Connor *et al.*, 1969; Dollahite *et al.*, 1963; Donnelly and Anthony, 1969; Drieger *et al.*, 1969; Fuller *et al.*, 1967; Joslyn and Glick, 1969; Pigeon *et al.*, 1962; Vohra *et al.*, 1966), while there were reports of beneficial effects of ingested tannin in some cases (Drieger *et al.*, 1969; Kendall, 1966).

The general significance of tannins in ecology received little attention until the last twenty years. However, the study by Goldstein and Swain in 1965, demonstrating that tannins prevented enzyme activity in *in vitro* tests, may have partly stimulated the rise in interest in the potential defensive roles for tannin that has been notable over the past twenty years. Key papers in this regard were by Feeny (1968, 1970), who suggested that oak tannins were important in the protection of foliage from caterpillar damage. Theories of plant defence invoking tannins as important players followed (Feeny, 1975, 1976; Fox, 1981; Rhoades, 1979; Rhoades and Cates, 1976; Swain, 1976). In the years since then there has been an explosion of work and new ideas, and there is a growing appreciation of the complexity of the topic—the diversity of chemicals with different effects, the diversity of chemical reactions involved, and the diversity of animal behavior and physiology with respect to tannins. This review is an update on the work of the last twenty years in particular, and of the current status of work on plant tannins and herbivores.

B. Distribution in Green Plants

The first major work on the distribution of tannins was by Bate-Smith (1957, 1962). He noticed that the presence of tannins coincided with the occurrence of a characteristic blue-black staining shown in sections of particular tissues and cells after histological processing involving production of iron salts. He examined over one thousand different species in 180 different families of dicotyledons, although he was not always able to use identical species in the comparisons. In a reanalysis of his data there is found an approximately 90% correspondance of histochemical evidence and gross chemical evidence of tannin. Tannin synthesis was clearly shown to be much more marked in some families than in others, with the strong tendency for herbaceous plants to lack them. Since the herbaceous habit is one of the characters attained more recently in plant evolution (Sporne, 1954), it is assumed that the capacity to synthesize tannins has decreased with advancement.

These interesting family differences have been investigated more widely by Harborne (1976) with particular reference to flavonoids in general and in particular the so-called condensed tannins. His data, obtained with more sophisticated chemical techniques, strongly support the conclusions of Bate-Smith. For example, of 300 species and 145 genera in the Umbelliferae, only one species contained condensed tannin, and it is considered a primitive species on other grounds. The woody Dilleniaceae by contrast gave positive results for all 107 species examined.

Bate-Smith (1974) and Swain (1976), in broader treatments of all green plants, suggested that condensed tannins evolved, perhaps in the Carboniferous when their presence became widespread, and that as Angiosperms evolved they first elaborated the hydrolyzable tannins, but then, later, families evolved that had neither.

Tannins may be present in any or all plant parts but are often at higher concentrations in woody, lignified tissues. In leaves they may decrease or increase with age of the leaf. In grasses, where only condensed tannins occur, the vegetative tissues have none while the grains of a few species such as *Sorghum* are tannin-rich (Eggum and Christensen, 1975; Haslam, 1979; Strumeyer and Malin, 1975; Watterson and Butler, 1983).

C. Correlative Evidence of Ecological Importance

Many studies have used correlative evidence for indicating the significance of tannins in the lives of animals. Distribution, abundance or species richness of animals can be variously correlated with amounts or general types of tannin. For example Feeny (1970) noted that more damage and more lepidopteran species occurred on oak in parts of Britain at times when tannin levels were relatively low. Levels of damage or performance of herbivores on different plants or plant parts can similarly be correlated with tannin levels. For example, Cooper Driver *et al.* (1977) found that acceptability of bracken fern to the locust *Schistocerca gregaria* fell off as tannin levels increased. There is, however, a particular danger in doing this with tannins, because their concentrations and types are often correlated with other factors known to be important for herbivores. For example, Coley (1988) found strong positive correlations between fiber and tannin content in leaves of rain forest trees, and since protein and water content are both negatively correlated with fiber content, high tannin leaves are often associated with very important nutrient deficiencies. It should be noted, however, that tannins interfere with fiber analysis (Reed, 1986). It has also become clear that correlations between insect performance and dietary tannin levels in their food plants may be positive or absent (Faeth, 1985; Fox and Macauley, 1977; Lawson *et al.*, 1984). Another complexity has emerged recently; single specific phenolic glycosides may be responsible for protection of particular plants from damage by particular herbivores, yet are themselves correlated with overall tannin content (Sunnerheim *et al.*, 1988). Finally, the within- and between-tree variation can be so profound that the question is raised about sampling error (Baldwin *et al.*, 1987). For these reasons it seems particularly inappropriate to use correlative evidence for estimating the effects of tannins.

Over the period since Feeny (1969) drew particular attention to the potential importance of tannins to phytophagous insects several attempts have been made to generalize about their defensive function. Best known are the theories of quantitative defense, which suggest that compounds such as tannins contrast with conventional toxins in that their effectiveness against herbivores is less potent, more graded, and considered to be more difficult to adapt to (Feeny, 1976; Rhoades and Cates, 1976). This was supported by the reasonable correlations noticed by these authors up to that time. The apparent similarity of tannins at least in their presumed primary mode of action as protein precipitants (and thereby digestibility reducers) was seen as evolutionarily convergent and biologically difficult to adapt to (Rhoades, 1979). Their widespread occurrence in plants presenting large quantity of foliage (trees) suggested some kind of generalized defence associated with vulnerability through being large and long lived ("apparent"). Much of this reasoning became less convincing when poor correlations and positive correlations were added to the list of tannin–herbivore interactions (Bernays, 1981a; Coley 1983; Faeth, 1985; Fox and Macauley, 1977).

Recent approaches have also addressed plant physiological constraints. Availability of soil nutrients has a profound effect on the production of phenolics including tannin in many plant species (Waterman and Mole, 1989). For example nitrogen fertilization of *Populus tremuloides* led to a reduction of tannins to less than one-fifth of the levels found in unfertilized control trees (Bryant *et al.*, 1987) and shortage of soil nutrients may commonly underly enhanced levels of carbon-based secondary compounds (Coley *et al.*, 1985). In addition, light levels profoundly influence foliar phenolic levels including tannins. For example, Mole *et al.* (1988) found very close positive correlations between light intensity and tannins in tropical rainforest leaves. Such findings point to the possibility that passive effects governed by abiotic factors may be more important than herbivores in determining concentrations of tannins in tannin-producing plants.

As with many biological topics, more detailed studies seem to be indicating that early generalizations do not continue to hold, though they may be relevant for particular groups of organisms. It is possible that vertebrates are affected by tannins more than arthropods (Section IVA,B). It is also possible that fungi and microorganisms have been the major selective agents for production and maintenance of tannins in plants (Section VB). Since herbivores create entry points for such pathogens, it is difficult always to draw clear lines between the relative importance of different types of attacking organisms.

We are concerned here particularly with documenting direct evidence of tannin–herbivore interactions, rather than using correlative data or speculating on what may be the selective pressures for tannin presence. Some of the papers concerning vertebrates discussed below rely heavily on correlations, however, and this is pointed out where possible.

III. CHEMICAL ASPECTS

A. Structure

Tannins are polyphenolic compounds with a molecular weight of between 300 and 3000 Daltons. Compounds that act as tannins in forming complexes with proteins and other macromolecules can be divided into four groups according to chemical structure, molecular weight, water solubility and tannin action (Swain, 1977). The two major structural groups of tannin are the condensed tannins (proanthocyanidins) and the hydrolyzable tannins (Haslam, 1981); the other two groups are the oxytannins and the *beta* tannins (Swain, 1979), although many do not consider these to be true tannins.

Condensed tannins are far and away the most widely distributed tannins in vascular plants. They are structurally related to flavonoids $(C_6–C_3–C_6)_n$ and consist of oligomers of two or more flavan-3-ols (such as catechin, epicatechin or the corresponding gallocatechin). Condensation to form oligomeric proanthocyanidins involves the addition of a flavan-3,4-diol to a flavan-3-ol or to an existing chain to form the "upper" unit. Interflavanoid bonds are generally between carbons 4 and 8 to form linear chains but branching can also occur due to 4–6 linkages (Stafford, 1988). The B ring of the flavan monomer is substituted with two or three *ortho* hydroxyl groups. The proanthocyanidins are so called because on treatment with hot mineral acid they yield small amounts (5–15%) of the corresponding anthocyanidins, namely cyanidin and delphinidin. In the past 15 years, due to the work of E. Haslam, L. J. Porter, H. Stafford and their colleagues, the structures of condensed tannins have become much better understood (Czochanska *et al.*, 1979, 1983; Fletcher *et al.*, 1977).

Hydrolyzable tannins are restricted to the dicotyledons. They are esters of glucose, or rarely other polyols, with gallic acid (3,4,5-trihydroxybenzoic acid), or hexahydroxydiphenic acids, or their derivatives (Swain, 1979). Hydrolyzable tannins can be readily hydrolyzed by hot mineral acid to yield the sugar and the constituent acids. Hydrolysis of hexahydroxydiphenic acid rapidly forms its dilactone, ellagic acid. Depending on the acids obtained from the hydrolysis, hydrolysable tannins are conventionally divided into gallotannins or ellagitannins. Chebulagic acid, dihydrodigallic acid, flavogallic acid and valoneic acid dilactone are defined as ellagitannins (Haslam, 1979).

B. Measurement of Tannins

Earlier studies attempted to define and measure tannins simply as polyphenols with a MW range of about 300–5000. Quantitative estimates of these tannins in plants have, however, proved to be more difficult to obtain than most other natural products, and as a result, inaccurate or inappropriate methods have flawed some of the published results. As knowledge of the variation in tannin structures and activities has improved, better or more specific methods of quantitative determination have been devised. There are now a variety of techniques which depend either on chemical analysis or on relative ability to bind with various proteins. The two are often not correlated (Martin and Martin, 1983) and the most suitable approach depends on whether tannin structures, amounts and properties are of interest or whether potential biological activity is to be estimated.

There have been several recent papers discussing the different techniques that are being used to measure tannins in ecological studies (Hagerman, 1987; Hagerman and Butler, 1989; Mole and Waterman, 1987c,d; Tempel, 1982; Wisdom *et al.*, 1987). Mole and Waterman (1987c,d) have made the most comprehensive detailed analysis of the problem of different results from different methodologies. They based their evaluation on an analysis of 17 plant extracts from fourteen different species and 12 different families. They used the methods of Folin-Denis (Ribereau-Gayon, 1972) and Hagerman and Butler (1978) for measuring total phenolics present in the leaves of these plants. They also estimated polymer lengths of condensed tannins (Butler *et al.*, 1984; Williams *et al.*, 1983), and used the vanillin (Burns, 1971) and proanthocyanidin methods (Swain and Hills, 1959 modified by Horowitz, 1970) for quantifying condensed tannins. Hydrolysable tannins were quantified by the iodate method for gallotannins (Bate-Smith, 1977) and by a newly developed method using ferric chloride (Mole and Waterman, 1987c). As a result, they concluded that whilst the overall levels of phenolics in extracts can be estimated with some confidence, the information imparted by more specific assays i.e. for condensed or hydrolysable tannins, is very dependent on the procedures employed, particularly when dealing with extracts from taxonomically highly diverse sources. At present most of the methods for estimating hydrolyzable tannins cannot be used with any degree of confidence. However, a new method of determining hydrolyzable tannin levels by measurement of gallic acid released by hydrolysis is now very promising (Inoue and Hagerman, 1988).

In contrast to the chemical methods, potential biological activity is thought to be related to the degree of protein precipitation that can be obtained. This has been used to define tannins in operational or ecologically functional terms (Martin and Martin, 1983). It is important to use appropriate proteins for the questions under study, however, as these vary in their affinity for tannins (Hagerman and Butler, 1981, Hagerman and Klucher, 1986). Protein precipitation methods have generally employed either hemoglobin (Bate-Smith, 1973; Schultz *et al.*, 1981) or bovine serum albumin (BSA) as the protein (Hagerman and Butler, 1978). Others have used the leaf protein, ribulose-1,5-bisphosphate carboxylase, RuBisco (Martin and

Martin, 1983) or *beta*-glucosidase (Becker and Martin, 1982). Recently Hagerman (1987) has developed a method that employs the formation of precipitin rings in protein-impregnated agarose gels by diffusion of tannin solutions.

Mole and Waterman (1987d) examined protein precipitating ability of tannins in extracts of the leaves of their 17 plant species. Tannins were extracted with 70% acetone (Foo and Porter, 1980) rather than 50% methanol or water (Bate-Smith, 1973). They used the method of Hagerman and Butler (1980) to the stage where a tannin-protein precipitate is generated. The protein precipitate was then analysed, following alkaline hydrolysis, using a ninhydrin assay for the amino acids generated by the hydrolysis (Marks *et al.*, 1985).

In all cases the problem of standards is critical because of the differences among different tannins and proteins. Wisdom *et al.* (1987) studied eight Sonoran desert species for total proanthocyanidins (condensed tannins). They carried out an evaluation of proanthocyanidin estimates using quebracho for the standard curve. Asquith and Butler (1985) had previously reported large differences in standard curves for tannins isolated from sorghum and quebracho. The efficiencies of precipitation by BSA by each different tannin, varied by up to a factor of fifty for tannins of different species. They indicate that standards need to be made from the plant species under study. It should be noted, however, that serious problems of technique in the paper by Wisdom *et al.* have been identified (Mole *et al.*, 1989). Hagerman and Butler (1989) suggest the use of a pure universal standard, which they could provide, and against which everything could be compared.

In order to measure the possible impact of tannins on digestive systems, the inhibition of cellulase activity by tannin extracts bound to the cellulose substrate and free in solution was studied by Bilgener (1988). Complexes between tannin and protein absorbed on a cellulose substrate are able to interfere with the digestion of that cellulose by cellulase enzymes. Such masking of cellulose was thought to potentially interfere with cellulolytic activities in the gut.

Combining the results of their two studies on tannin chemistry (Mole and Waterman, 1987c) and biochemistry (Mole and Waterman, 1987d), Mole and Waterman concluded that there was little correlation between methods used. First, there was not an absolute positive correlation between the level of protein precipitation and the incorporation of tannin in the tannin–protein precipitate. For example, as relative protein concentrations increase, the proportion of the tannin bound in the precipitate decreases, leading to less stable precipitates. Some basic amino acids will precipitate with tannins and therefore could potentially influence the amino acid balance of the diet. By measurements of total phenolics and condensed tannins it is impossible to make a reliable prediction of protein precipitation ability of the crude extracts of the leaves under study. The presence of hydrolysable tannins (or any other specific types of phenolic) can in no way be inferred from estimates of total phenolics and condensed tannins. There are difficulties in obtaining universal or even accurate measurements due to the general co-occurrence of many chemically distinct variants of a tannin in a plant. Hagerman and Butler (1980) suggest combining the results of tannin concentration and binding efficiencies, and expressing them as a measure of the protein precipitating potential, or "specific activity" of a leaf. They feel that this is a more realistic parameter for between-species comparisons than tannin concentration or binding efficiencies alone.

Finally, the extraction techniques used will influence the results obtained (Hagerman, 1988). Thus, at all levels, there are severe difficulties in obtaining the measures ecologists require. Questions must be asked about the importance and relevance of crude or precise assays, and the level of accuracy required, because in most cases attempting a useful analysis of tannins is a major undertaking. The most useful concise practical guide for dealing with tannin extraction and quantification now is by Hagerman and Butler (1989), while a guide to ecological tannin assays may be found in Mole *et al.* (1989). In summary, they make the following points:

(a) A check must be made for different types of tannin because of their different solubility properties.
(b) While it is desirable to use tannins from the plant under study for the standards, this is usually impractical. A purified and characterized standard is required.
(c) A tannin–protein precipitation reaction is the way to assay for tannins in the broadest operational sense. The "specific activity" of Hagerman and Butler (1980) is most appropriate.
(d) Ecologists need to follow standard procedures and become aware of the limitations of existing methods.
(e) Hydrolyzable tannins are chemically difficult to quantify and crude extracts are unsuitable. The method of Inoue and Hagerman (1988) should be used.
(f) Vanillin tests are appropriate as chemical assays for condensed tannins.

C. Mode of Action

Tannins complex with natural polymers, such as dietary proteins, digestive enzymes, polysaccharides (starch, cellulose, hemicellulose etc) (Loomis, 1974; Mole and Waterman, 1987d; Price *et al.*, 1980), fats and nucleic acids (Takechi and Tanaka, 1987) and amino acids (Mole and Waterman, 1987b, Takechi and Tanaka, 1987). Asquith and Butler (1985) demonstrated that the tannin–protein interactions may be very specific for tannins having different molecular structures. They suggested than tannins were rather specific in their interactions with proteins.

There have been several studies concerning the relative protein-binding efficiencies of condensed tannins versus tannic acid (a hydrolyzable tannin), often with conflicting results. Bate-Smith (1975) developed the technique of hemoglobin precipitation from hemolyzed blood solutions. Tannins were judged in terms of "relative astringency" (RA) or protein precipitating efficiency of a particular tannin preparation. He found that tannic acid was the most efficient precipitating agent with an RA of 1·0; chebulagic acid was next with an RA of 0·76; procyanidin dimers 0·1; trimers 0·2–0·3 and oligomers 0·4–0·5. Porter and Woodruffe (1984) extended this study and measured protein precipitating ability with hemoglobin at constant pH 5·0 using pure polymeric proanthocyanidins. They found that the predominant factor influencing RA is molecular weight. Provided the molecular weight of the polymer is approximately 2,500, the condensed tannins and tannic acid (MW 1250 ± 60) have a very similar RA. In their study the stereochemistry and B-ring oxidation of the procyanidins appeared to have little effect.

Their finding is in contrast to that of Bate-Smith (1975), Haslam (1974) and McManus *et al.* (1981), who found that condensed tannins having monohydroxy-B-ring structures are less efficient in binding with proteins than condensed tannins having additional vicinal hydroxyl groups. The amount of protein precipitated by condensed tannins, purified from sorghum grain, and commercially purified tannic acid was directly determined using radioiodinated BSA. Various amounts of tannin were added to the BSA and the amount of protein precipitated was determined by counting radioactivity in the supernatant after centrifugation. The results of Hagerman and Klucher (1986) suggest that condensed tannin, on a molar or a weight basis, is a more effective protein precipitating agent than chromatographically purified commercial tannic acid.

Every tannin–protein complexing system displays unique kinetics depending on its molecular structure, pH of the medium (Hagerman and Butler, 1981; Takechi and Tanaka, 1987; Van Sumere *et al.*, 1975), and the presence of other chemicals (Asquith and Butler, 1985; Beart *et al.*, 1985a,b; Hagerman and Butler, 1980, 1981; Haslam, 1974; McManus *et al.*, 1981; McManus *et al.*, 1985; Takechi and Tanaka, 1987). This is partly because of the variety of chemical interactions between tannins and other macromolecules have been postulated: covalent interactions, ionic interactions, hydrogen bonding interactions or hydrophobic interactions (Hagerman and Klucher, 1986). The most common mode of interaction between tannin and protein involves hydrogen bond formation between the protein amide carbonyl and the phenolic hydroxyl (Hagerman and Butler, 1980). This interaction is pH dependent.

The effects of protein structure on the interaction of tannin with proteins was determined with a competitive binding assay (Hagerman and Butler, 1981). Proteins, especially proline-rich proteins, have a very high affinity for either condensed or hydrolyzable tannin. For example, proline-rich proteins are bound several orders of magnitude more strongly than small compact proteins. Proteins that are small and tightly folded have low affinity for either condensed or hydrolyzable tannins. Some highly glycosylated proteins have low affinity for tannins (Strumeyer and Malin, 1970) but glycosylation of other proteins enhances affinity for tannins (Asquith *et al.*, 1987).

The importance of molecular structures in hydrolysable tannin–protein interactions is relatively well understood (Beart *et al.*, 1985; Haslam, 1974; McManus *et al.*, 1981). Haslam and Lilley (1986) summarize recent work in Haslam's laboratory on the interactions, both covalent and "non-bonding," that occur between naturally occurring polyphenols (*syn* vegetable tannins) and other substances, proteins and polysaccharides. They investigated the interaction of biosynthetic intermediates in gallic acid metabolism in plants (Haslam, 1974; Haddock *et al.*, 1982; McManus *et al.*, 1985) with BSA. For each system studied the number of moles of phenolic material bound per mole of protein were obtained as a function of unbound phenolic concentrations. The molecular size of the polyphenolic is important; interaction with BSA increases approximately exponentially in the galloyl-D-glucose series (tri-tetra-penta). Both molecular size and increased flexibility cause the molecule to bind more readily. Therefore the degree of binding between tannin and protein (bovine serum albumin, BSA) is greatly influenced by molecular size of the tannins and their biosynthetic intermediates, and the conformational flexibility and mobility of the polyphenols.

The precipitates formed between BSA and polyphenolics are reversible; however, under certain conditions irreversible associations occur through the formation of covalent bonds between the protein and the polyphenolic substrate. This tends to occur under oxidative conditions and at relatively high pH when the phenols are transformed into quinones that react with nucleophilic groups such as –SH, $-NH_2$ on the protein. Beart *et al.* (1985) investigated the hydrolytic decomposition of a range of procyanidins, as a function of pH and temperature, and from the results obtained suggest that the decomposition is specific acid catalysed. It is postulated that in the presence of protein the carbocations produced are captured by e.g. –SH groups on the protein and the modification in the protein induced by this reaction leads ultimately to aggregation and then to precipitation. It is possible that analogous interactions to give covalently bound polyphenolics may occur in naturally occurring ecological situations, such as, for example, the microbial decomposition of plant tissues and the acid-mediated breakdown of polyphenolic compounds in the stomachs of ruminants. This has not been demonstrated, although simple phenolics will covalently bind to free amino acids (Pierpoint, 1969a) and the BSA (Pierpoint, 1969b).

Relatively subtle changes in polyphenol structure also lead to marked changes in affinities for the dextran gels (polysaccharides). Once again, for the simple galloyl-D-glucose series, molecular size and flexibility are the two major features that moderate binding. There is the possibility that phenolic residues can be occluded into cavities in polysaccharide structures (Haslam and Lilley, 1986).

It has been believed that once tannin–protein complexes form they are insoluble and eventually precipitate from the media. In an excess of either proteins or tannins, cross linking between separate protein molecules via tannins is not completed and soluble tannin–protein complexes can occur without precipitation (Hagerman and Robbins, 1987; Mole and Waterman, 1985). Mole and Waterman (1987b) further analysed the effects of conformational changes in proteins caused by tannin binding. Takechi and Tanaka (1987) commented that the specificity of tannin-protein binding was not as marked as that of antigen-antibody interactions. They suggested that tannin-protein interactions should not be investigated by competitive binding assays such as used by Hagerman and Butler (1981).

The kinetics of digestive enzyme inhibition by tannins, with pepsin, a pancreatic protease, a bacterial protease, alpha amylase and hemicellulase was studied by Bilgener (1988). These experiments showed quite clearly that the mode of enzyme inhibition by tannins is dependent on the chemical structures of the tannins used, pH values of the reaction mixture and also the presence of other food polymers (such as cellulose) in the media (Bilgener, 1988).

The biologically and ecologically important properties of tannins were thought to depend on their generalized complexation with digestive enzymes as well as proteins and other components in herbivore diets with the consequent reduction in digestion (see Sections II and IV). Zucker (1983) argues for a specific rather than a generalized effect and, due to the great structural diversity, tannin complexes may involve very stereospecific interactions, as opposed to generalized interactions. Zucker presents the case for regarding tannins not as generalized defense compounds but rather as specific inhibitors targeting a variety of digestive processes that involve different enzymes. Although the antidigestive functions probably do not occur *in vivo*, these arguments may be relevant for effects on epithelial cell membrane components (Section V). Stafford (1988) suggests that in addition to the proposed functions of tannins, the great variety of structures could imply informational functions. She cites the potential information available in the sequence of flavonoid units of the oligomeric chains of proanthocyanidins, since frequently both 2,3-*trans* and 2,3-*cis* isomers as well as diphenols and triphenolic B-rings are involved.

In summary, potential modes of action relate largely to the manner in which tannins bind with biologically important compounds, especially macromolecules. The sites of action could be any parts of any cellular structure. The vulnerable tissue is the gut epithelium since movement through this cell layer is unlikely. Binding processes in the gut lumen from the time of chewing and throughout the time the materials are in the gut *could* influence digestion and absorption in a wide variety of ways. These are discussed in Section IV.

IV. BIOLOGICAL ACTIVITY IN PRACTICE

A. Vertebrates

1. Behavioral responses

Separation of behavioral deterrence from postingestional effects is impossible in many studies. In addition, much of the evidence for inhibition of feeding is correlative. Therefore the appraisal presented here is a mixture of types of data that overall invoke tannins as feeding deterrents, and the treatment is illustrative rather than comprehensive. Unlike the majority of insects, vertebrate species ingest a wide variety of plant species, albeit with some degree of specialization on herbaceous plants (grazers), trees and shrubs (browsers), seeds or fruits. Browsers will naturally ingest more tannins than grazers (Section IIB), but the availability of choices and the ingestion of a mixed diet may differentially affect the selectivity of animals at different times, making it difficult to obtain precise answers.

Few studies have been carried out with phytophagous reptiles but tortoises have clearly been shown to be deterred by both condensed and hydrolysable tannins (Swain, 1976). Most of the work with birds has involved domestic poultry where usually there seems to be reduction in feeding in the presence of both types of tannins (e.g. Connor *et al.*, 1969; Martin-Tanguy *et al.*, 1977).

Most studies have been on mammals where some degree of deterrence seems to be common, the extent varying, as might be expected, with the normal exposure the animals have to dietary tannins. Joslyn and Glick (1969) showed that condensed and hydrolyzable tannins can reduce feeding by rats at relatively high concentrations in artificial diets. On the other hand, 5% tannic acid in dry powdered diet did not appear to reduce feeding in mice (Freeland *et al.*, 1985b).

Among ruminants there seem to be some effects of tannins on palatability. However, work with domestic animals is generally governed by a need to know whether weight gain is affected, rather than a need to study mechanisms, and in most cases, feeding inhibition cannot be separated from poor growth caused by tannin ingestion. Cooper *et al.* (1988) studied the seasonal selection of plants by hand-reared kudus and impalas in savanna vegetation in northern Transvaal, South Africa. Based on an *a priori* palatability classification, discriminant function analysis separated relatively palatable species from unpalatable species in terms of a linear combination of protein and condensed tannin concentrations. In an earlier paper, Cooper and Owen-Smith (1985) suggested that condensed tannins had a deterrent effect if above a threshold concentration of 5% of dry mass based on data for the late wet season. No such threshold effect was evident in the seasonally more comprehensive data set.

There have been a number of recent behavioral studies on selection of food resources by primates in the field. These include studies on the feeding behavior of howler monkeys (*Alouatta palliata*) (Bilgener, 1988; Choo *et al.*, 1981; Glander, 1978, 1981, 1982; Milton, 1979, 1980; Milton *et al.*, 1980); black colobus monkeys (*Colobus satanus*) (McKey and Waterman, 1982; McKey *et al.*, 1981); black and white colobus (*Colobus quereza*) (Oates *et al.*, 1977); the leaf monkey, *Presbytis johnii* (Oates *et al.*, 1980); rhesus monkeys (Marks *et al.*, 1987); vervet monkeys (*Cercopithicus aethiops*) (Wrangham and Waterman, 1981); chimpanzees (*Pan troglodytes*) (Wrangham and Waterman, 1983); gorillas (*Gorilla gorilla*) (Calvert, 1985); and Malagasy primate lemurs (Ganzhorn, 1988).

Although it is hard to interpret field observations on feeding of whole plants, owing to differing environmental parameters and the presence of other dietary components, these types of studies are more ecologically realistic and some general trends do emerge. Many primates appear to select food plants with high levels of protein (Bilgener, 1988; Hladik, 1978; McKey *et al.*, 1981; Milton, 1979; Nagy and Milton, 1979; Oates *et al.*, 1980). However, when protein is not limiting, tannins may be an important factor in food selection. For example, in the howler monkeys, phenolics seem unimportant compared to proteins and fiber, and only become important in food choice when protein levels are high (Bilgener, 1988). Tannins appear to be an important factor in the food choice of leaf-eating monkeys with ruminant-like digestion (Marks *et al.*, 1987; McKey *et al.*, 1981; Oates *et al.*, 1977; Oates *et al.*, 1980), and selection appears to be for leaves low in fiber and tannins. Rhesus monkeys selected plants which were less astringent and contained lower levels of hydrolysable tannins (Marks *et al.*, 1987). Similarly, the concentration of condensed tannins may influence feeding by chimpanzees on fruits (Wrangham and Waterman, 1983) and may influence food partitioning among lemurs (Ganzhorn, 1988).

2. *Postingestional effects that are negative*

That many animals tend to reject very astringent food is not in doubt, but what they avoid by doing so is still something of an open question (Mole and Waterman, 1987). Because tannins precipitate proteins under certain conditions and reduce the activity of many enzymes, it has been assumed that tannins act by reducing the availability of proteins in the diet and the activity of digestive enzymes; that is, they are digestibility-reducing compounds.

Such processes have been widely considered to reduce the nutritional value of tannin-containing plant material to vertebrate herbivores. Whatever the mechanisms, measurable effects of dietary tannin have included reduction of weight gain in muscovy ducklings (*Cairana moschata*) (Martin-Tanguy *et al.*, 1977); diminished weight gains and poor feeding efficiency in birds (Blair and Mitaru, 1983); chicks and rats (Featherston and Rogler, 1975); hamsters (Mehansho *et al.*, 1985a,b); mice (Asquith *et al.*, 1985); livestock (Donnelly, 1983a,b); bullfinches (Wilson and Blunden, 1984; Wilson, 1983); reduced egg laying in hens (Davidson, 1972); and reduction of trypsin and amylase activities in rats (Griffiths and Moseley, 1980). Mole and Waterman (1987a) summarize the overall effects of tannins from 38 studies of vertebrates, representing work on reptiles, birds, and both marsupial and eutherian mammals. Many of these studies indicate low palatability, reduced consumption of high tannin diets, depressed growth rates and faecal nitrogen increasing with high levels of tannin in the diet. Increased faecal nitrogen is now well established as an effect of tannins in several very thorough studies. For example the tannin-rich diet of howler monkeys increases the faecal excretion of nitrogen (Milton *et al.*, 1980), which could be due to complexation of tannins and proteins (Bilgener, 1988). Overall evidence that tannins can produce deleterious effects on vertebrate herbivores is beyond doubt, but these effects are not necessarily symptomatic of digestibility-reducing agents, and the assumption that the sole or primary effects of dietary tannins are due to their inhibition of digestion is now being questioned (Butler *et al.*, 1986; Mole and Waterman, 1987b).

The effects of dietary tannins may be less in ruminants than in non-ruminants presumably because of the effectiveness of microbial fermentation in dealing with a wide range of dietary components (Sandanandan and Arora, 1979; Kumar and Singh, 1984). Cooper and Owen-Smith (1985) suggested that the main nutritional effect of condensed tannins was inhibition of microbial breakdown of plant cell walls. This means that carbohydrate availability is limited although it results in greater post-ruminal digestion and greater net absorption of nitrogen by the animals. Such an effect has been found for sheep (Barry and Manley, 1984, 1986). However Robbins *et al.* (1987b) suggest that browsing mammals such as deer may avoid this effect by deactivation of tannins (Section IV.A.3) and that browsing ruminants are less susceptible to effects of tannins than are grazers.

Because of the emphasis on reduction in digestibility it is useful to examine some of the studies in this area, although other luminal modes of action are possible and, as discussed below, systemic effects as a result of uptake of tannin breakdown products from the digestive tract can also occur (Butler *et al.*, 1986). *In vitro* studies have shown that digestive enzymes such as trypsin and amylase are inhibited by tannins (Griffith and Moseley, 1980) as is the zymogen, enterokinase (Oh and Hoff, 1986). Although tannins inhibit most enzymes *in vitro*, it is more difficult to demonstrate that this inhibition is significant *in vivo*.

Blytt *et al.* (1988) investigated the inhibition by condensed tannins from sorghum seeds and quebracho on two enzymes, alkaline phosphatase and 5′-nucleotide phosphodiesterase, solubilized from bovine intestinal mucosa, purified to homogeneity. In their purified soluble form, these enzymes are strongly inhibited by purified condensed tannins.

In freshly prepared homogenates of intestinal mucosas, both these enzymes occur predominantly in a particulate membrane-bound form. Blytt *et al.* (1988) found that these enzymes, in the washed particulate fraction, proved to be less susceptible than the soluble form to inhibition by tannins at comparable tannin concentrations. They suggest that in the particulate form other proteins may compete for binding the tannins. Also the presence of phospholipids in the membrane may diminish tannin inhibition. Thus, as some digestive enzymes occur in membrane-bound forms, these may be less susceptible to inhibition by tannins than the soluble forms purified and tested *in vitro*. The protection of digestive enzymes from tannins by membrane components may be quite common. Dietary tannins have opportunities to complex with a wide variety of dietary proteins and other proteins of the digestive tract that have a high affinity for protein (Mehansho *et al.*, 1983) before being exposed to the major digestive enzymes. Thus, in addition to diminished susceptibility to tannin of digestive enzymes in their predominantly *in vivo* form, under physiological conditions dietary tannin is probably not accessible to digestive enzymes. Dietary proteins (Butler *et al.*, 1984) or specialized tannin binding proteins of the saliva (Mehansho *et al.*, 1983) are available to form complexes with dietary tannin before it is exposed to most digestive enzymes.

Even if digestive enzymes are subject to inhibition by dietary tannins, the effect is likely to be prevented or reversed by surfactants (detergents) such as bile acids (Mole and Waterman, 1985) or by various tannin binding materials encountered in the digestive tracts (Oh and Hoff, 1986). Mole and Waterman (1985) demonstrated that the presence of glycocholic acid (a bile salt) in the tryptic hydrolysis of proteins prevented tannic acid complex formation with trypsin *in vitro*. It is probable that such bile components, produced for a variety of other purposes, serve an additional role in preventing tannin–protein complexation.

Blytt *et al.* (1988) conclude that the antinutritional effects of dietary tannins are probably not due to binding and inhibition of digestive enzymes by tannins, especially those that predominantly occur in a membrane-bound form. The relative insensitivity to tannin of membrane-bound enzymes also suggests that other membrane associated processes such as absorption may be little affected by dietary tannins. The anti-nutritional effects of dietary tannins that result in reduced protein digestibility may be due to the formation of less-digestible complexes of dietary proteins with tannins (Butler *et al.*, 1984). Other possible antinutritional effects are unrelated to digestibility (Rogler *et al.*, 1985, Mehansho *et al.*, 1985b).

Mole and Waterman (1985) stress the importance of tannin–protein interactions in solution. Although tannins do form insoluble precipitates with proteins, they found that tannins and proteins can be present in soluble systems at reactant concentrations that might be reasonably expected to occur in the digestive system. In these soluble systems they found substantial inhibition of the tryptic degradation of proteins in the presence of high levels of tannins. In contrast, increased levels of proteolysis were observed when low levels of tannin were incorporated in a system containing BSA as the substrate for trypsin; i.e. tannins can denature protein structure when they form complexes with them. The rate-enhanced tryptic proteolysis observed with BSA and relatively low levels of tannin may also be of *in vivo* importance. They conclude that any model for tannin–protein interactions must include more than pH, concentrations of the buffer, enzyme, substrate and tannin. It must also take into account the actual enzyme–substrate system and be nutritionally realistic, while the *in vivo* activity of gastrointestinal mucoproteins and bile surfactants should also be included in the equation.

Some studies on the toxicity of hydrolyzable tannins have shown ulceration of parts of the gut and severe effects on liver and kidneys probably due to the effects of absorption of phenolic products of tannic acid hydrolysis in the digestive tract (McLeod, 1974; Robbins *et al.*, 1987a; Singleton and Kratzer, 1969). The precise modes of action are not known although complexes of many generalized kinds could occur with different cell components. Preventing significant interaction of tannin and tannin products with the gut epithelium may be the most important requirement, since these materials will certainly act as irritants. Indeed, erosion of the intestinal mucosa could well be the main effect of tannin itself.

One of the main lines of evidence that digestion is diminished by dietary tannin comes from the observation that faecal nitrogen is increased as tannin levels increase (see above). The situation is, however, much more complex than was previously realized. Faecal nitrogen increase in rats and pigs was found to be associated with increases in proline (proline-rich proteins) rather than just poorly digested dietary protein (Eggum and Christensen, 1975; Mitaru *et al.*, 1984). Thus, while nitrogen retention is reduced, the reason must be related to endogenous nitrogen losses rather than reduced digestion of protein. The production of proline-rich proteins is probably a protective mechanism on the part of animals and is discussed in Section IVA3. Other studies indicate that high concentrations of nitrogen in the faeces are also due to a combination of increased mucus and intestinal cell debris caused by tannin erosion (Mitjavila *et al.*, 1977; Freeland *et al.*, 1985b). The studies indicate that high faecal nitrogen is not an indication of lowered digestion.

Mineral balance may be disturbed by ingestion of tannins, and in particular there may be sodium depletion, caused by the increased secretion of mucus and cell damage following ingestion of tannin (Freeland *et al.*, 1985)—a potentially very serious problem with many physiological repercussions. This problem may be particularly important in vertebrates, which often suffer from the inherent low sodium status of many plants even without the presence of tannin. Tannins can also chelate minerals such as iron and cause deficiency of them (Freeland *et al.*, 1985a,b).

3. How animals reduce the effects

Animals that regularly utilize tannin-containing plant materials for food may be expected to have developed defensive mechanisms against the potentially detrimental effects of dietary tannins. Tannins are no different from other classes of deterrent plant chemicals in that herbivores can adapt to their presence in the diet. As indicated in Section IVA2, tannins probably have minimal effects on digestive processes directly. Whether this is due to specific adaptations or to fortuitous effects of the various lumen constituents cannot be determined. It is clear that moderate levels of tannin do not interfere with digestion because of the combination of surfactants in the form of bile salts and the overall nature of the ionic milieu (Mole and Waterman, 1985) (Section IVA2). At very high levels of dietary tannin and low levels of protein, effects on digestion are probable but likely to be subsidiary to effects on the digestive tract itself.

A general increase in mucus production along the digestive tract is perhaps the simplest line of defense against the irritant properties of tannins (Freeland *et al.*, 1985b). Production of substances that preferentially bind with tannin is another way in which animals gain protection. Rats fed high levels of tannins in the diet have been shown to respond by producing salivary fluid containing proteins rich in proline (Freeland *et al.*, 1985b; Mehansho *et al.*, 1983) that bind readily to tannins. Salivary proline-rich proteins are induced in the parotid glands of mice (Mehansho *et al.*, 1985b) and a variety of other animals (Robbins *et al.*, 1987b) as a response to dietary tannins. Parotid salivary glands in ruminants (per unit body mass) are three times larger in browsers than in grazers, while they are intermediate in mixed feeders (Kay *et al.*, 1980). While large glands are induced by tannins in some species they are constitutive in deer. Proline-rich proteins are reported to comprise about 70% of the parotid gland secretions in humans (Bennick,

1982). The topic of salivary proline-rich proteins and their role has been reviewed by Mehansho *et al.* (1987).

The importance of the proline-rich proteins is becoming clear. These proteins preferentially bind tannins so that they are not free to act in any other more harmful capacity. Initially this was thought to be a means of preventing antidigestive effects since the efficiency of the binding is many times greater than with most other proteins so that net loss of protein would be reduced (Hagerman and Robbins, 1987). In fact the significance may relate rather to the protection of the gut mucosa (Freeland *et al.*, 1985a,b; Robbins *et al.*, 1987b), which would not be affected by the complex, and to perhaps prevent any other deleterious processes such as chelation of essential minerals. In addition, preventing breakdown of tannins to smaller phenolic products that can be absorbed into the blood may prevent the development of toxic effects in the liver and kidneys.

The formation of protein–tannin complexes within the animal may be modified by the presence of other chemicals in the diet (Freeland *et al.*, 1985a; Martin and Martin, 1984, 1985). Freeland *et al.* (1985a) demonstrated that simultaneous consumption of tannins and saponin by mice may result in interactions that reduce the detrimental effects of each. Clearly, behavioral mixing of diet components is an important mechanism for reducing the overall intake of tannin, and for obtaining a mix that promotes beneficial interactions.

4. Positive effects

There are few studies that show tannins having a positive effect on vertebrates. The presence of tannins may influence the effect of other secondary metabolites in the herbivores' diet. Goldstein and Spencer (1985) found that cyanogenesis in *Carica papaya* was inhibited by the presence of tannins. There is also some evidence that tannins may reduce the physiological effects of alkaloids by preventing the absorption of alkaloids into the bloodstream and alkaloids may be passed out in the faeces without any harmful effects to the animal's physiology (Freeland and Janzen, 1974).

Tannins may reduce the detrimental effects of saponins. This has been shown in mice (Freeland *et al.*, 1985a). In ruminants, tannins reduce the likelihood of bloat. This production of a stable foam caused by certain legumes is prevented by moderate quantities of dietary tannin (Jones and Lyttleton, 1971; Reid *et al.*, 1974).

Low levels of tannin may improve digestion by denaturing proteins (Mole and Waterman, 1985) while there are examples of increased protein retention in ruminants caused by some tannins in the diet (Driedger and Hatfield, 1972; Jones and Mangan, 1977; Mangan *et al.*, 1974). The possible reason is that some complexing of tannins and protein in the rumen could reduce deamination by microorganisms and thus loss of ammonia nitrogen. Dissociation of the complexes in the duodenum subsequently allows digestion and absorption to proceed. Thus high tannin levels that inhibit rumen carbohydrate digestion may favour amino acid retention by the animals (Barry and Manley, 1986), and the ingestion of tannins may have value under conditions of high protein requirement, rather than high energy requirement.

The overall conclusion must be that effects are very different under different circumstances of dose, overall diet, species of vertebrate, and whether the tannin is hydrolyzable or not.

B. Arthropods

1. Behavioral responses

There are few actual comparative observations of arthropod responses to a substrate with or without tannin. Negative impact over time without any behavioral measurement has tended to be associated with post-ingestional effects rather than feeding deterrence. In addition, compensatory feeding may occur if nutrient availability is reduced (Simpson and Abisgold, 1985) while food aversion learning is also possible if there are deleterious post-ingestional effects (Lee and Bernays, 1988). In spite of these complications, it is clear that tannins can be phagostimulants or deterrents for different insect species, or may be stimulating at low to moderate concentrations and either have no effect or be deterrent at high concentrations.

Some authors report reduced feeding on artificial diets when a tannin is incorporated. However, in these moist warm environments most tannins alter within hours and become darker-colored oxidation products variously bound to dietary components, and fresh diet is provided at most once a day. Since the quinones produced are themselves more reactive products and generally quite deterrent (Chapman, 1974), these assays unfortunately tell us little that is relevant to a natural situation. Table 1 summarizes work on the behavioural responses of insects to tannins, omitting those papers where there is likely to have been deterrence due to such changes.

Even from the limited information available, it appears that species that normally feed on plants that do not contain tannin are more likely to be deterred by tannin than those that feed habitually on tannin-rich plants (e.g. Bernays and Chamberlain, 1982). In the latter case, phagostimulation is also common. Behaviorally, tannins influence herbivorous insects in ways that parallel other secondary compounds in plants. Low molecular weight tannins appear to be more deterrent than high molecular weight polymers (Marini Bettolo *et al.*, 1986).

Table 1
Effects of tannins on feeding behavior of insects

Insect species	Tannin	Concentration (approx %)	Effect	Reference
A. Species normally feeding on tannin-containing plants				
Euproctis chrysorrhoea	tannic acid	?1	++	1, 2
Lymantria dispar	tannic acid	?1	++	2
	tannic acid	0·2	+	3
Orgyia antiqua	tannic acid	?1	++	2
Operophtera brumata	tannic acid	?1	++	2
Acronycta aceras	tannic acid	?1	++	2
Phalera bucephala	tannic acid	?1	++	2
Anacridium melanorhodon	tannic acid	10	+	4
Schistocerca gregaria	tannic acid	5	+	5, 15
	tannic acid	10–40	0	6, 15
	quebracho	5–10	0	6, 15
	sorghum tannin	5	0	4
	chestnut tannin	5	0	4
Heliothis zea	cotton cond. tannin	5	0	11
Costelytra zealandica	*Lotus* cond. tannin	1	0	12
Atta cephalotes	hydrolyzable tannin	?	+	13
	condensed tannins	?	0	13
B. Species normally feeding on plants without tannins or with low tannin concentrations				
Pieris brassicae	tannic acid	0·5	−	7
Hypera postica	tannic acid	?1	−	8
Dysdercus koenigii	tannic acid	0·01M	−	9
Myzus persicae	tannic acid	0·01M	−	10
Locusta migratoria	tannic acid	10	0	4, 15
Chortoicetes terminifera	tannic acid	10	−	4, 6
Acrida conica	tannic acid	10	−	4, 6
Locustana pardalina	tannic acid	10	−	4, 6
Gastrimargus africanus	tannic acid	10	−	4, 6
Locusta migratoria	quebracho	10	−	15
Chortoicetes terminifera	quebracho	10	−	4
Acrida conica	quebracho	10	−	4
Locustana pardalina	quebracho	10	−	4
Gastrimargus africanus	quebracho	10	−	4
Hemileuca oliviae	cond. tannin	1	−	14

1. Grevillius, 1905; 2. Gornitz, 1954; 3. Meisner and Skatulla, 1975; 4. Bernays unpub. 5. Bernays and Chamberlain, 1980; 6. Bernays *et al.*, 1980, 1981; 7. Jones and Firn, 1979; 8. Bennett, 1965; 9. Schoonhoven and Derksen-Koppers, 1973; 10. Schoonhoven and Derksen-Koppers, 1976; 11. Klocke and Chan, 1982; 12. Sutherland *et al.*, 1982; 13. Howard, 1987; 14. Roehrig and Capinera, 1983; 15. Bernays and Chamberlain, 1982.
+ = stimulates feeding, 0 = no effect, − = inhibits feeding

2. Negative postingestional effects

The affinity of tannins for proteins has been highlighted in the literature, and it seems likely that this property is at least partly responsible for some of the detrimental effects seen. The major effect cited in the literature and found in current textbooks concerns the putative "antidigestive effects" of tannins, predicted because of the likelihood that dietary protein and digestive enzymes will be bound by tannins and thereby lose their value (and discussed further in Section IVA). We wish to emphasize however, that there is no proven case of this occurring in insect herbivores *in vivo*, and that this particular possibility is probably negated in a variety of different ways (see IVA3). This is not to say that there are no negative effects—but they do not concern direct effects on digestion.

Perhaps the first clue to alternative mechanisms was in the work of Bernays on graminivorous grasshoppers (Bernays *et al.*, 1980), where midgut lesions were noted in tannin-fed insects. Histology of species differentially affected by tannic acid showed that phenolics had accumulated in the midgut cells of grassfeeders, but not in polyphagous species that normally ingest tannin in their diets. Post-mortems on insects that had died in the tannin treatments showed that invariably there were lesions severe enough to allow leakage of gut contents into the hemolymph. Similar findings were reported by Berenbaum (1984) and Steinly and Berenbaum (1985) in papilionid caterpillars that normally do not feed on tannin-containing plants. It seems that insects habitually feeding on tree foliage and other tannin-rich plants produce a relatively thick protective chitin/protein peritrophic membrane from the midgut epithelium (Adang and Spense, 1982) and among grasshopper species there is a positive correlation between relative mass of peritrophic membrane and quantities of tannins in the diet (Bernays and Simpson, 1989). This would agree with the finding that these membranes appear to selectively adsorb tannins from the gut lumen as well as providing a barrier that protects the midgut epithelium from direct effects of tannins (Bernays, 1978; Bernays and Chamberlain, 1980; Bernays *et al.*, 1981).

Midgut lesions produced by tannin ingestion in sensitive insects are preceded by sloughing of midgut epidermis, while in less sensitive species there may additionally be formation of thicker than usual peritrophic membranes. These two factors are not without cost, and will tend to cause loss of protein nitrogen in the faeces. The result of this is that in gravimetric studies of nutritional indices, such insects could appear to digest less of the dietary protein. It is possibly partly due to this that measures of digestion appear low, and the perception of "antidigestive effects" is enhanced. In the most extreme case so far examined, the grasshopper *Bootettix argentatis*, which feeds on the very highly resinous plant *Larrea tridentata*, was estimated to use approximately 10% of the dietary protein in the production of an exceptionally stout peritrophic membrane (Bernays, unpublished). It must be noted, however, that this protein has been through the metabolic pool, and is not lost due to poor digestive processes. In terms of nutritional indices, a low digestibility may be measured when in fact it is a low efficiency of use of digested protein for growth.

Tannin effects on the midgut epithelium are not unprecedented and occur also in vertebrates (IVA2). They probably represent the well-known ability of tannins to combine with proteins and other macromolecules. In the case of hydrolyzable tannins, hydrolysis does occur in the gut of insects. Thus in the grasshoppers *Anacridium melanorhodon* and *Schistocerca gregaria*, which both feed on trees and shrubs, hydrolysis of tannic acid was found to be over 50% (Bernays *et al.*, 1980). In the former, the released gallic acid was found to be utilizable by the insect (Bernays and Woodhead, 1982a,b) (IVB4), but it may be that in other cases the hydrolysis products are harmful, because they pass through the peritrophic membrane, are absorbed, and would have to be detoxified and excreted. Nothing is known of the potential or actual detrimental effects of tannin hydrolysis products on insects, but this is worthy of further study, particularly since hydrolysis products that may be absorbed have been shown to be more important than tannins themselves in some mammalian studies (Freeland *et al.*, 1985b; Robbins *et al.*, 1987a,b).

A potential effect of tannins on insects is via negative effects on symbiotic microorganisms. Little is known of this, however, and in most insects feeding on green plants, gut microorganisms are not thought to be particularly important. There are possibilities in the case of wood feeders such as termites but these have not been explored. However, some work has been done on leaf cutting ants, which have fungus gardens and must select leaves suitable for fungal growth. Seaman (1984) found that low concentrations of tannic acid and other phenolics stimulated growth of the fungus, while high levels were inhibitory.

3. Reducing potential negative effects

Since the most publicised effects of tannins are expected to be antidigestive, and realistic work with insects shows little or no effect on digestion, it is reasonable to ask why not. In the case of grasshoppers, it was proposed that the peritrophic membrane preferentially adsorbed tannins, thereby making them ineffective for binding with important dietary or digestive enzyme proteins (Bernays *et al.*, 1980). It was found that certain species, at least, had special regions of the gastric caeca that appeared to sequester noxious macromolecules including tannins (Bernays, 1981b). Preferential binding to proline-rich proteins is well established (Hagerman and Butler, 1981) and it is possible that peritrophic membrane proteins or glycoproteins are proline-rich.

In caterpillars, Feeny (1970) and Berenbaum (1980) noted that tree feeders, which may be assumed to be ingesting higher quantities of tannin, had, in general, much higher gut pH. Since high pH conditions reduce protein-tannin binding, it was assumed that the high pH was an adaptive phenomenon, reducing the likelihood of antidigestive effects of tannins in caterpillars with high tannin diets. Whether it is specifically adaptive is not known, but it is probable that the high pH does contribute to the ineffectiveness of tannins as protein precipitants in the gut lumen. Recently, Schultz and Lechowicz (1986) further showed that luminal pH first decreases and then increases after feeding. This is partly due to the fact that the pH of freshly chewed leaf tissue from hardwood trees is initially low and ranges from about 4 to 6. Fourth instar larvae of *Lymantria dispar* fed on *Quercus robus* show an increase in gut pH from 8·5 to 11 several hours after feeding. These authors suggest that the question may be one of protein-tannin associations first forming and then dissociating. Such a process may even stimulate protein digestibility (Mole and Waterman, 1985). More or less neutral gut pH is characteristic of other plant-feeding insect orders, whether or not tannins occur in the diet.

Other factors are emerging. Martin and Martin have shown that gut lumenal surfactants have a critical role to play in preventing formation of tannin–protein complexes. Lysolecithin and linoleoylglycine were shown to be important surfactants in *Manduca sexta* (Martin and Martin, 1984), while they also review the limited literature in the study of insect gut detergents and find that such materials are of widespread occurrence at levels that would protect them from antidigestive effects of tannin, even in species that would never ingest such compounds. Detergents are of particular value in insects with near neutral gut pH, and further work (Martin *et al.*, 1985, 1987) has demonstrated that they are of varied chemistry, are influenced by the diet and gut ionic strength, and are very effective in preventing any detrimental effects of tannin in the lumen. The locust *Schistocerca gregaria* had a greater variety of different surfactant materials than caterpillars.

4. Positive effects of tannins

A positive effect of tannic acid was demonstrated with the tree locust *Anacridium melanorhodon*. Addition of tannic acid to low phenol foliage resulted in increased efficiency of food utilization and a resultant 15% increase in growth rate (Bernays *et al.*, 1980). The tannin is hydrolyzed and the resulting gallic acid is absorbed and deposited in the cuticle where it is believed to be utilized in cuticular sclerotization in lieu of the usual DOPA produced from aromatic amino acids (Bernays and Woodhead, 1982a,b). This process conserves protein by the amount that would be needed to extract the appropriate quantities of phenylalanine and/or tyrosine. Schopf *et al.* (1982) report growth benefits relating to the presence of phenolics including tannins, and perhaps it is to be expected that insects feeding on woody plants low in protein and rich in phenolics, including tannins, might commonly utilize the phenolics. There is even evidence that certain insects may be able to make phenylalanine and tyrosine from the phenolic ring (Sin *et al.*, 1984). The sclerotized exoskeletons of insects may make up a very high proportion of the body weight—40% in grasshoppers (Bernays and Woodhead, 1984)—and the special nutritional needs of insects for this material require further study.

There is a possibility that tannins benefit herbivores in indirect ways. Faeth and Bultman (1986) painted tannins on leaves inhabited by leaf miners and found reduced levels of mortality from diseases. Taper *et al.* (1986) and Taper and Case (1987) similarly report benefits to cynipid gall wasps from tannins. High tannin levels of oaks are correlated with species richness and reduced mortality from fungal attack, while similar reduction in mortality can be shown by artificial application of tannic acid to leaves with cynipid galls on them.

Protection from disease has been little explored in externally feeding insect herbivores, although phenolics of various kinds have potential value (Koike *et al.*, 1979). Different host plants of the gypsy moth (*Lymantria dispar*) have differential effects on mortality due to nuclear polyhedrosis virus (Keating and Yendol, 1987) and it is possible that tannins in the leaves of trees may be important. Antiviral effects of tannins are known and could occur in the guts of insects feeding on tannin-rich plants, thereby influencing infectivity of pathogens (Keating *et al.*, 1988). However, although such possibilities exist, further work is required to demonstrate their occurrence.

V. OVERVIEW AND SIGNIFICANCE

A. Comparisons Among Animals

From the point of view of evolutionary ecology there is considerable interest in whether or not animals can have provided a selection pressure for tannin production in plants, and by examination of the effects of tannins on herbivores there is the hope of obtaining insights on this theme. We should at least discover the degree of vulnerability of them, and the costs to them of tannin ingestion. In addition, differences among animal groups with respect to adverse effects of tannins on them may suggest possible differences in their selection pressure on plants. Alternatively, differences may simply reflect taxonomic differences in relation to the potential target systems.

Deterrence is widespread but not universal. Even high levels of tannins may have no measurable effects on food acceptability to certain vertebrates and arthropods that habitually feed on tannin-containing foods.

Presumably, such animals are not severely affected by them if they are ingested. Behavioral sensitivity to tannins may or may not be a measure of their likely detrimental effects following ingestion, though it is clear that extreme deterrence in both arthropods and vertebrates is exhibited when serious negative effects occur. Low to moderate concentrations of tannins may have stimulatory effects in species of animals from any taxon, humans and insects included.

In dealing with tannins after ingestion there appear to be some common themes but also possible important differences among animal taxa. First of all it now seems that although tannin–protein associations are important, the effects are not directly antidigestive in the sense of preventing digestion by complex formation with dietary proteins or digestive enzymes. This is due to various aspects of the gut milieu. The major serious negative effect is on digestive epithelia. Adaptations include thicker peritrophic membranes in the case of insects and increased mucus production in vertebrates. In addition, in some mammals, hypertrophy of salivary glands occurs as these secrete proline-rich proteins preferentially binding the dietary tannins. If these adaptations are overwhelmed by increases in dietary tannins, irritation of gut epithelial cells occurs, and eventually cells are sloughed off into the lumen. This can result in protein loss in the faeces. The mucus and cellular debris can thus result in falsely low estimates of protein digestibility. Yet much of the faecal N has been through the metabolic pool and is not therefore undigested protein. It is likely that all cases of "reduced digestibility" are due to this phenomenon and unfortunately it is very much more difficult to determine the origin of the faecal nitrogen than to simply measure N-levels. Important progress would ensue if we could determine the origin of faecal N.

Secondary effects of damage to gut epithelium, increased mucus production and increased salivary secretion can include changes in ion balance. In mammals sodium depletion (and marked salt appetite) occurs when dietary tannins are very high. This, in turn, may cause a suite of tertiary problems, including hypertrophy of the adrenal glands (Freeland *et al.*, 1985b). No studies have been undertaken on mineral depletion in arthropods, and the situation may be somewhat different since sodium is replaced by potassium to a large extent as the major cation in body fluid of phytophagous insects.

In vertebrates and arthropods having important symbiotes—gut flora and fauna and fungus gardens—there are conflicting results but in general the symbiotes seem well adapted to tannins at normal or moderate levels, though there is some controversy as to the effects of tannins on the microbial flora of monogastric mammals.

Finally, absorption of the phenolic products of digested hydrolysable tannins can have detrimental effects that are as yet not clearly defined but indicate overloading of and damage to the liver and kidneys in vertebrates when detoxification and excretory processes do not keep up. Seed feeding birds may be more sensitive than mammals, perhaps in relation to the drier nature of their diets. Analogous problems may occur in insects but have not yet been studied.

B. Other Potential Roles

1. Internal plant roles

By and large these are unexplored and generally not considered likely. Bu'lock (1980) suggests that among microorganisms many secondary compounds may have roles in maintaining basic metabolism when circumstances are not propitious for growth, such as in cases of extreme nutrient imbalance, and possible physiological roles of secondary compounds in green plants may be worth further investigation. High levels of phenolics including tannins are associated with nutrient poor soils (Bryant *et al.*, 1983; Coley *et al.*, 1985) and slow growth rates when plants are grown in poor soil have now been suggested as a *cause* of high levels of tannin (Bryant *et al.*, 1985). Also, nitrogen fertilization of particular plant species has been shown to cause a decrease in their production of tannins (Bryant *et al.*, 1987). Any tannin-producing species has higher levels of all phenolics in the sun relative to shade (Waterman and Mole, 1989). Such environmentally mediated differences in resource allocation have been largely discussed in terms of defense (Coley *et al.*, 1985) without serious consideration of plant physiological needs and constraints that remain possible but unexplored. Coley (1988) has shown very clearly with 41 neotropical tree species that tannin concentration is significantly correlated with growth rate, but what is cause and what is effect remains confused. Correlations found between concentrations of tannins, total phenolics and lignin are well known (Barry and Manley, 1986) and may not be without physiological significance. Recently Haslam (1988) has indicated that in tannin-producing species, tannin involvement in lignification could be very important. This would be consistent with the fact that woody plants tend to be those containing tannin (see Section IIB).

2. Defense against fungi and microorganisms

Tannins, along with a wide variety of other phenolics, will certainly influence growth of fungi and microorganisms on or in the plant. This has been discussed by various authors (e.g. Swain, 1979; Zucker, 1983). As with insects there is correlative evidence concerning tannin levels and relative vulnerability to certain pathogens. The histological distribution of tannins indicates that epidermis is the most common site of intense tannin deposition in leaves and especially the outer epidermal cell wall (Bate Smith, 1957; McKey, 1979). This may be circumstantial evidence of their importance in preventing penetration of disease forming or saprophytic organisms including fungi. Protective roles of tannins against disease organisms may be difficult to separate from protective roles against insects, which provide the entry points for pathogens in many cases. It is possible that the risk of diseases is much more important than minimal levels of herbivory; yet if they go together, they must also be considered in concert. As a working hypothesis, we suggest that tannins are more important as defenses against pathogens than against herbivores, and that other secondary compounds may play the more important roles in herbivory.

3. *Effects on soil*

Effects on soil and soil production have been examined to some extent. For example, tannins in leaf litter decrease rates of decomposition of leaf litter by soil arthropods, fungi and microorganisms (Anderson, 1978; Cameron and LaPoint, 1978; Harrison, 1971), so influencing humus levels and nutrient availability. In addition tannins leached from fallen leaves can directly affect soil pH and cation exchange properties (Davies, 1971) and there is the possibility of allelopathic effects. In addition, tannins appear to reduce mineralization processes generally (Baldwin *et al.*, 1983). These processes will impact many organisms, but whether or not they are adaptive for the producers is generally unknown. In some cases alteration of soil properties by tannin has been shown to favor the producer by increasing nutrient acquisition (R. Northup, pers. com.). This area requires further work.

C. Tannins as defenses

Almost all of the biological literature refers to tannins as defensive compounds. Insofar as the presence of tannins does protect the plants possessing them from damage by certain organisms, this can be accepted. However, it is important to recognize that the label "defensive" carries with it the idea of production and maintainance under selective pressure by appropriate herbivores and pathogens. Yet the evidence for this being so is hard to come by, and there is a real possibility that tannins have a wide variety of roles, with effects on herbivores sometimes being quite incidental. The poor correlations between astringency (protein precipitation potential) and probable production cost of various tannins (Beart *et al.*, 1985) is not necessarily supportive of antiherbivore defense arguments, while the large amount of carbon tied up in tannins may be no loss, if other elements are limiting (Fox, 1981). A recent reappraisal by Haslam (1988) once again suggests that basic plant physiological requirements are dominant factors.

To obtain a more comprehensive understanding of the evolutionary ecology of tannins a variety of different approaches is essential. It is important to study a few plant species in depth. What are the various potential roles of tannin for a particular plant? How many of these appear to be realized? Can the importance of different roles be weighted in respect of protection from damage by abiotic and various biotic factors? Are roles involving soil processes, competition and allelopathy significant relative to protective values?

Another important way forward will be the study of fitness parameters of individuals with genetically different potential for tannin production. In what way are these genotypes different in dealing with the array of specific requirements for survival and maximal reproductive output? Only the work of Coley (1986) on intraspecific differences in tannin production is suggestive so far. A good negative correlation between tannin level and herbivory was demonstrated, though the possibility of other factors being causative was not ruled out. Even careful, in-depth studies will require supplementary experimental techniques using pure tannins, since naturally occurring variation in tannins will be accompanied by associated variation in simple phenolics.

Clues may reside in a multitude of life history details. Where tannins are elaborated immediately prior to leaf fall (Feeny, 1970), perhaps soil effects are most important. Where tannins are at their highest in young lush leaves (Coley, 1983) perhaps protection from herbivory may be indicated. If slow-growing species have higher levels of tannins than fast-growing species, are physiological factors significant? If differences in tannin levels are engendered among species or in a single genotype by soil and climatic differences, do they have particular adaptive significance for nutrient acquisition?

We suggest that research will be most revealing in this difficult subject if interdisciplinary teams attack the outstanding problems together, and that studies on multiple effects should be given priority over studies using single species interactions, which have predominated so far.

ACKNOWLEDGEMENTS

We are grateful to Ann Hagerman for very helpful comments on an earlier draft. Part of the funding for this work was provided by NSF Grant BSR-8705014 to EAB.

REFERENCES

Adang, M. J. and Spence, K. D. (1982). Biochemical comparisons of the peritrophic membrane of the lepidopterans *Orgyia pseudotsugata* and *Manduca sexta*. *Comp. Biochem. Physiol.* **73B**, 645–649.

Anderson, J. M. (1978). Inter- and intra-habitat relationships between woodland *Cryptostigmata* species diversity and the diversity of soil and litter microhabitats. *J. Anim. Ecol.* **44**, 475–596.

Asquith, T. N. and Butler, L. G. (1985). Use of dye-labeled protein as spectrophotometric assay for protein precipitants such as tannin. *J. Chem. Ecol.* **11**, 1533–1544.

Asquith, T., Mehansho, H., Rogler, J., Butler, L. G. and Carlson, D. M. (1985). Induction of proline-rich protein biosynthesis in salivary glands by tannins. *Fed. Proc.* **44**, 1097.

Asquith, T. N., Uhlig, J., Mehansho, H., Putnam, L., Carlson, D. M. and Butler, L. (1987). Binding of condensed tannins to salivary proline-rich glycoproteins: The role of carbohydrate. *J. Agric. Food Chem.* **35**, 331–334.

Baldwin, I. T., Olson, R. K. and Reiners, W. A. (1983). Protein binding phenolics and the inhibition of nitrification in subalpine balsam fir soils. *Soil Biol. Biochem.* **15**, 419–423.

Baldwin, I. T., Schultz, J. and Ward, D. (1987). Patterns and sources of leaf tannin variation in yellow birch (*Betula allegheniensis*) and sugar maple (*Acer saccharum*). *J. Chem. Ecol.* **13**, 1069–1078.

Barry, T. N. and Manley, T. R. (1984). The role of condensed tannins in the nutritional value of *Lotus pedunculatus* for sheep. 2. Quantitative digestion of carbohydrates and proteins. *Brit. J. Nutr.* **51**, 493–504.

Barry, T. N. and Manley, T. R. (1986). Interrelationships between the concentrations of total condensed tannin, free condensed tannin and lignin in *Lotus* sp. and their possible consequences in ruminant nutrition. *J. Sci. Food Agric.* **37**, 248–254.

Bate-Smith, E. A. (1957). Leucoanthocyanins 3. The nature and systematic distribution of tannins in dicotyledonous plants. *J. Linn. Soc. (Bot.)* **55**, 669–705.

Bate-Smith, E. C. (1962). The phenolic constituents of plants and their taxonomic significance: I Dicotyledons. *J. Linn. Soc. (Bot.)* **58**, 95–173.

Bate-Smith, E. C. (1973). Haemanalysis of tannins: the concept of relative astringency. *Phytochemistry* **12**, 907–912.

Bate-Smith, E. C. (1974). The systematic distribution of ellagitannins in relation to the phylogeny and classification of the Angiosperms. In: Chemistry in Botanical Classification (Ed. G. Bendz and J. Santesson), pp. 93–102. Nobel Foundation; Stockholm.

Bate-Smith, E. C. (1975). Phytochemistry of proanthocyanidins. *Phytochemistry* **14**, 1107–1113.

Bate-Smith, E. C. (1977). Astringent tannins of *Acer* species. *Phytochemistry* **16**, 1421–1426.

Beart, J. E., Lilley, T. H. and Haslam, E. (1985a). Plant polyphenols—secondary metabolism and chemical defense: some observations. *Phytochemistry* **24**, 33–38.

Beart, J. E., Lilley, T. H. and Haslam, E. (1985b). Polyphenol interactions. Part II. Covalent binding of procyanidins to proteins during acid-catalyzed decomposition: observations on some polymeric proanthocyanidins. *J. Chem. Soc. Perkin Trans.* **11**, 1439–1443.

Becker, P. and Martin, J. S. (1982). Protein-binding capacity of tannins in *Shorea* (Dipterocarpaceae) seedling leaves. *J. Chem. Ecol.* **8**, 1353–1367.

Bennett, S. E. (1965). Tannic acid as a repellent and toxicant to alfalfa weevil larvae. *J. Econ. Ent.* **58**, 372.

Bennick, A. (1982). Salivary proline-rich proteins. *Mol. Cell Biochem.* **45**, 83–99.

Berenbaum, M. (1980). Adaptive significance of midgut pH in larval Lepidoptera. *Amer. Nat.* **115**, 138–146.

Berenbaum, M. (1984). Effects of tannins on growth and digestion in two species of papilionids. *Ent. Exp. Appl.* **34**, 245–250.

Bernays, E. A. (1978). Tannins: an alternative viewpoint. *Ent. Exp. Appl.* **24**, 244–253.

Bernays, E. A. (1981a). Plant tannins and insect herbivores: an appraisal. *Ecol. Ent.* **6**, 353–360.

Bernays, E. A. (1981b). A specialized region of the gastric caeca in the locust, *Schistocerca gregaria. Physiol. Ent.* **6**, 1–6.

Bernays, E. A. and Chamberlain, D. (1980). A study of tolerance of ingested tannin in *Schistocerca gregaria. J. Insect Physiol.* **26**, 415–420.

Bernays, E. A. and Chamberlain, D. (1982). The significance of dietary tannin for locusts and grasshoppers. *J. Nat. Hist.* **16**, 261–266.

Bernays, E. A., Chamberlain, D. and McCarthy, P. (1980). The differential effects of ingested tannic acid on different species of Acridoidea. *Ent. Exp. Appl.* **28**, 158–166.

Bernays, E. A., Chamberlain, D. and Leather, E. M. (1981). Tolerance of acridids to ingested condensed tannin. *J. Chem. Ecol.* **7**, 247–256.

Bernays, E. A. and Simpson, S. J. (1989). Feeding and nutrition. In: *Biology of Grasshoppers*, Eds R. F. Chapman and A. Joern, Wiley: Chichester. (in press).

Bernays, E. A. and Woodhead, S. (1982a). Plant phenols utilized as nutrients by a phytophagous insect. *Science* **216**, 201–203.

Bernays, E. A. and Woodhead, S. (1982b). Incorporation of dietary phenols into the cuticle in the tree locust *Anacridium melanorhodon. J. Insect Physiol.* **28**, 601–606.

Bernays, E. A. and Woodhead, S. (1984). The need for high levels of phenylalanine in the diet of *Schistocerca gregaria* nymphs. *J. Insect Physiol.* **30**, 489–493.

Bilgener, M. (1988). Chemical components of howler monkeys (*Alouatta palliata*) food choice and kinetics of tannin binding with natural polymers. PhD Dissertation, Boston University.

Blair, R. and Mitaru, B. N. (1983). New information on the role of tannins in the utilization of feedstuffs by growing birds. Proc. Florida Nutrit. Conf. Univ. Florida, Gainsville. pp. 139–149.

Blytt, H. J., Guscar, T. K. and Butler, L. G. (1988). Antinutritional effects and ecological significance of dietary condensed tannins may not be due to binding and inhibiting digestive enzymes. *J. Chem. Ecol.* **14**, 1455–1465.

Booth, A. N. and Bell, T. A. (1968). Physiological effects of *Sericea* tannin in rats. *Proc. Soc. Exp. Biol. Med.* **128**, 800–803.

Bryant, J. P., Chapin, F. S. and Klein, D. R. (1983). Carbon/nutrient balance of boreal plants in relation to vertebrate herbivory. *Oikos* **40**, 357–368.

Bryant, J. P., Chapin, F. S., Reichardt, P. and Clausen, T. (1985). Adaptations to resource availability as a determinant of chemical defense strategies in woody plants. In: *Chemically Mediated Interactions between Plants and other Organisms* (Ed. G. Cooper Driver, T. Swain, and E. E. Conn), Plenum: New York.

Bryant, J. P., Clausen, T. P., Reichardt, P. B., McCarthky, M. C. and Werner, R. A. (1987). Effect of nitrogen fertilization upon the secondary chemistry and nutritional value of quaking aspen (*Populus tremuloides* michx.) leaves for the large aspen tortrix (*Choristoneura conflictana* (Walker)). *Oecologia* **73**, 513–517.

Burns, R. E. (1971). Method for estimation of tannin in grain *Sorghum. Agron. J.* **63**, 511–512.

Butler, L. G., Price, M. C. and Brotherton, J. E. (1982). Vanillin assay for proanthocyanidin (condensed tannin): modification of the solvent for estimation of the degree of polymerization. *J. Agric. Food Chem.* **30**, 1087–1089.

Butler, L. G., Reidl, D. J., Lebryk, D. G. and Blytt, H. J. (1984). Interaction of proteins with sorghum tannin: mechanism, specificity and significance. *J. Am. Oil Chem. Soc.* **61**, 916–920.

Butler, L. G., Rogler, C., Mehansho, H. and Carlson, D. M. (1986). Dietary effects of tannins. In: *Plant Flavonoids in Biology and Medicine: Biochemical, Pharmacological and Structure-Activity Relationships*. Alan R. Liss: New York. pp. 141–156.

Bu'lock, J. D. (1980). Mycotoxins In: *The Biosynthesis of Mycotoxins*, Ed by P. S. Steyn, pp. 1–16, Academic Press: London.

Calvert, J. J. (1985). Food selection by western gorillas in relation to food chemistry. *Oecologia* **65**, 236–246.

Cameron, G. N. and LaPoint, T. W. (1978). Effects of tannins on the decomposition of Chinese tallow leaves by terrestrial and aquatic invertebrates. *Oecologia* **32**, 349–366.

Chang, S. I. and Fuller, H. L. (1964). Effect of tannin content of grain sorghums on their feeding value for growing chicks. *Poultry Sci.* **43**, 30–36.

Chapman, R. F. (1974). The chemical inhibition of feeding by phytophagous insects; a review. *Bull. Ent. Res.* **64**, 339–363.

Choo, G. M., Waterman, P. G., McKey, D. B. and Gartlan, J. S. (1981). A simple enzyme assay for dry matter digestibility and its value in studying food selection by generalist herbivores. *Oecologia* **49**, 170–178.

Coley, P. D. (1983). Herbivory and defensive characteristic of tree species in a lowland tropical forest. *Ecological Monographs* **53**, 209–229.

Coley, P. D. (1986). Costs and benefits of defense by tannins in a neotropical tree. *Oecologia* **74**, 531–536.

Coley, P. D. (1988). Effects of plant growth rate and leaf lifetime on the amount and type of anti-herbivore defense. *Oecologia* **74**, 531–536.

Coley, P. D., Bryant, J. P. and Chapin, S. (1985). Resource availability and plant antiherbivore defense. *Science* **230**, 895–899.

Connor, J. K., Hurwood, I. S., Barton, H. W. and Fuelling, D. E. (1969). Some nutritional aspects of feeding sorghum grain of high tannin content to growing chickens. *Austr. J. Exp. Agric. Anim. Husb.* **9**, 497–501.

Cooper, S. M. and Owen-Smith, N. (1985). Condensed tannins deter feeding by browsing ungulates in a South African savanna. *Oecologia* **67**, 142–146.

Cooper, S. M., Owen-Smith, N. and Bryant, J. P. (1988). Foliage acceptibility to browsing ruminants in relation to seasonal changes in the leaf chemistry of woody plants in a South African savanna. *Oecologia* **75**, 336–342.

Cooper-Driver, G., Finch, S., Swain, T. and Bernays, E. A. (1977). Seasonal variation in secondary plant compounds in relation to the palatability of *Pteridium aquilinum*. *Biochem. Syst. Ecol.* **5**, 177–183.

Czochanska, Z., Foo, L. Y., Newman, R. H., Porter, L. J. and Thomas, W. A. (1979). Direct proof of a homogeneous polyflavan-3-ol structure for polymeric proanthocyanidins. *J.C.S. Chem. Commun.* 375–377.

Czochanska, Z., Foo, L. Y., Newman, R. H. and Porter, L. J. (1983). Polymeric proanthocyanidins: stereochemistry, structural units of molecular weight. *J.Chem.Soc. (Perkin)*., 2278–2286.

Davidson, J. (1972). The nutritive value of field beans (*Vicia faba*) for laying hens. *Proc. Nutr. Soc.* **31**, 50.

Davies, R. I. (1971). Relation of polyphenols to decomposition of organic matter and to pedogenetic processes. *Soil Sci.* **111**, 80–85.

Dollahite, J. W., Housholder, G. T. and Camp, B. J. (1963). Calcium hydroxide, a possible antidote for shin oak (*Quercus havardi*) poisoning in cattle. *Southwestern Vet.* **16**, 115–117.

Donnelly, E. D. (1983a). Breeding low-tannin sericea. 1. Selecting for resistance to *Rhizoctonia* species. *Crop Sci.* **23**, 14–16.

Donnelly, E. D. (1983b). Breeding low-tannin sericea. 11. Effects of *Rhizoctonia* foliar blight on vigor and seed yields, *Crop Sci.* **23**, 17–19.

Donnelly, E. D. and Anthony, W. B. (1969). Relationship of tannin, dry matter digestibility, and crude protein in *Sericea lespedeza. Crop Sci.* **9**, 361–362.

Drieger, A. and Hatfield, E. E. (1972). Influence of tannins on the nutritive value of soybean meal for ruminants. *J. Anim. Sci.* **34**, 464–468.

Drieger, A. and Hatfield, E. E. and Garrigus, U. S. (1969). Effect of tannic acid treated soybean meal on growth and nitrogen balance. *J. Anim. Sci.* **29**, 156–157.

Eggum, B. E. and Christensen, K. D. (1975). Influence of tannin on protein utilization in feedstuffs with special reference to barley. In: *Breeding for seed protein improvement using nuclear techniques*, IAEA-PL-570/10 Vienna, pp. 135–143.

Faeth, S. H. (1985). Quantitative defense theory and patterns of feeding by oak insects. *Oecologia* **68**, 34–40.

Faeth, S. S. H. and Bultman, T. L. (1986). Interacting effects of increased tannin levels on leaf-mining insects. *Ent. Exp. Appl.* **40**, 297–302.

Featherston, W. R. and Rogler, J. C. (1975). Influence of tannins on the utilization of sorghum grain in rats and chicks. *Nutr. Rep. Intl.* **11**, 491–497.

Feeny, P. (1968). Effect of oak leaf tannins on larval growth of the winter moth *Operophtera brumata. J. Insect Physiol.* **14**, 805–817.

Feeny, P. (1969). Inhibitory effect of oak leaf tannins on the hydrolysis of proteins by trypsin. *Phytochemistry* **8**, 2119–2126.

Feeny, P. (1970). Seasonal changes in oak leaf tannins and nutrients as a cause of spring feeding by winter moth caterpillars. *Ecology* **51**, 565–581.

Feeny, P. (1975). Biochemical coevolution between plants and their insect herbivores. In: *Coevolution of Plants and Animals*, Eds L. E. Gilbert and P. H. Raven, Texas University Press, pp. 3–19.

Feeny, P. (1976). Plant apparancy and chemical defence. In: *Biochemical Interactions between Plants and Insects*. Eds J. W. Wallace and R. L. Mansell, Plenum Press, New York, pp. 1–40.

Fletcher, A. C., Porter, L. J. and Haslam, E. (1977). Plant proanthocyanidins. 111. Conformational and configurational studies of natural procyanidins. *J. Chem. Soc. (Perkin)* 1628–1643.

Foo, L. Y. and Porter, L. J. (1980). The phytochemistry of proanthocyanidin polymers. *Phytochemistry* **19**, 1747–1754.

Fox, L. R. (1981). Defense and dynamics in plant-herbivore systems. *Amer. Zool.* **21**, 853–864.

Fox, L. R. and Macauley, B. J. (1977). Insect grazing on *Eucalyptus* in response to variation in leaf tannins and nitrogen. *Oecologia* **29**, 145–162.

Freeland, W. J. and Janzen, D. H. (1974). Strategies in herbivory by mammals: the role of plant secondary compounds. *Am. Natur.* **108**, 268–289.

Freeland, W. J., Calcott, P. H. and Anderson, L. R. (1985a). Tannins and saponin: interaction in herbivore diets. *Biochem. System Ecol.* **13**, 189–193.

Freeland, W. J., Calcott, P. H. and Geiss, D. P. (1985b). Allelochemicals, minerals and herbivore population size. *Biochem. Syst Ecol.* **13**, 195–206.

Fuller, H. L., Chang, S. I. and Potter, D. K. (1967). Detoxication of dietary tannic acids by chicks. *J. Nutrit.* **91**, 477–481.

Ganzhorn, J. U. (1988). Food partitioning among Malagasy primates. *Oecologia* **75**, 436–450.

Glander, K. E. (1978). Howling monkey's feeding behavior and plant secondary compounds: a study of strategies. In: *The Ecology of Arboreal Folivores*. Ed. G. G. Montgomery, Smithsonian: Washington.

Glander, K. E. (1981). Feeding patterns in mantled howling monkeys. In: *Foraging Behavior: Ecological, Ethological, and Physiological Approaches.* Eds A. C. Kamil and T. Sargent, Garland Publishing: New York.

Glander, K. E. (1982). The impact of plant secondary compounds on primate feeding behavior. *Yearbook of Physical Anthropology* **25**, 1–18.

Goldstein, J. L. and Swain, T. (1965). The inhibition of enzymes by tannins. *Phytochemistry* **4**, 185–192.

Goldstein, W. and Spencer, K. C. (1985). Inhibition of cyanogenesis by tannins. *J. Chem. Ecol.* **11**, 847–858.

Gornitz, K. von (1954). Frassauslosende Stoffe fur polyphagen Holzgewachsen fressenden Raupen. *Verhand. Deutch. Gesellschaft Angew. Ent. E.V.* **6**, 38–47.

Grevillius, A. Y. (1905). Zur Kenntnis der Biologie des Goldafters (*Euproctis chrysorrhoea*) L. und der durch denselben verusachten Beschadigungen. *Botanisches Zentralblatt* **18**, 222–322.

Griffith, D. W. and Moseley, G. (1980). The effects of diets containing field beans of high or low polyphenolic content on the activity of digestive enzymes in the intestine of rats. *J. Sci. Food Agric.* **31**, 255–249.

Haberlandt, G. (1914). *Physiological Plant Anatomy.* Macmillan: London.

Haddock, E. H., Gupta, R. K., Al-Sharifi, S. M. K., Layden, K., Magnolato, D. and Haslam, E. (1982). The metabolism of gallic acid and hexahydroxydiphenic acid in plants: biogenetic and molecular taxonomic considerations. *Phytochemistry* **21**, 1049–1062.

Hagerman, A. E. (1987). Radial diffusion method for determining tannin in plant extracts. *J. Chem. Ecol.* **13**, 437–449.

Hagerman, A. E. (1988). Extraction of tannin from fresh and preserved leaves. *J. Chem. Ecol.* **14**, 453–461.

Hagerman, A. E. and Butler, L. G. (1978). Protein precipitation method for the quantitative determination of tannins. *J. Agric. Food Chem.* **26**, 809–812.

Hagerman, A. E. and Butler, L. G. (1980). Determination of protein in tannin–protein precipitates. *J. Agric. Food Chem.* **28**, 944–947.

Hagerman, A. E. and Butler, L. G. (1981). The specificity of proanthocyanidin-protein interactions. *J. Biol. Chem.* **256**, 4494–4497.

Hagerman, A. E. and Butler, L. G. (1989). Choosing appropriate methods and standards for analyzing tannins. *J. Chem. Ecol.* (in the press).

Hagerman, A. E. and Klucher, K. M. (1986). Tannin-protein interactions. In: *Plant Flavonoids in Biology and Medicine: Biochemical, Pharmacological and Structure-Activity Relationships*, pp. 67–76. Alan R. Liss: New York.

Hagerman, A. E. and Robbins, C. (1987). Implications of soluble tannin-protein complexes for tannin analysis and plant defense mechanisms. *J. Chem. Ecol.* **13**, 1243–1253.

Harborne, J. B. (1976). Flavonoids and the evolution of the Angiosperms. In: *Secondary Metabolism and Coevolution.* Eds M. Luckner, K. Mothes, and L. Nover. Deutsche Akademie der Naturforscher Leopoldina: Halle, pp. 563–604.

Harrison, A. F. (1971). The inhibitory effect of oak leaf litter tannins on the growth of fungi in relation to litter decomposition. *Soil Biol. Biochem.* **3**, 167–172.

Haslam, E. (1974). Polyphenol-protein interactions. *Biochem. J.* **139**, 285–288.

Haslam, E. (1979). Vegetable tannins. In: *Vegetable Tannins.* Ed. by T. Swain, J. B. Harborne and C. F. Van Sumere. *Rec. Adv. Phytochem.* **12**, 475–524.

Haslam, E. (1981). Vegetable tannins. In (Conn, E. E. ed): *The Biochemistry of Plants.* Volume 7. New York: Academic Press. pp. 527–544.

Haslam, E. (1988). Plant polyphenols (syn. vegetable tannins) and chemical defense—a reappraisal. *J. Chem. Ecol.* **14**, 1789–1806.

Haslam, E. and Lilley, T. E. (1986). Interactions of natural phenols with macromolecules. In: *Plant Flavonoids in Biology and Medicine: Biochemical, Pharmacological, and Structure-Activity Relationships.* Alan R. Liss, New York. pp. 53–65.

Hladik, C. M. (1978). Adaptive strategies of primates in relation to leaf eating. In: *The Ecology of Arboreal Folivores.* Ed. G. G. Montgomery, Smithsonian Institute Press, Washington D.C., pp. 373–395.

Horowitz, R. H. (1970). Official methods of analysis of the Association of Official Analytical Chemists (11th Edn), AOAC, Washington, D.C.

Howard, J. J. (1987). Leafcutting and diet selection: the role of nutrients, water, and secondary chemistry. *Ecology* **68**, 503–515.

Inoue, K. H. and Hagerman, A. E. (1988). Determination of gallotannin with rhodanine. *Anal. Biochem.* **169**, 363–369.

Jones, C. G. and Firn, R. D. (1979). Some allelochemicals of *Pteridium aquilinum* and their involvement in resistance to *Pieris brassicae*. *Biochem. Syst. Ecol.* **7**, 187–192.

Jones, W. T. and Lyttleton, J. W. (1971). Bloat in cattle XXXIV. A survey of legume forages that do and do not produce bloat. *New Zeal. J. Agric. Res.* **14**, 101–107.

Jones, W. T. and Mangan, J. L. (1977). Complexes of the condensed tannins of sainfoin (*Onobrychis viciifolia*) with fraction 1 leaf protein and with submaxillary mucoprotein, and their reversal by polyethylene glycol and pH. *J. Sci. Fd. Agric.* **28**, 126–136.

Joslyn, M. A. and Glick, Z. (1969). Comparative effects of gallotannic acid and related phenolics on the growth of rats. *J. Nutrit.* **98**, 119–126.

Kay, R. N. B., Engelhardt, W. V. and White, R. G. (1980). The digestive physiology of wild ruminants. In: *Digestive physiology and metabolism in ruminants.* Eds Y. Ruckebusch and P. Thivend, MTP Press: Lancaster, pp. 743–761.

Keating, S. T. and Yendol, W. G. (1987). Influence of selected host plants on gypsy moth (*Lepidoptera: Lymantriidae*) larval mortality caused by a baculovirus. *Environ. Ent.* **16**, 459–462.

Keating, S. T., Yendol, W. G. and Schultz, J. C. (1988). Relationship between susceptibility of gypsy moth larvae (*Lepidoptera: Lymantriidae*) to a baculovirus and host plant foliage constituents. *Environ. Ent.* **17**, 952–958.

Kendall, W. A. (1966). Factors affecting foams with forage legumes. *Crop Sci.* **6**, 487–492.

Klocke, J. A. and Chan, B. G. (1982). Effects of cotton condensed tannin on feeding and digestion in the cotton pest, *Heliothis zea*. *J. Insect Physiol.* **28**, 911–916.

Koike, S., Iazuka, T. and Mizutani, J. (1979). Determination of caffeic acid in the digestive juice of silkworm larvae and its antibacterial activity against the pathogenic *Streptococcus faecalis* AD-4. *Agric. Biol. Chem.* **43**, 1727–1731.

Kumar, R. and Singh, M. (1984). Tannins: their adverse role in ruminant nutrition. *J. Agric. Food Chem.* **32**, 447–453.

Lagerheim, G. (1900). Zur Frage der Schutzmittel der Pflantzen gegen Raupenfrass. *Entomolo. Tidskrift* **21**, 209–232.

Lawson, D. L., Merritt, R. W., Martin, M. M., Martin, J. S. and Kukor, J. J. (1984). The nutritional ecology of larvae of *Alsophila pometaria* and *Anisota senatoria* feeding on early- and late-season oak foliage. *Ent. Exp. Appl.* **35**, 105-114.

Lee, J. and Bernays, E. A. (1988). Declining acceptability of a food plant for the polyphagous grasshopper *Schistocerca americana:* the role of food aversion learning. *Physiol. Ent.* **13**, 291–301.

Loomis, W. D. (1974). Overcoming problems of phenolics and quinones in the isolation of plant enzymes and organelles. *Methods Enzymol.* **31**, 528–544.

Mangan, J. L., West, J. and Jordan, D. J. (1974). The effect of condensed tannins of sainfoin *Onobrychis viciifolia* on protein degradation in the rumen. Report 1972–3, Inst. Anim. Physiol. Babraham, Cambridge: Agric. Res. Council.

Marini Bettolo, G. B., Marta, M., Pomponi, M. and Bernays, E. A. (1986). Flavan oxygenation pattern and insect feeding deterrence. *Biochem. Syst. Ecol.* **14**, 249–250.

Marks, D. L., Buchsbaum, R. and Swain, T. (1985). Measurement of total phenolics in plant samples in the presence of tannins. *Anal. Biochem.* **147**, 136–143.

Marks, D. L., Swain, T., Goldstein, S., Richard, A. and Leighton, M. (1987). Chemical correlates of Rhesus monkey food choice: the influence of hydrolyzable tannins. *J. Chem. Ecol.* **14**, 213–235.

Marsh, C. D., Clawson, A. B. and Marsh, H. (1918). Oak poisoning of livestock. Report A.I.32. U. S. Government Printing Office, Washington D.C.

Martin, J. S., Martin, M. M. and Bernays, E. A. (1987). Failure of tannic acid to inhibit digestion or reduce digestibility of plant protein in gut fluids of insect herbivores: implications for theories of plant defense. *J. Chem. Ecol.* **13**, 605–621.

Martin, M. M. and Martin, J. S. (1983). Tannin assays in ecological studies: precipitation of ribulose-1,5-biphosphate carboxylase/oxygenase by tannin acid, quebracho and oak foliage extracts. *J. Chem. Ecol.* **9**, 285–294.

Martin, M. M. and Martin, J. S. (1984). Surfactants: their role in preventing the precipition of proteins by tannins in insect guts. *Oecologia* **61**, 342–345.

Martin., M. M., Rockholm, D. C. and Martin, J. S. (1985). Effects of surfactants, pH, and certain cations on precipitation of proteins by tannins. *J. Chem. Ecol.* **11**, 485–494.

Martin-Tanguy, J., Guillaume, J. and Kossa, A. (1977). Condensed tannins in horse bean seeds: chemical structure and apparent effect on poultry. *J. Sci. Fd. Agric.* **28**, 757–765.

McKey, D. B. (1979). The distribution of secondary compounds within plants. In: *Herbivores: Their Interaction with Secondary Plant Metabolites.* Eds G. A. Rosenthal and D. H. Janzen, London: Academic Press, pp. 56–133.

McKey, D. B., Gartlan, J. S., Waterman, P. G. and Choo, G. M. (1981). Food selection by black colobus monkeys (*Colobus satanus*) in relation to food chemistry. *Biol. J. Linn. Soc.* **16**, 115–146.

McKey, D. B. and Waterman, P. G. (1982). Ranging behavior of a group of black colobus (*Colobus satanus*) in the Doula-Edea Forest Reserve, Cameroon. *Folia Primatol.* **39**, 264–304.

McLeod, M. N. (1974). Plant tannins—their role in forage quality. *Nutrition Abstr. Rev* **44**, 803–815.

McManus, J. P., Davis, K. G., Beart, J. E., Gaffney, S. H., Lilley, T. H. and Haslam, E. (1985). Polyphenol interactions. Part I. Introduction: Some observations on the reversible complexation of polyphenols with proteins and polysaccharides. *J. Chem. Soc. Perkin Trans. II* 1429–1438.

McManus, J. P., Davis, K. G., Lilley, T. H. and Haslam, E. (1981). The association of proteins with polyphenols. *J. Chem. Soc. Chem. Commun.* **7**, 309–311.

Mehansho, H., Butler, L. G. and Carlson, D. M. (1987). Dietary tannins and salivary proline-rich proteins: Interactions, induction and defense mechanisms. *Ann. Rev. Nutr.* **7**, 423–440.

Mehansho, H., Clements, S., Sheares, B. T., Smith, S. and Carlson, D. M. (1985b). Induction of proline-rich glycoprotein synthesis in mouse salivary glands by isoproterenol and by tannins. *J. Biol. Chem.* **260**, 4418–4423.

Mehansho, H., Hagerman, A., Clements, S., Butler, L., Rogler, J. and Carlson, D. M. (1983). Studies on environmental effects of gene expression: modulation of proline-rich protein biosynthesis in rat parotid glands by sorghums with high tannin levels. *Proc. Natl Acad. USA* **80**, 3948–3952.

Mehansho, H., Rogler, J., Butler, L. G., Carlson, D. M. (1985a). An unusual growth inhibiting effect of tannins on hamsters. *Fed. Proc.* **44**, 1960.

Meisner, J. and Skatulla, U. (1975). Phagostimulation and phagodeterrency in the larva of the gypsy moth *Porthetria dispar. Phytoparasitica* **3**, 19–26.

Milton, K. (1979). Factors influencing leaf choice by howler monkeys: a test of some hypotheses of food selection by generalist herbivores. *Amer. Nat.* **114**, 362–378.

Milton, K. (1980). *The foraging strategy of Howler Monkeys. A study in Primate Economics.* Columbia University Press: New York.

Milton, K., Van Soest, J. and Robertson J. B. (1980). Digestive efficiencies of wild howler monkeys. *Physiol. Zool.* **53**, 402–409.

Mitaru, B. N., Reichert, R. D. and Blair, R. (1984). The binding of dietary protein by sorghum tannins in the digestive tract of pigs. *J. Nutr.* **114**, 1787–1796.

Mitjavila, S., Lacombe, C., Carrera, G. and Derache, R. (1977). Tannic acid and oxidised tannic acid on the functional state of rat intestinal epithelium. *J. Nutrit.* **197**, 2113–2118.

Mole, S., Butler, L. G., Hagerman, P. G. and Waterman, P. G. (1989). Ecological tannin assays: a critique. *Oecologia* **78**, 93–96.

Mole, S., Ross, J. A. M. and Waterman, P. G. (1988). Light-induced variation in phenolic levels in foliage of rain-forest plants. 1. Chemical changes. *J. Chem. Ecol.* **14**, 1–22.

Mole, S. and Waterman, P. G. (1985). Stimulatory effects of tannins and cholic acid on tryptic hydrolysis of proteins: ecological implications. *J. Chem. Ecol.* **11**, 1323–1332.

Mole, S. and Waterman, P. G. (1987a). Tannins as antifeedents to mammalian herbivores—still an open question? In: *Allelochemicals: Role in Agriculture and Forestry.* Ed. by G. R. Waller. ACS Symposium Series 330. pp. 572–587. A.C.S. Washington, D.C.

Mole, S. and Waterman, P. G. (1987b). Tannin acid and proteolytic enzymes: enzyme inhibition or substrate deprivation? *Phytochemistry* **26**, 99–102.

Mole, S, and Waterman, P. G. (1987c). A critical analysis of techniques for measuring tannins in ecological studies. 1. Techniques for chemically defining tannins. *Oecologia* **72**, 137–147.

Mole, S. and Waterman, P. G. (1987d). A critical analysis of techniques for measuring tannins in ecological studies 11. Techniques for biochemically defining tannins. *Oecologia* **72**, 148–156.

Nagy, K. A. and Milton, K. (1979). Aspects of dietary quality nutrient assimilation, and water balance in wild howler monkeys (*Alouatta palliata*). *Oecologia* **39**, 249–258.

Nierenstein, M. (1934). *The Natural Organic Tannins: History; Chemistry; Distribution.* J. and A. Churchill: London.

Oates, F. J., Swain, T. and Zantovska, J. (1977). Secondary compounds and food selection by colobus monkeys. *Biochem. Syst. Ecol.* **5**, 317–321.

Oates, F., Waterman, P. G. and Choo, G. M. (1980). Food selection by the South Indian leaf-monkey, *Presbytis johnii*, in relation to leaf chemistry. *Oecologia* **45**, 45–46.

Oh, H. I. and Hoff, J. E. (1986). Effect of condensed grape tannins on the in vitro activity of digestive proteases and activation of their zymogens. *J. Food Sci.* **51**, 577–580.

Pierpoint, W. S. (1969a). O-quinones formed in plant extracts. *Biochem. J.* **112**, 609–618.

Pierpoint, W. S. (1969b). O-quinones formed in plant extracts. Their reactions with bovine serum albumin. *Biochem. J.* **112**, 619–629.

Pigeon, R. F., Camp, B. J. and Dollahite, J. W. (1962). Oral toxicity and polyhydroxyphenol moiety of tannin isolated from *Quercus havardi* (shin oak). *Amer. J. Vet. Res.* **23**, 1268–1270.

Porter, L. J. and Woodruffe, J. (1984). Haemanalysis: the relative astringency of proanthocyanidin polymers. *Phytochemistry* **23**, 1255–1256.

Price, M. L., Hagerman, A. E. and Butler, L. G. (1980). Tannin in sorghum grain: effects of cooking on chemical assays and on antinutritional properties in rats. Nutritional Reports International p. 21.

Reed, J. D. (1986). Relationships among soluble phenolics, insoluble proanthocyanidins and fiber in East African browse species. *J. Range Management* **39**, 5–7.

Reid, C. S. W., Ulyatt, M. J. and Wilson, J. M. (1974). Plant tannins, bloat, and nutritive value. *Proc. New Zeal. Soc. Anim. Prod.* **34**, 82–93.

Rhoades, D. F. (1979). Evolution of plant chemical defence against herbivores. In: *Herbivores: Their Interaction with Secondary Plant Metabolites.* Eds G. A. Rosenthal and D. H. Janzen. Academic Press: New York, pp. 4–55.

Rhoades, D. F. and Cates, R. G. (1976). Toward a general theory of plant antiherbivore chemistry. *Rec. Adv. Phytochem.* **10**, 168–213.

Ribereau-Gayon, P. (1972). *Plant Phenolics.* Hafner: New York.

Robbins, C. T., Hanley, T. A., Hagerman, A. E., Hjeljiord, O., Baker, D. L., Schwartz, C. C. and Mautz, W. W. (1987a). Role of tannins in defending plants against ruminants: reduction in protein availability. *Ecology* **68**, 98–107.

Robbins, C. T., Mole, S., Hagerman, A. E., and Hanley, T. A. (1987b). Role of tannins in defending plants against ruminants: reduction in dry matter digestion? *Ecology* **68**, 1606–1615.

Roehrig, N. E. and Capinera, J. L. (1983). Behavioral and developmental responses of range caterpillar larvae, *Hemileuca oliviae*, to condensed tannin. *J. Insect Physiol.* **29**, 901–906.

Rogler, J. C., Ganduglia, H. R. R. and Elkin, R. G. (1985). Effects of nitrogen source and level on the performance of chicks and rats fed low and high tannin sorghum. *Nutr. Res.* **5**, 1143–1151.

Sandanandan, K. P. and Arora, S. P. (1979). Influence of tannins on rumen metabolism. *J. Nucl. Agric. Biol.* **7**, 118–121.

Seaman, F. C. (1984). The effects of tannic acid and other phenolics on the growth of the fungus cultivated by the leaf-cutting ant, *Myrmicocrypta buenzlii. Biochem. Syst. Ecol.* **12**, 155–158.

Schoonhoven, L. M. and Dersken-Koppers, I. (1973). Effects of secondary plant substances on drinking behaviour in some Heteroptera. *Ent. Exp. Appl.* **16**, 141–145.

Schoonhoven, L. M. and Dersken-Koopers, I. (1976). Effects on some allelochemics on food uptake and survival of a polyphagous aphid *Myzus persicae. Ent. Exp. Appl.* **19**, 52–56.

Schopf, R., Mignat, C. and Hedden, P. (1982). As to the food quality of spruce needles for forest-damaging insects. *Z. Angew. Ent.* **93**, 217–220.

Schultz, J. C., Baldwin, I. T. and Northnagle, P. J. (1981). Haemoglobin as a binding substrate in the quantitative analysis of plant tannins. *J. Agric. Food Sci.* **29**, 823–826.

Schultz, J. C. and Lechowicz, M. J. (1986). Hostplant, larval age, and feeding behavior influence midgut pH in the gypsy moth (*Lymantria dispar*). *Oecologia* **71**, 133–137.

Simpson, S. J. and Abisgold, J. (1985). Compensation by locusts for changes in dietary nutrients: behavioural mechanisms. *Physiol. Ent.* **10**, 443–452.

Sin, E. Y., Brunet, P. and Sin, I. L. (1984). Metabolism of hydroxybenzoic acids in *Periplaneta americana*. *Comp. Biochem. Physiol.* **77B**, 791–798.

Singleton, V. L. and Kratzer, F. H. (1969). Toxicity and related physiological activity of phenolic substances of plant origin. *J. Agr. Food. Chem.* **17**, 497–512.

Smith, H. A. (1959). The diagnosis of oak poisoning. *Southwestern Vet.* **13**, 34–37.

Sporne, K. R. (1954). Statistics and the evolution of dicotyledons. *Evolution* **8**, 55–66.

Stafford, H. A. (1988). Proanthocyanidins and the lignin connection. *Phytochemistry* **27**, 1–6.

Steinly, B. A. and Berenbaum, M. (1985). Histopathological effects of tannins on the midgut epithelium of *Papilio polyxenes* and *Papilio glaucus*. *Ent. Exp. Appl.* **39**, 3–9.

Strumeyer, D. H. and Malin, M. J. (1970). Resistance of extracellular yeast invertase and other glycoproteins to denaturation by tannins. *Biochem. J.* **118**, 899.

Strumeyer, D. H. and Malin, M. J. (1975). Condensed tannins in grain sorghum: Isolation, fractionation, and characterization. *J. Agric. Food Chem.* **23**, 909–914.

Sunnerheim, K., Palo, R. T., Theander, O. and Knutson, P. (1988). Chemical defense in birch: platyphylloside: a phenol from *Betula pendula* inhibiting digestibility. *J. Chem. Ecol.* **14**, 549–560.

Sutherland, O. R. W., Hutchins, R. F. N. and Greenfield, W. J. (1982). Effect of lucerne saponins and *Lotus* condensed tannins on survival of grass grub, *Costelytra zealandica*. *New Zealand J. Zool.* **9**, 511–514.

Swain, T. (1976). Reptile-Angiosperm co-evolution. In: *Secondary Metabolism and Coevolution*. Eds. M. Luckner, K. Mothes and L. Nover. Deutche Akademie der Naturforscher Leopoldina: Halle, pp. 551–561.

Swain, T. (1977). Secondary compounds as protective agents. *Ann. Rev. Plant Physiol.* **28**, 479–501.

Swain, T. (1979). Tannins and lignins. In: (G. A. Rosenthal and D. H. Janzen, eds.) *Herbivores*. Academic Press: New York pp. 657–682.

Swain, T. and Hillis, W. E. (1959). Phenolic constituents of *prunus domestica* 1. The quantitative analysis of phenolic constituents. *J. Agric. Food Chem.* **10**, 63–68.

Takechi, M. and Tanaka, Y. (1987). Binding of 1,2,3,4,6-pentagalloyl glucose to proteins, lipids, nucleic acids and sugars. *Phytochemistry* **26**, 94–97.

Taper, M. L. and Case, T. J. (1987). Interaction between oak tannins and parasite community structure: unexpected benefits of tannins to cynipid gall-wasps. *Oecologia* **71**, 254–261.

Taper, M. L., Zimmerman, E. M. and Case, T. J. (1986). Sources of mortality for a cynipid gall-wasp (*Dryocosmus dubiosus* Hymenoptera: Cynipidae): the importance of the tannin/fungus interaction. *Oecologia* **68**, 437–495.

Tempel, A. S. (1982). Tannin measuring techniques. *J. Chem. Ecol.* **8**, 1289–1298.

Van Sumere, C. F., Albrecht, J., De Donder, A., De Pooter, H. and Pe, I. (1975). Plant phenolics. In: *The Chemistry and Biochemistry of Plant Proteins*, Ed. by J. B. Harborne and C. F. Van Sumere, Academic Press, New York.

Vohra, P., Kratzer, F. H. and Joslyn, M. A. (1966). The growth depressing and toxic effects of tannins to chicks. *Poultry Sci.* **45**, 135–142.

Waterman, P. G. (1984). Food acquisition and processing as a function of plant chemistry. In: *Food Acquisition and Processing in Primates*, Ed. by D. J. Chivers, B. A. Wood and A. Bilsborough, Plenum Press: New York, pp. 177–211.

Waterman, P. and Mole, S. (1989). Factors influencing the shikimate pathway in plants. Ed. E. A. Bernays, CRC Press, Boca Raton, Florida. In: Focus on Plant Insect Interactions Vol 1.

Watterson, J. J. and Butler, L. G. (1983). Occurrence of an unusual leucoanthocyanidin and absence of proanthocyanidins in sorghum leaves. *J. Agric. Food Chem.* **31**, 41–45.

Williams, V. M., Porter, L. J. and Hemingway, R. W. (1983). Molecular weight profiles of proanthocyanidin polymers. *Phytochemistry* **22**, 569–572.

Wilson, M. S. (1983). Comparisons of tannin levels in developing fruit buds of two orchard pear varieties using two techniques, Folin Dennis and protein precipitation assay. *J. Chem. Ecol.* **10**, 493–498.

Wilson, M. S. and Blunden, J. (1984). Changes in the levels of polyphenols in three pear varieties during bud development. *J. Sci. Food Agric.* **34**, 973–978.

Wisdom, C. S., Gonzalez-Coloma, A. and Rundel, P. W. (1987). Ecological tannin assays. Evaluation of proanthocyanidins, protein binding assays and protein precipitating potential. *Oecologia* **72**, 395–401.

Wrangham, R. W. and Waterman, P. G. (1981). Feeding behavior of vervet monkeys on *Acacia tortilis* and *Acacia xanthophloea*: with special reference to reproductive strategies and tannin production. *J. Anim. Ecol.* **50**, 715–731.

Wrangham, R. W. and Waterman, P. G. (1983). Condensed tannins in the fruits eaten by chimpanzees. *Biotropica* **15**, 217–222.

Zucker, W. V. (1983). Tannins: does structure determine function? An ecological perspective. *Am. Nat.* **121**, 335–365.

General Ecological Principles Which are Illustrated by Population Studies of Uropodid Mites

F. ATHIAS-BINCHE

ADVANCES IN ECOLOGICAL RESEARCH Vol. 19
ISBN 0-12-013919-7

I. INTRODUCTION

Ecology of soil mites, and especially demography, is not very well known because of the difficulty of direct observation—the animals being usually small-sized and the entire life cycle being endogenous—and the problems of reproducing natural habitat factors under laboratory conditions.

In fact, studies of population dynamics are strictly limited to the use of indirect methods, the best of these being derived from field observations, possibly supported by data derived from reared material. In this case the results have only a statistical value, and depend on the local environmental factors.

For these reasons, the sampling programme is very important. One can find various sampling methods used for the study of the demography and the population dynamics in Southwood (1978). Among these methods, the techniques relative to the study of soil Arthropods are characterized by the following features:

The samples are generally made at random in a homogeneous plot (about 10–20 m^2 in the case of the Acari up to about 100–500 m^2 in the case of larger Arthropods as Myriapoda).

The sampling periodicity has to be short, about once a week or a fortnight in the case of small Arthropods (Acari, Collembola) and once a month in larger forms.

The sampling period depends on the life duration; it is preferable to be able to study the data relative to one to two complete generations (from egg to the death of old adults) and not only an annual phenology. The sampling duration ranges from about one year in the case of small Acari or Collembolla, characterized by a rapid turn-over time, to two years for Acari and more for large Myriapoda.

The absolute population estimates are obtained by sampling a unit of habitat with a soil corer. In general, the larger the animal and the sparser its population, the bigger the sample unit (about 20 cm^2 for small Arthropods to about 1/4 m^2 for Myriapoda). The depth to which it is necessary to sample varies with the animal and the soil, but also with the climate, from 8 to 10 cm deep in temperate forest ecosystems to 15–40 cm in mediterranean or tropical open habitats. The number of sample units should be rather high in order to catch the instar possessing a low probability of capture because the susceptibility of an animal to being caught may alter with age (short-living instars like Acari larvae) or with immobile stage (e.g. pre-ecdysial stage). The minimal number of sample units may be calculated by the Haley method, with $t = x\ Vn/\sigma$ (x = average number of animals per sample, σ = standard deviation, t = Student's t test of standard statistical tables at the 5% level and n = expected number of samples). The parameters of the individual/area curve may be also used (Athias-Binche, 1981) or some other similar methods

(Southwood, 1978). In fact, the density being estimated by counting dead animals, the statistical validity of the quantitative data is very important.

Some other methods of sampling may be used, especially in the case of large, active, surface dwelling Arthropods (spiders, Myripoda), e.g. direct hand sorting *in situ* or using pitfall traps, but they are little improvement on relative estimates of the population density. Emergence traps are used in Insects, especially Diptera.

The fauna is usually extracted by dynamic methods: large Berlese funnels, Tullgren funnels or Macfadyen high-gradient extractors (Macfadyen, 1968; Bieri *et al.*, 1978). These are selected according to their efficiency (see the reviews of Lasebikan, 1975; Lasebikan *et al.*, 1978). Obviously, immobile stages, especially eggs, are not extracted by behavioural methods, this obliges one to use methods of extrapolation for estimating their density. Egg number could be estimated from culture observations, or counting mature eggs in cleared females (Athias-Binche, 1985). Flotation methods allow one to collect immature stages. However this is not an absolute method but it gives good results in some special kinds of substrates (peat, deep mineral soil, see Block, 1967). Soil section methods, especially a gelatine-embedding technique, may be used in substrates rich in plant debris and organic matter (litter, moorland, lichens), but the group studied must be very abundant (Anderson and Healy, 1970; Pande and Berthet, 1973). However these non-dynamic methods require many manipulations, and they are mostly used as a complement of the classical extractors.

If possible, reared populations should be studied, even if the total life cycle is not obtained. These cultures allow one to estimate the life duration and the fecundity, and to observe the behaviour and some other features like the pre-ecdysial stage, as will be seen in the case of the Uropodina. The usual method used is the culture cell (Petri dish with a wet mixture of charcoal-plaster of Paris or blotting paper).

The papers devoted to the soil population dynamics and demography—not simply to descriptive studies of seasonal phenology—are rather rare. One can refer to Athias-Henriot (1976) and Hihm and Chang (1976) for the Gamasina; Lebrun (1977, 1984), Schatz (1983, 1985), Schrenker (1986) and Stamou (1986) for the Oribatida; Kaliszewski *et al.* (1984) and Lindquist (1986) for the Tarsonemida; Gregoire-Wibo (1980), Gregoire-Wibo and Snider (1978), Hutson (1981), Longstaff (1976, 1977), Petersen (1980) and van Straalen (1985a,b) for the Collembola; Blower and Miller (1977) and Geoffroy (1981) for the Myriapoda. One can find some methods used in hemi-edaphic groups (Diptera) in Deleporte (1986).

This chapter is not an exhaustive study of the methods used in edaphic population dynamics, but the example of the Uropodina may be applied to various soil-living groups, such as Acari, Collembola, Pseudoscorpionida and some Myriapoda. The cases for which the methods used for the

Uropodina are not convenient for other Acari are indicated in the text.

Uropodid mites constitute a very homogeneous cohort within the Gamasida (= Mesostigmata), which belong to the Anactinotrichida. Among the other mites (= Actinotrichida), the Oribatida are the most numerous in soil; some other groups, Actinedida, Acaridida and Tarsonemida, these later being very minute mites, are also present in soil. Most of the Uropodina are edaphic forms, whilst some lineages have colonized specialized microhabitats.

A series of papers has been devoted to the study of field ecology of Uropodid mites in temperate and mediterranean soils (Athias-Binche 1981a,b,c, 1982a,b, 1983a,b, 1985). The main results showed that Uropodina constitute a very sensitive indicator of the soil biological activity, especially at the litter/soil interface in forest ecosystems. The present study, devoted to demography, including survivorship, fecundity and productivity, attempts to complete this ecological study and to propose precise methods for studying population dynamics of soil dwelling mites. The purely descriptive study of growth and phenology is completed by an evaluation of demographic parameters, productivity devoted to growth and reproduction, and P/B ratio. This work ends with an analysis of demographic strategies and their relationships with the way of life and the degree of ecological evolution of the habitat.

II. METHODS AND TECHNIQUES, STUDY PLOTS

A. Sampling

The study plot is a slope on acid brown soil situated in the valley of the Massane river (Albères Mountains, Eastern Pyrénées, France). This is the RG plot described in detail in a previous paper (Athias-Binche, 1981c). Seven quantitative samples (20 cm^2 each) were made every week; the results were calculated from the mean of two successive sets of samples. In order to obtain abundant material, the sampling was supplemented every week by two large semi-quantitative samples (250 cm^2 each). The litter was taken separately from soil, the latter being sampled on a depth of 2 cm. Animals were extracted by means of Berlese-Tullgren funnels. During the two year sampling period (June 1975–June 1977), 585 quantitative samples and 197 semi-quantitative samples were taken (Athias-Binche, 1985).

B. Observation of Animals, Body Measurements

The mites were cleared in lactic acid, then observed on an open slide under a light microscope (phase contrast). Juvenile instars, eggs visible in females and about 50% of the adults have been measured (length, width). The material (especially immature stages) has not to be stored in lactic acid more than 5–6 days in order to avoid swelling which could introduce measurement errors. A permanent slide method is not convenient because it induces deformation or flattening of the body, especially in the case of juvenile instars.

Juvenile instars are measured for determining pre-ecdysial periods. This method was previously used for the demographic analysis of an *Allodinychus flagelliger* population in dead wood (Athias-Binche, 1978, 1979a,b), and the method will be herein only briefly recapitulated. During the week before the moult, the immature stages swell with hemolymph during the internal reorganization of organs. A simple graphical method (Harding, 1949) may be used for distinguishing the pre-ecdysial period which marks the change from one instar to the successive one. Cumulative percentages of each class of length frequencies of the same instars are plotted on "probit" probability paper; they lie on a straight line in the case of normal (= gaussian) distribution. On the contrary, if the curve shows inflexions, the sample contains several statistical populations of different size. The different size classes may be discriminated, and then their statistical parameters re-calculated. The largest class represents the oldest individuals which are in the pre-moulting period.

It is to be noticed that the pre-moulting swelling occurs only in the Gamasina and Uropodina. In the other Acari, the only means to detect the pre-ecdysial period is to seek for individuals in which the following instar is visible inside the body.

C. Individual Biomass

The animals are collected in large samples (*ca* 300 cm^2) and extracted in a Berlese funnel on wet plaster of Paris cells. This living material is afterwards anaesthetized in ethyl acetate vapour, and weighed using a microbalance (0·1 µg). The anaesthetic must be used in the case of very active animals (Gamasida, Myriapoda, Collembola), it could be avoided in slow moving mites, such as Oribatida.

Some animals were reared in the laboratory, but the complete developmental cycle was never observed (soil living Uropodina being more sensitive to laboratory conditions than forms adapted to specialized non-edaphic microhabitats, see Athias-Binche, 1981b, 1985). However some other soil Arthropoda are less sensitive to rearing conditions and the life cycle may be observed in the laboratory (e.g. Oribatida, Acaridida, some Gamasida, Collembola, see Lebrun, 1977; Mitchell, 1977; Schatz, 1985; Longstaff, 1976; Gregoire-Wibo and Snider, 1978).

D. Climate of the Massane Beechwood

The detailed study of the Massane forest climate was published in a previous paper (Athias-Binche, 1981c). The Massane forest belongs to the mediterranean beech-woods (Thiebaut, 1982). The locality is situated in the perhumid cool climates according to the Emberger nomenclature. The Bagnouls and Gaussen's ombrothermic diagram for the two years studied indicates that the year 1975 was rather wet in summer, but arid in October (Fig. 1). The year 1976 was more close to mediterranean conditions, with a dry period in July, but August was very rainy. Autumn and winter were wetter than the previous year. January, May and June 1977 were very rainy.

III. BRIEF RÉSUMÉ OF UROPODID MITE BIOLOGY

The biology of Uropodina was described in detail in a previous paper (Athias-Binche, 1981b).

Uropodina are medium-sized mites, their body length ranging from 450 to 1200 mμ, as in Gamasida or Oribatida. They feed on living organic substances. They are either liquid-feeder (fungal hyphae, slow moving or feeble prey) or microphagous, eating small particles such as yeast or unicellular algae. Thus, they are not to be regarded as true decomposers—feeding more or less directly on dead plant material, such as most Oribatida—they are in fact usually situated near the end of the trophic chains in the soil sub-system, and they are heavily dependent on the efficiency of the matter/energy

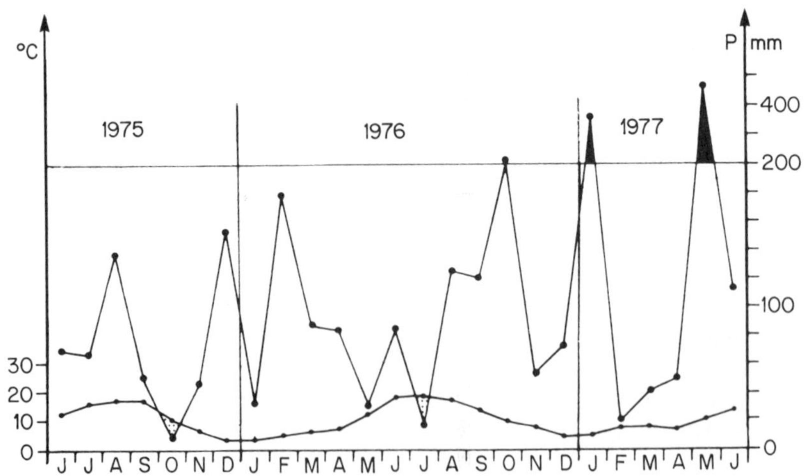

Fig. 1. Massane forest. Ombrothermic diagram of the studied period. °C: average monthly temperature, P mm: monthly rainfall (mm) (1°C = 2 P mm).

exchanges in the decomposing processes within their substrate (Athias-Binche, 1981b). This is the reason why these mites are more sensitive than detritivores to any disturbance of the edaphic energy pathways. For these reasons, Uropodina are mainly encountered in biologically active habitats, and the soil-living forms are the most numerous in the fragmentation layer of the litter in temperate deciduous forests, or in the various ecological niches of tropical rain forests, where their maximum number and species diversity are found. In the species colonizing specialized microhabitats, their biotopes are also always rich in decomposing organic matter (rotten wood, manure, guano, carcasses, tidal debris, stored food, insect and mammal nests). These species are obligatory phoretic (Athias-Binche, 1984a).

Uropodina are slow-moving mites, with a low energy budget, compared to that of the Gamasid mites, of which most are active predators (Wood and Lawton, 1973).

The reproductive behaviour is highly elaborate within the mites and involves generally a courtship period, mating behaviour and low egg production. In bisexual forms, the fertilization is external; the male produces a bag-shaped spermatophore which is passed to the female during the mating courtship. The female introduces the spermatophore into the endogyne whilst the male remains clasping the female. This general scheme may be complicated by various behaviours which depend on the species and its habitat. This pseudo-mating, a very complex behaviour for such small invertebrates, serves to increase the probability of reproductive success, while in the majority of soil Actinotrichida, the spermatophore is laid at random by the male in the substrate, thus the probability of reproduction success is lesser than in the case of Uropodid mating behaviour.

As for many other Acari, thelytoky (production of females by parthenogenesis) is very frequent in Uropodina. The phoretic species are usually bisexual but, as in the Gamasina, arrhenotoky (haploid males produced without fertilization) may be very common; this question needs further research. Parthenogenesis might limit the occurrences of a few competitive genes (Suomalainen *et al.*, 1976); from an ecological point of view, it ensures the production of immature stages, even in the case of a very low population density, for instance when the soil substrate is poor.

Eggs are telolecythic and usually of large size (Table 3). The gravid female bears generally two to four pre-mature eggs, it lays one egg per clutch, which occurs about every 24 h. From laboratory data, the female may produce a maximum of 30 to 60 eggs in its life.

Postembryonic development comprises usually four instars in soil living forms: larva, protonymph, deutonymph and adults. Development may be abbreviated in species colonizing specialized microhabitats (Athias-Binche, 1987a). The total life duration occupies approximately one year in edaphic Uropodina. Depending on rearing conditions and species, hatching occurs 8 to 26 days after egg laying, the larva moults after 13 to 39 days, the protonymph duration may be 2 weeks to 2 months. The deutonymphal longevity is very variable, requiring about 3 to 5 months; the last moult can be also delayed for several months in unfavourable conditions. The deutonymph plays the main part in the persistence of the species, because of its resistance to adverse environmental factors. It may be obligatory or facultatively phoretic, depending on the specificity and the characteristics of the habitat (Athias-Binche, 1984a,b). Adults live from 2 to 5 months. The total duration of development requires at least 3 months, thus the total life cycle takes about one year. The total life duration is similar in the other soil Acari under temperate conditions, except in the small forms (Tarsonemida, some oribatid families) in which the generations range from about several days to few months.

IV. RESULTS

A. Species Collected and Their Main Characteristics

The following list contains the soil uropodid mites collected in the Massane forest (body dimensions are given in Table 2). In addition, the present study includes an obligatory phoretic species, *Allodinychus flagelliger* (Berlese 1910), living in dead wood, of which the ecology and demography were shown to be different from that of edaphic species (Athias-Binche 1978, 1979a,b).

Polyaspididoidea sensu *Athias-Binche and Evans 1981*

Trachytes aegrota (Koch 1841). Palearctic species, probably on its southern geographical limits in the Massane forest. Very common in temperate soils, scarce in the Massane soil (less than 3 individuals/m^2). Telytokous, non phoretic, strictly litter living and stenotopic in the Massane forest (Table 1).

T. lamda Berlese 1903 (Fig. 1b). Southern and central Europe, not frequent in eumediterraneous region. Rather few in the Massane forest (25 ind/m^2). Telytokous, non phoretic. Litter living, rather stenotopic in the Massane conditions.

T. sp aff. *T. baloghi* Hirschmann and Z.-Nicol 1969. Telytokous, non phoretic. Scarce in the Massane beech-wood (1 ind/m^2), but more abundant in eumediterranean evergreen forests in the Eastern Pyrénées. Humicolous.

Polyaspinus quadrangularis Athias-Binche 1981 (Fig. 1a). One of the rare Uropodina with marked predatory behaviour. Rather abundant in the Massane forest (33 ind/m^2), where this species is humicolous, seems to be rare in eumediterraneous xeric ecosystems. Non phoretic.

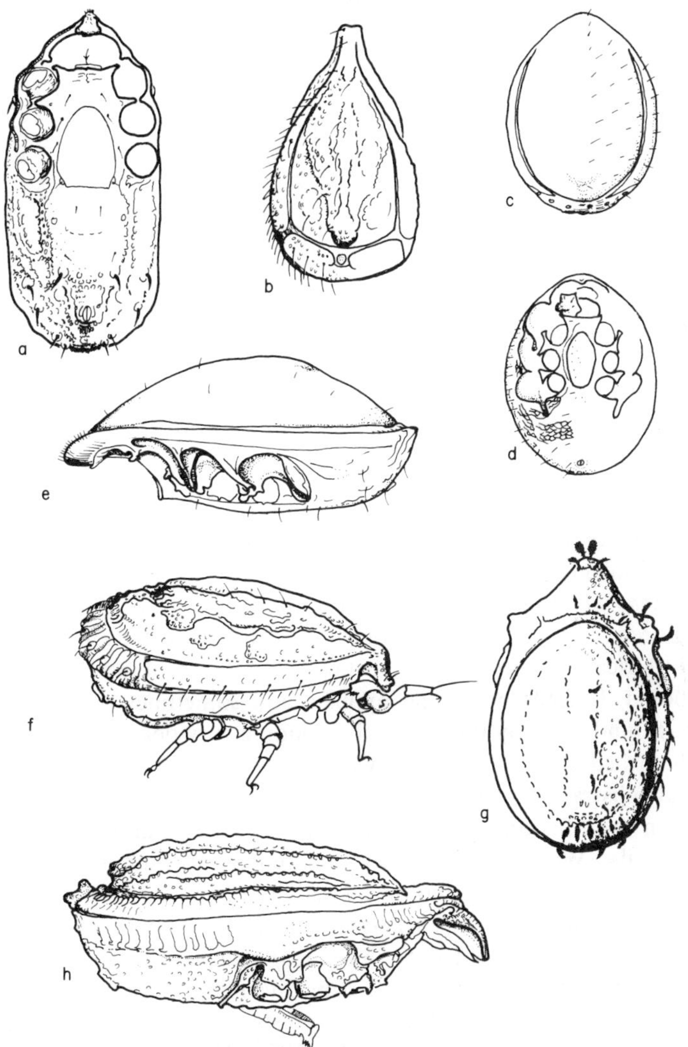

Fig. 1. Some soil Uropodid mites of the Massane forest. a: *P. quadrangularis*, female, ventral view, gnathosoma and legs removed, b: *T. lamda*, female, dorsal view, c, d: *O. minimus*, female, dorsal and ventral view, gnathosoma and legs removed, e: *C. erlangensis*, female, lateral view, gnathosoma and legs removed, f: *N. catalonica*, female, habitus, g: *A. coriacea*, female, dorsal view, h: *U. carinatus*, female, lateral view, gnathosoma and legs removed (a–b: Polyaspidisoidea, c–h: Uropodoidea).

Table 1
Vertical distribution and stenotopy of the soil Uropodina in the Massane forest (from Athias-Binche, 1982b).

	L%	ST		L%	ST
T. aegrota	99·8	27·3	*O. minimus*	38·5	11·3
O. alveolus	95·0	48·9	*P. quadrangularis*	36·6	23·3
T. lamda	75·7	25·5	*A. coriacea*	32·6	12·8
U. carinatus	72·0	15	*C. erlangensis*	28·6	19·7
N. catalonica	48·3	14·2	*T. cf baloghi*	23·8	77·1

L%: percentage of individual numbers in the litter, ST: stenotopy in the Massane RG plot, difference between maximum and minimum frequency along the 22 levels of the transect; ST is the highest in the most stenotopic species.

Uropodoidea sensu *Athias-Binche and Evans 1981*

Cilliba erlangensis (Hirschmann and Z.-Nicol 1969) = *Colliba massanae* (Athias-Binche 1981) (Fig. 1e). Temperate european soils, rather scarce in the Massane forest (20 ind/m^2), more rare in eumediterranean forests. Telytokous, facultatively phoretic.

Olodiscus minimus (Berlese 1910) (Fig. 1c,d). This telytokous and facultatively phoretic species is notable for its very wide distribution in Europe and North Africa. Its ecological plasticity is very striking for an edaphic Uropodina. It feeds on fungal hyphae, and sometimes on chlorophyllous plants. Found in many types of soil, including very degraded mediterranean soils disturbed after wild fires, may be observed in ant nests, able to colonize artificial substrates in soil (Athias and Mignolet, 1979). The most abundant species in the Massane soils (227 ind/m^2), hyperdominant in degraded and eroded soils. Without a marked preference for humus or litter layer, eurytopic, less sensitive to biological activity of soils.

Neodiscopoma catalonica Athias-Binche 1981 (Fig. 1f). Mediterraneous species, very closely related to *Neodiscopoma cosmogyna* (Berlese 1916) described from Italy. Supra- and eumediterreanean climatic areas in the Pyrénées-Orientales, but restricted to forests. In the Massane beech-wood, holds the second rank after *O. minimus* (190 ind/m^2) and it is a competitor of this species in the active brown mulls rich in litter. No layer preference, rather eurytopic.

Urodinychus carinatus (Berlese 1888) (Fig. 1h). Italy, Southern France, Algeria. Large litter-living species, non phoretic, restricted to the least xeric forests. Moderately abundant in the Massane forest (28 ind/m^2).

Oodinychus alveolus Athias-Binche 1981. Large litter dwelling species, only in the wettest forests in the Pyrénées-Orientales. Non phoretic. Rare in the Massane beech-wood (1 ind/m^2).

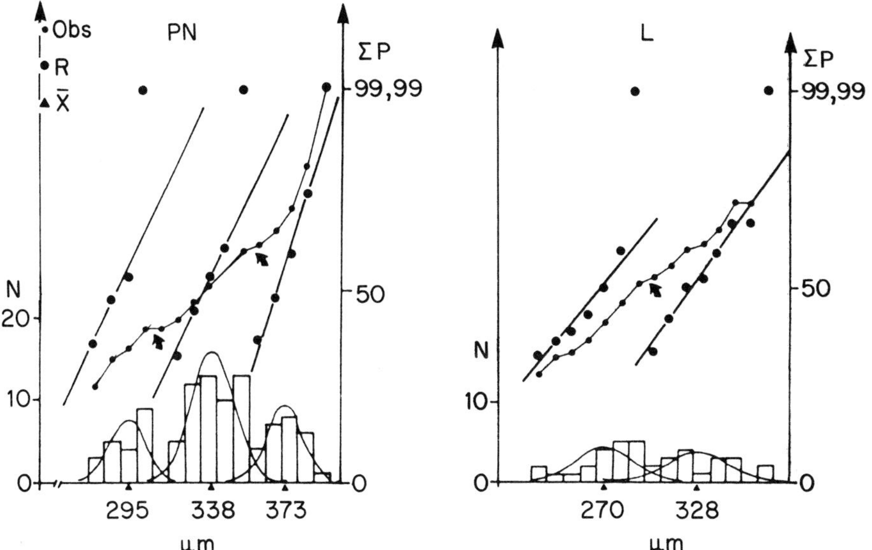

Fig. 2. Use of probability paper for graphical analysis of polymodal frequency size distribution: histogram of size frequencies, examples of size partition of larvae (L) and protonymphs (PN) in *O. minimus*. P: cumulative percentages of frequencies ("probit" scale); N: frequency for each 10 µm interval; obs: observed probits; R: re-aligning of the probits after partition of the statistical populations and fitting to the normal distribution; x: mean of the new size classes.

Armaturopoda coriacea Athias-Binche 1981 (Fig. 1g). Small humicolous species, non phoretic. At present, only known from the Massane forest, where it is rather abundant (67 ind/m^2).

The demographic analysis will take into account only the six most abundant populations.

B. Dimensions and Biomasses

The Harding method distinguishes two to three size classes in the immature stages (Fig. 2, Table 2). In the deutonymphal instars, the pre-ecdysial period is well marked and the distribution of frequencies is often bimodal; on the contrary, the standard deviation is higher in the other immature stages and only the Harding's method can help for separating the different statistical populations. The *t*-test was also used to discriminate differences in length between males and females in bisexual forms. The adult body lengths are significantly different in *N. catalonica* and *U. carinatus* (Table 2).

Mature eggs are large (Table 3), their length represents 30 to 49% of female body size; this ratio may exceed 60% in *O. minimus*, which is a small

Table 2
Body length (μm) of the 4 instars and size classes discriminated by the Harding's method.

Species	♀	♂	DN 1	DN 2	PN 1	PN 2	PN 3	L 1	L 2	t	S
O. minimus	442·5		395·5	419·0	295·2	338·1	373·8	270	328·8		
CV	2·8		1·4	1·4	3·0	3·1	2·3	7	8·4		
N	180		58	84	21	53	26	20	15		
N. catalonica	572·0	586·4	485·1	520·3	300·6	373·0	436·9	294·8	356·6	8·13	99·9%
CV	2·3	2·0	2·7	5·8	4·41	6·1	4·5	0·2	9·1		
N	99	101	82	112	12	22	32	27	9		
O. carinatus	864·4	880·7	699·5	792·4	503·6	590·3	664·0	312·8	409·2	2·68	99%
CV	3·4	3·1	4·0	4·1	6·6	4·1	0·8	8·4	7·8		
N	44	45	13	24	22	19	4	10	24		
C. erlangensis	528·2			477·3		287·4		220·4			
CV	2·5			1·8		11·8		16·2			
N	24			11		7		13			
A. coriacea	458·8	460·4	420·0	464·8	294·8		375·2	257·1	304·2	0·88	
CV	2·7	2·2	3·4	3·6	10·8		5·0	7·4	5·7		
N	67	92	42	27	18		24	7	7		

O. alveolus	864·8	848·0		681·6			560·1			433·6			
CV	3·3	5·5		4·8			8·8			13·1			
N	5	8		5			7			7			
T. aegrota	642·0			548·6			315·2			301·2			
CV	3·0			5·8			6·3			16·2			
N	8			10			10			5			
T. lamda	585·0		486·4		530·1	386·0		422·7	251·2		295·8		
CV	1·5		1·3		2·4	10·4		2·4	4·1		5·9		
N	72		11		29	16		13	8		12		
P. quadrangularis	614·8	614·8	520·7		578·0	364·4		440·0		365·1		0	—
CV	3·0	2·7	3·1		1·3	3·7		4·4		17·7			
N	24	50	13		8	9		6		7			
T. cf. baloghi	637·2			566·4									
N	9			2									

DN1, 2, PN1, 2, 3, L1, 2: resp. size classes of deuto- and protonymphs, and larvae, CV: variation coefficient (100 δ/x), N: number of measured individuals, t: t-test on the difference between male and female body length, S: significance of t on the difference of the average lengths.

Table 3
Minimum and maximum size of eggs measured in the females.

	Min.		Max.		N	%L/♀
	L	W	L	W		
O. minimus	40	40	280	200	87	63
N. catalonica	132	92	230	180	30	40
U. carinatus	276	100	318	192	6	36
C. erlangensis	156	104	258	198	14	48
A. coriacea	120	110	200	140	9	43
Tr. lamda	140	100	260	200	27	44
P. quadrang.			229	163	4	48
O. alveolus			262	172	3	30
Tr. cf. bal.			208	322	4	32

(Min, Max, μm). L: length, w: width, N: number of individuals measured, %L: percentage of the ratio egg max. length/female length.

species. The egg width is larger than the female genital aperture, but its chorionic flexibility allows some deformation during egg-laying.

The fresh weights are indicated in Table 4. For the same size, members of the superfamily Uropodoidea are heavier than the Polyaspididioidea because of their more sclerotized integuments. The size/mass relationships are illustrated by a scatter diagram (Fig. 3) including species found in the Massane forest and five other species, collected in the Meerdael forest (Belgium), and studied in previous papers (Athias and Mignolet, 1979; Athias-Binche 1982b). The correlation coefficient between length and mass reaches 0·928 in Uropodoidea (significant at the 99·9% level for N = 9 paired data) and 0·800 (significant at the 90% level for N = 6). The differences are mainly marked for the slopes, the regression equations being resp. $Y = 0{\cdot}203X - 83{\cdot}7$ and $Y = 0{\cdot}032X + 1{\cdot}37$; thus, mass increases more rapidly in relation of size in the Uropodoidea. From an ecological point of view, these equations demonstrate that the Uropodoidea devote an important part of their production to developing a sclerotized strong carapace, inert and without metabolic function. The ratio of production allocated to reproduction is consequently lower in Uropodoidea.

C. Graphical Determination of the Generations

In order to distinguish the beginning and the end of each generation, I used a simple graphical method, rather rough but efficient for locating modal classes.

On a graph with time as abscissa, numbers as ordinate, one marks the

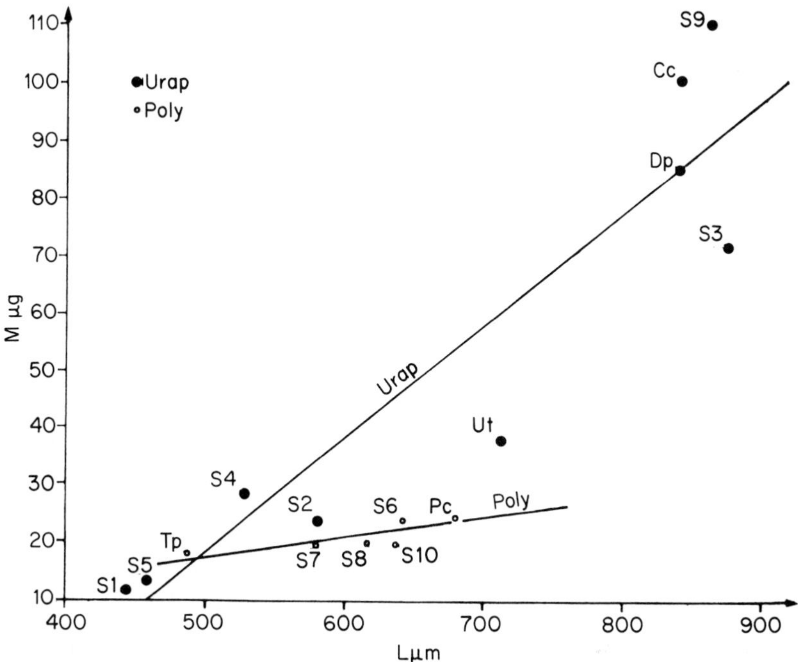

Fig. 3. Size/mass regression lines (L µm/M µg) in Uropodiodea and Polyaspididoidea adults. Massane species, Uropodoidea (Urap) S1: *O. minimus*, S2: *N. catalonica*, S3: *U. carinatus*, S4: *C. erlangensis*, S5: *A. coriacea*, S9: *O. alveolus*; Polyaspididoidea (Poly): S6: *T. aegrota*, S7: *T. lamda*, S8: *P. quadrangularis*, S10: *T. cf baloghi*. Other species (Meerdael Quercocarpinetum, Belgium), Uropodoidea: Cc: *C. cassidea*, Dp: *Dinychus perforatus*, Ut: *Urodiaspis tecta*, Polyaspididoidea Pc: *Polyaspinus cylindricus*, Tp: *Trachytes pi*.

Table 4

Individual body weight (µg, fresh mass) of the 4 instars in the Uropodina of the Massane forest (from Athias-Binche, 1982b).

Species	A	DN	PN	L
O. minimus	12·0	8·1	5·7	3·4
N. catalonica	24·0	12·0	9·2	4·9
U. carinatus	77·0	49·8	20·9	8·7
C. erlangensis	29·0	20·7	10·5	5·4
A. coriacea	13·5	10·4	6·1	3·6
T. aegrota	24·2	19·7	10·0	5·0
T. lamda	19·0	14·1	7·8	4·3
P. quadrangularis	20·0	14·8	8·1	4·4
O. alveolus	112·0	69·8	27·1	11·5
T. cf. baloghi	19·0	14·1	7·8	4·3

A, DN, PN, L: resp. adults, deutonymph, protonymph. larva.

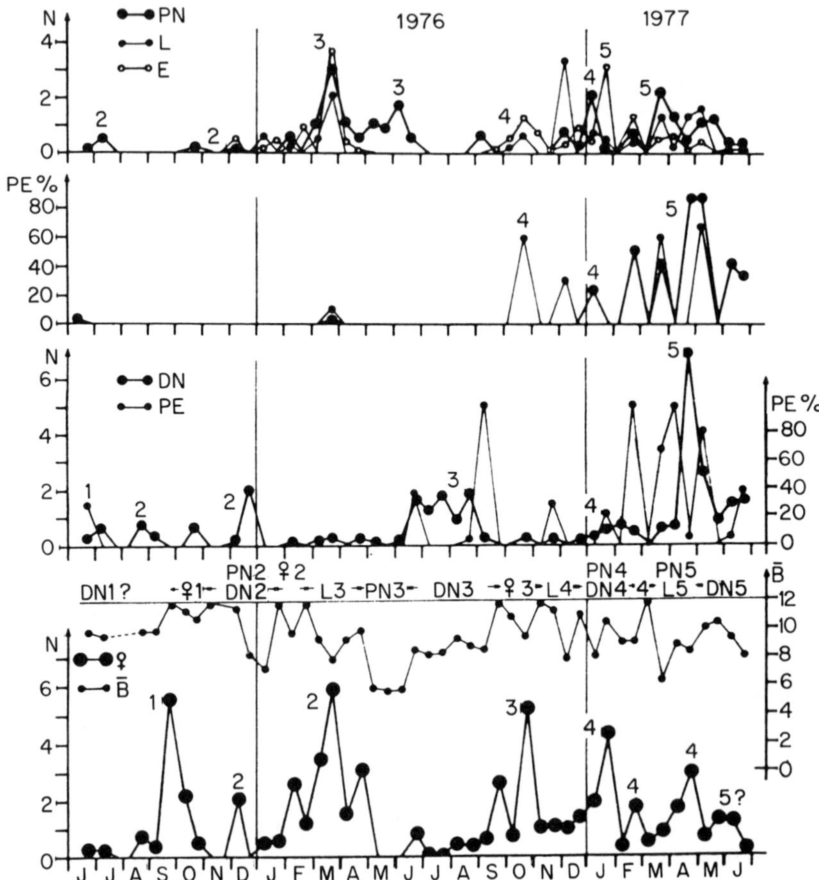

Fig. 4. Phenology of the Massane population of *O. minimus*, cohort partition. Ordinates: N: number of individuals (N.10^{-2}/m^2), PE%: percentage of immatures in pre-ecdysial period, B: mean population biomass (mg/m^2); lines: PN: protonymphs, L: larvae, E: eggs (counted inside females), DN: deutonymphs, PE: percentage of deutonymphs in pre-ecdysial period, ♀: females, bold type numbers mark the different generations.

numbers of the four instars (by means of separate symbols) at each sampling date. The main recruitment period is indicated by a mode; the decreasing of this curve, associated with an increase of the frequency of pre-ecdysial immatures, indicates the end of the occurrence of the instar. The average biomass of the population ($B = b_i/N$, in which b_i = the total biomass of the population at the time i and N = the density) is a parameter of the population structure. The maximum average biomass equals the adult mass and the minimal value is the larval weight. When the average biomass decreases, the population is in a recruitment period (numerous immature stages); on the contrary, when the average biomass tends towards its maximum, the population is dominated by adults. Finally, the count of mature eggs in females indicates the beginning of a new generation.

This method has several limitations, of which many are common to all field demographic studies of Invertebrates. Firstly, the probability of capture of one instar is proportional to its life duration: some transient stages, such as larvae or immatures in the pre-ecdysial period have a low probability of capture; thus their number is under-estimated and the evaluation of the density needs correction. On the other hand, in a population with a slow turn-over and weak seasonal fluctuations, as in *A. coriacea* and *P. quadrangularis*, the successive generations overlap each other and it is rather difficult to distinguish the different cohorts. Despite these disadvantages, analysis of the average biomass usually permits a good discrimination of the main events of the biological cycle and it may be adapted to the study of many soil Arthropods.

Figure 4 gives an example of this method for *O. minimus*. Females collected in the year 1975 belonged to the generation no. 1 which would die out in November. At the end of 1975, only some eggs and some protonymphs of the second generation were collected. On the contrary, the long living deutonymphs were more numerous. The larval moult might be very rapid and, by the way, their probability of capture was then very low. The average biomass reached its maximum in September and October, marking the peak of adults of the first generation (Fig. 4: ♀); the biomass decreased afterwards with the arrival of the young stages of the second generation (Fig. 4: PN2, DN2), thus the larval hatching may have taken place in September and October. The deutonymphs "2" moult rapidly, generating young females: the main biomass reached its maximum in February and March (Fig. 4: ♀2). The eggs were produced and the larvae of generation no. 3 hatched, with a maximum in March 1976, marked by a concomitant fall of the average biomass (Fig. 4: L3). Protonymphs "3" were caught from March to June, then deutonymphs "3" were collected from June to August. The imaginal moults were heavy in September 1976 during which 100% of the deutonymphs were collected in the pre-ecdysial stage. The biomass increased afterwards with the appearance of females of the third generation (Fig. 4: ♀3). The young females laid their eggs and the larval hatchings of generation no. 4 began in October, reaching their maximum in December 1976 (Fig. 4: L4). The successive moults were very rapid: the protonymphs "4" appeared in December–January and the deutonymphs produced immediately the young females "4"; in fact these deutonymphs "4" were mostly collected as young whitish females. All the immature stages moulted from March 1977 and the biomass quickly reached its maximum with the appearance of the females "4". Egg-laying followed immediately and hatchings were marked by a sudden fall of the biomass at the end of March (Fig. 4: L5). The protonymphs "4" were observed at the beginning of April and they began to moult at the end of this month, given the first deutonymphs of generation no. 5. At the end of the sampling period, in May–June 1977, one could observe the first imaginal moults and perhaps the arrival of young females "5".

The limits of the different generations and cohorts being known, it is possible to draw the cumulative curves of egg and immature numbers, the final plateau marking the maximum recruitment and the change to the following instar. By joining the successive cohort limits, one can visualise the different generations (Figs 5 to 8). Egg-laying and mainly hatching were mostly observed during rainy periods, as in November–December 1975, in spring and autumn 1977, then in winter and spring 1977. The deutonymphs moult more rapidly during wet periods (cf deutonymphs 2 and 4). On the contrary, in summer, especially when it was rather dry as in 1976, deutonymphs entered an estivation period and the moulting was delayed; so their life duration was longer than that of the deutonymphs of wet periods. In autumn 1975, which was rather dry, the last moult was put back until January–February. It is clear that the cycle of female appearance also depends on rainfall, because imaginal moults occurred mostly during wet periods, and also because the animals migrate down in deeper layers during summer in response to the increase of evaporation. During these migratory movements, population activity is reduced because of poor trophic resources in the deep layers of soil.

Generation length depended therefore on season and climatic conditions. Some data from other populations of *O. minimus* from a drier locality (cork-oak stand of Valmy, Argelès-sur-Mer, *ca* 5 km east of the Massane forest), indicated that the estivation period was much longer than in the Massane forest, especially in an open habitat where litter and plant cover was destroyed by fire (Athias-Binche and Saulnier, 1986; Athias-Binche, 1987b). This ecological plasticity constitutes the main characteristic of *O. minimus*, an opportunistic and euryecic species; the other Uropodina encountered in the Massane beech-wood are more specialized and less able to adapt their life cycle to habitat fluctuations.

D. Examples of Generation Analysis

Four examples of generation analysis will be given (Figs 5 to 8).

As was seen, in *O. minimus* reproduction occurred mainly during wet periods. During the 2 year study, one could observe five cohorts of females, the first and the fifth being incomplete (Fig. 5, for details of seasonal phenology, see comments in Section C).

In *N. catalonica*, egg-laying and larval hatching were more regular and seemed to be less dependent on the climatic parameters (Fig. 6). Winter eggs

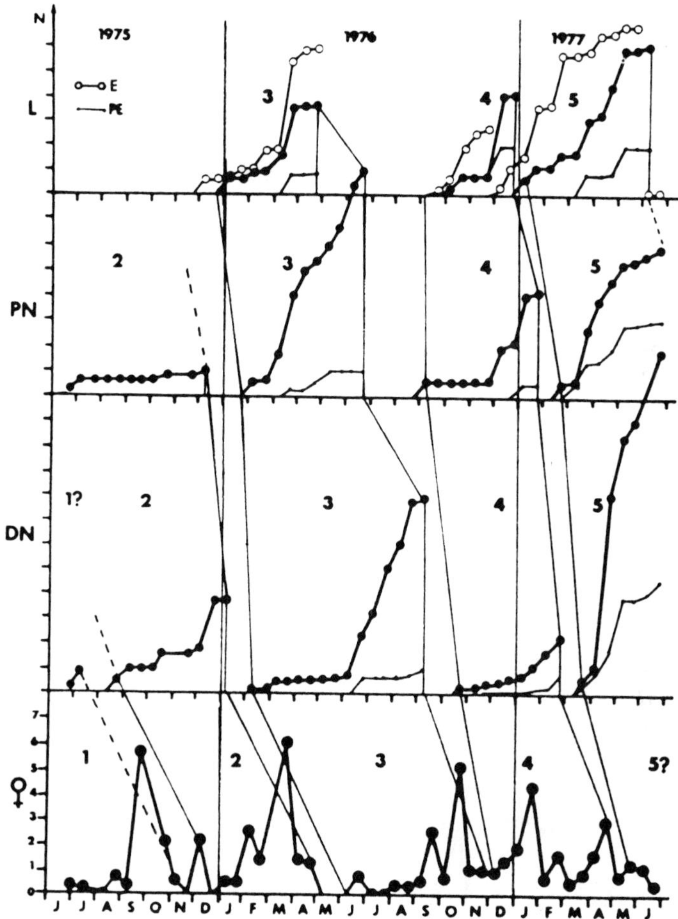

Fig. 5. Cohorts and generations in *O. minimus*, average number by fortnights ($N.10^{-2}/m^2$), cumulative data for immature stages and eggs. E, L, PN, DN, ♀: resp. eggs, larvae, proto- and deutonymphs, females, PE: immature stages in pre-ecdysial period. Generations no. 3 and 4 are complete.

rapidly develop into spring larvae; then some delay can be observed in summer between egg production by the females, egg-laying and autumnal larval hatching. Protonymphs were mostly abundant in summer. This fact might explain why *N. catalonica* is restricted to the most shady and wet forests of the mediterranean climatic area because the protonymphal instar is less dryness-resistant than the deutonymph and could not withstand severe estivation (Athias-Binche, 1981c). Protonymphs were abundant in January. Deutonymphs were long lived, they were collected all through the year, but their peak was mainly seen in summer and spring. Adults could be observed

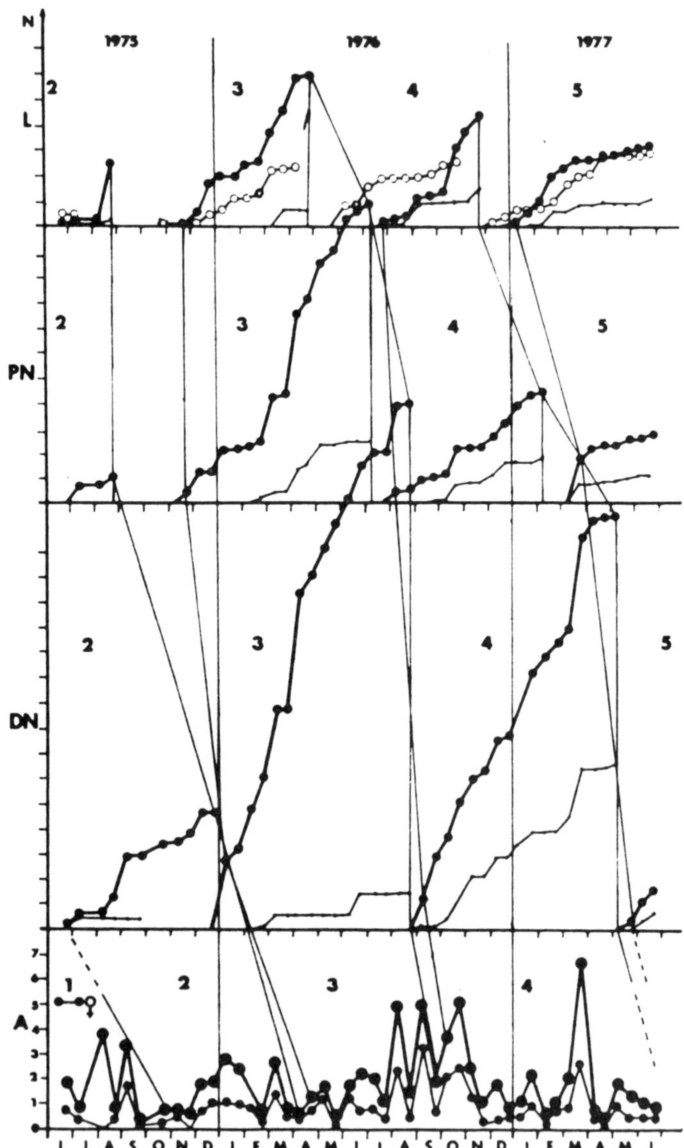

Fig. 6. *N. catalonica*, same explanation as for Fig. 5. Only generation no. 3 is complete.

all the year, their seasonal fluctuations being less strong than for *O. minimus*. Unlike the *O. minimus* population, females may hatch their eggs during summer and some larvae were produced, as in August 1976, however the most abundant hatching occurred mostly in autumn after the first rains.

During the two years of sampling, only four generations were observed, thus the longevity was higher than that of *O. minimus*.

In *U. carinatus*, appearances of immature stages were very regular (Fig. 7); production of young seemed to be less than in the two previous populations; adults were also less numerous. Larvae were born in autumn and in spring; the winter and summer cohorts of protonymphs and deutonymphs were well separated. If the appearance of juveniles was well marked and comparable from one year to one another, imaginal moults were regularly spread over all the seasons, and adult numbers did not fluctuate greatly. Thus, it is rather difficult to distinguish the main periods of recruitment of young adults. One can observe some differences between adult and immature densities, the adult mortality may have been precocious and have occurred immediately after reproduction. Phenology appeared to be independent of climatic conditions. In addition, this large litter-living species cannot easily move to deep layers during dry periods; which may be the reason why this species is mostly observed in the wettest mediterranean forests. Furthermore, as with many litter-living Uropodina, this species is more sensitive than *N. catalonica* and *O. minimus* to the soil biological activity (Athias-Binche, 1982a), and it is mostly found in habitats characterized by hight decomposing activity.

The different female cohorts were easily distinguishable in *T. lamda* (Fig. 8). Egg production and hatching took place once a year, but births were spread over several months, usually from autumn to spring. Egg laying and

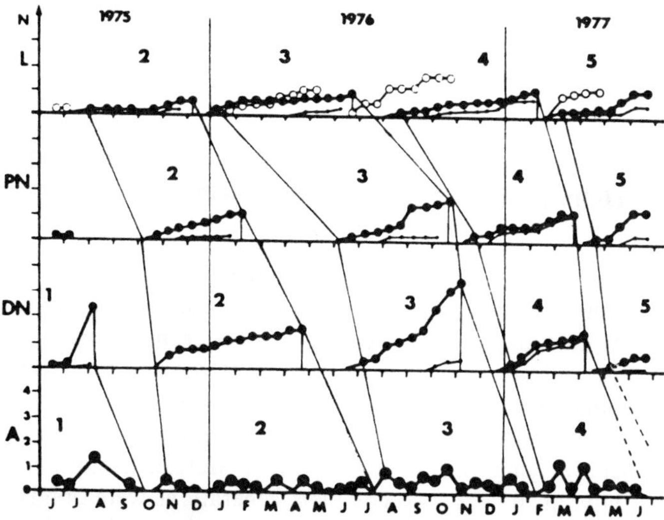

Fig. 7. *U. carninatus*, same explanation as for Fig. 5. Generation no. 2 incomplete (eggs and some larvae), generation 3 complete, adult extinctions probably missing in generation no. 4.

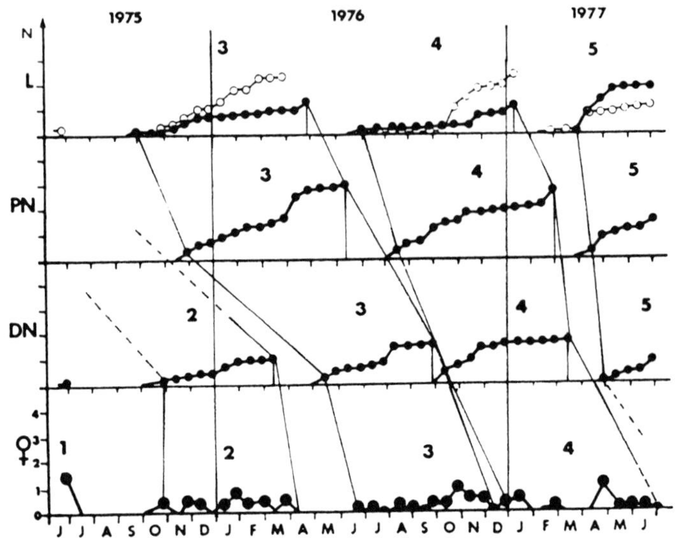

Fig. 8. *T. lamda*, same explanation as for Fig. 5. Generation no. 3 complete, last females probably missing in generation no. 4.

larval hatching seemed to be more frequent during wet seasons (November 1975 to April 1976, October 1976 to January 1977 and early egg-laying in the wet spring in May–June 1977). As in the case of *O. minimus*, imaginal moults seem to be favoured during rainy periods. Females may be present during summer.

E. Generation and Cohort Duration, Some Remarks About Arthropod Demographic Studies in the Field

Figures 5 to 8 indicate that one generation from egg to adult required at least six months in the populations studied. These life durations appeared to be longer than in data obtained by several authors from laboratory experiments, the observed postembryonic development ranging from 120 to 190 days (4–6 months) (Vallee, 1954; Krasinskaya, 1961; Radinovsky, 1965), but it is to be noticed that temperatures in their experiments were higher than those of the field, varying from 15 to 30°C. In the Massane conditions, during a two year study, at the best two complete generations were observed. The first one is often cut out because the cumulative calculated numbers of the young instars represented only a part of immatures produced during the period preceding the beginning of the sampling. In the same way, the end of the last generation was not known because the final adults could die after the end of the study.

These comments illustrate some methods of demographic study of soil Arthropoda, small animals, living under the litter and of which rearing is often difficult. Firstly, a perfect systematic and biological knowledge of the material is essential, and all the juvenile instars of the concerned species must be precisely identified. On an other hand, the generation duration being *a priori* unknown, a two year sampling period has to be considered as a minimum, if two complete generations must be observed, especially in a temperate climate where life duration of soil mites usually varies from three months to about one year. Thus, a simple one-year study cannot give a good appraisal of the demography, but only a simple description of seasonal phenology. In a tropical climate, generation time might be shorter, whilst in cold conditions, the life duration may be considerably lengthened (as an example, some Oribatid mites may live for four years in cold Canadian soils, Mitchell, 1977). In addition, in soil mite populations, which live in a very stable habitat, without any disruptive factors, demographic balances do not change from one year to one another. On the contrary, in mite populations living in temporary or limited microhabitats, the dynamics may fluctuate greatly, depending on annual or seasonal characteristics, or colonization/emigration movements. In these cases, sampling could take several years and growth models may be used (Athias-Binche 1979b).

For evaluating the life duration of the different instars and generations, it will be enough to locate on the diagrams (Figs 5 to 8) the number of two-week periods which separate the first larval birth from the disappearance of the last adult. By this means, one can obtain a theoretical maximal life duration (in the field), which should be observed if the first hatched larva could survive a long enough time to produce the last disappearing adult. This method is the same for each instar: as an example, the maximal life duration of a larva will correspond to the case for which the first hatching larva would be also the last to moult, giving a protonymph.

To evaluate the average life duration of each instar, a parameter more closely related to the natural conditions, all modal points are joined together (corresponding to the maximum number of each instar), starting from the egg numbers. This method postulates that gravid females lay their eggs immediately (this could be inaccurate because a delay may exist between the imaginal moult and the first clutch, see Fernandez and Athias-Binche, 1986). Then, data are collected and life durations may be calculated for an average generation (average of 2 to 3 complete field generations). These evaluations are rough and overall because the exact age of each individual is never known in field invertebrate populations.

In the soil-dwelling Uropodid mites, the average total life duration never exceeds one year and varies from 8 months, in *O. minimus*, to 11 months in *T. lamda* (Table 5). Adults are short lived, seldom exceeding 1 month in field (the life duration might be longer under laboratory conditions). The most part of the soil Uropodid life consequently occurs as an immature stage (as in the case of many other Arthropods), a period which covered 6 to 9 months in the Massane forest conditions. Postembryonic development extended from a minimum of 76% in *O. minimus* (short living species reproducing early) to a maximum of 90% in *A. coriacea* (Table 5). These average data might vary with the season, especially in *O. minimus*. It is to be noticed that these values are only valid for soil Uropodid mites or other edaphic mites. In relation to the habitat and the correlated adaptative strategies, the relative duration of the postembryonic development may be considerably reduced, some juvenile instars could even be eliminated by ovolarviparity as in Gamasids or Uropodids colonizating fluctuating microhabitats (Athias-Binche, 1987) or some parasitoid mites, like Tarsonemida (Lindquist, 1986).

Life duration is longer in the dead wood-living species *A. flagelliger*, with a maximum life duration of 22 months and an average generation time reaching 12 months. This length is due to the period of the phoretic travel of the deutonymph, which emigrates during spring and returns in autumn. The post-embryonic development is interrupted during the phoretic period and the imaginal moulting is delayed (Athias-Binche, 1979, 1984a,b). The relative duration of the postembryonic development occupies 75%, i.e. slightly less than for *O. minimus*.

Table 5

Life duration of the different instars (months) and total life duration of an average generation of 6 populations of soil Uropodina, and *A. flagelliger*, phoretic species living in dead wood.

		L	PN	DN	A	+	DA
A. flagelliger	MAX	9·5	10·0	11	11·5	22·5	9
	X	3·6	5·1	9·2	9·7	12·5	2·5
O. minimus	MAX	3·0	5·2	7·0	8·2	10·2	2·0
	X	1·5	4·1	6·1	7·6	8·0	0·4
N. catalonica	MAX	4·7	7·2	9·5	9·7	12·5	2·8
	X	2·3	6·0	7·7	9·6	10	0·4
U. carinatus	MAX	5·3	8·0	9·1	9·4	12·7	3·3
	X	2·6	6·6	8·5	8·8	9·9	1·1
V. coriacea	MAX	2·5	8·0	10·0	10·5	11·5	1·0
	X	1·2	5·2	9·0	9·5	10·0	0·5
T. lamda	MAX	6·7	8·2	10·2	11·2	14·2	3·0
	X	3·8	7·5	9·2	10·7	11·5	1·3
P. quadrangularis	MAX	4·5	7·7	9·5	10·5	12·0	1·5
	X	3·0	6·1	8·6	9·5	10·0	0·5

MAX: theoretical maximum duration. X: average durations; L: larvae; PN: protonymphs; DN: deutonymph; A: adult; +: end of the cohort; DA: life duration at the adult stage.

F. Growth in Size and Weight

The average age of each instar being known, it is possible to draw size/weight growth curves (Fig. 9). Different methods are possible (cf Leveque *et al.*, 1977), herein, results are expressed as percentage of the total life duration (pivotal ages, 100% corresponding to the death of the last adults) plotted against body length or mass, 100% corresponding to the adult length or mass.

In terms of lengths, the larval size is proportionally larger in small species; the larval length represents more than 60% of adult length in *O. minimus* and *A. coriacea.* Moreover, in these two species, the growth curve is sub-linear from larva to adult. In the other species, of which the larval body length represents less than 60% of the adult size, one can observe an inflexion at the change from protonymph to deutonymph, which marks a sudden size increase. In *U. carinatus*, a large species with pronounced cuticular sculpturing, this threshold is more delayed, occuring at the deutonymph to adult moult.

As regards weight, in the smallest species, the larvae represent more than 25% of the adult mass. The growth in weight is linear from larva to deutonymph, the imaginal moult is marked by a sudden mass increase, more or less marked, depending on the species, which is mainly due to the cuticular sclerotization in the adult. In the other species, the weight increase is earlier. In *U. carinatus*, in which the larva represents only 11% of the adult mass, the weight increase speeds up from the protonymph, and the change from deutonymph to adult (which is relatively very rapid) is marked by a very important gain of mass, which is due to the development of sclerotized and ornamented integuments. Thus, the growth curve is not linear, but more related to a parabolic model. One can notice very similar models of growth in the two members of the Polyaspididoidea, *T. lamda* and *P. quadrangularis*.

These data demonstrate that the small species devote a large part of their productivity to reproduction, whilst on the contrary, the largest species use the major part of their production for growth. These observations may be generalized to many other Arthropod species.

V. DEMOGRAPHIC ASSESSMENTS, PRODUCTION, P/B RATIO, DEMOGRAPHIC STRATEGIES

A. Methods

Demographic parameters were calculated for an average generation (mean of two to three successive generations) and for a one year (January to December 1976) period during which all populations presented at least one complete generation.

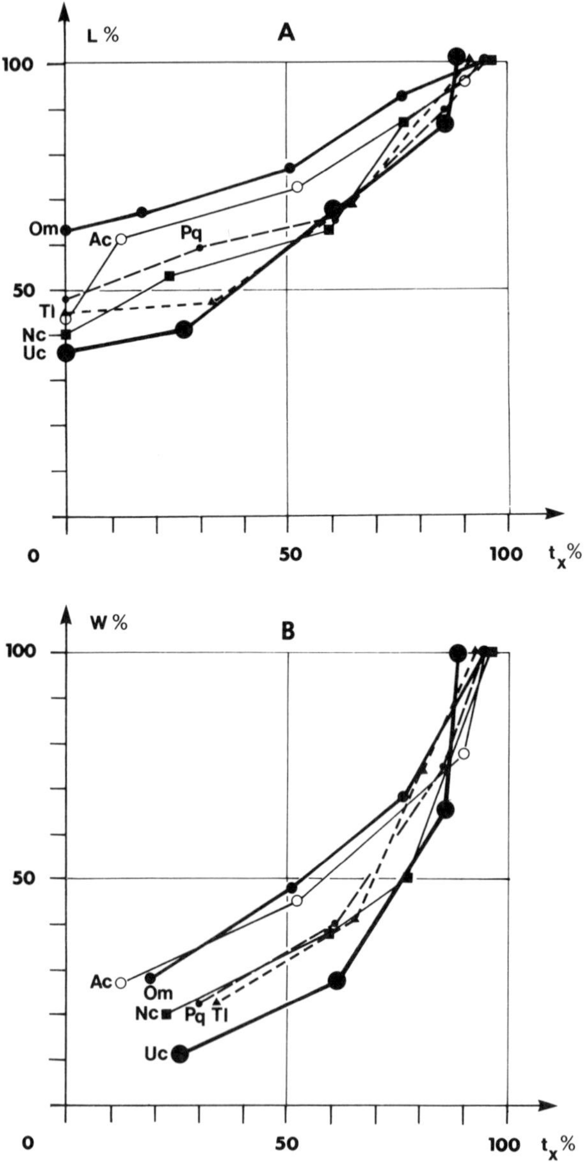

Fig. 9. Growth in size and weight in soil Uropodina. A. Size-age curves from egg to adult. L%: percentage of the adult body length, t_x%: percentage of the total life duration. B. Weight/age curves from larva to adult. W%: percentage of the adult weight. Ac: *A. coriacea*, Nc: *N. catalonica*, Om: *O. minimus*, Pq: *P. quadrangularis*, Tl: *T. lamda*, Uc: *U. carinatus*.

Total egg laying is equivalent to the number of mature eggs (just about to be laid) observed in females. This number is added to the cumulative abundance of larvae. The result allows one to evaluate the abundance of produced, then hatched eggs; it represents a minimal estimation, because death rate between laying and larval hatching is not known.

The immature stage recruitment is estimated by the cumulative number of larvae which were born during one generation or one year. These numbers have to be corrected when the fundamental equation of population dynamics is not checked. In fact, larvae are very sensitive instars which could die during the funnel extraction; in addition some of them might moult between two sampling dates. The fundamental equation (cf Williamson, 1972) is written as follows: $n_{x+1} = n_x + B - D$, where n = the total abundance of the population resp. at the x + 1 and x fortnights, B = the birth number and D = the deaths (supposing no immigration/emigration movements in edaphic populations). The number of births, B, has to be corrected when $n_{x+1} > n_x$ in order to balance the equation. The corrected value of B is often similar to the protonymph abundances, because some larval moults may occur between two successive samplings.

The number of adults actually hatched is equal to the observed maximum abundance of the adults of each generation (cf Figs 5 to 8).

The fecundity of one generation is equal to the total number of eggs produced during one generation divided by the number of the parent adults (EGG/A). This ratio is a minimum estimation provided by field observation; in laboratory conditions, a female could produce 30 to 60 eggs during its life because of a lower death rate and an artificially prolonged life duration. Annual fecundity is evaluated by the total egg/adult ratio.

The larval recruitment allows one to evaluate the births. As for fecundity, it is a minimum estimation because the mortality between hatching and the larval/protonymphal moulting is not known. Birth rate is equal to the recruitment divided by the average density of the population, either during one generation or one year. The annual birth rate is frequently higher than the natality for one generation, mostly in early mature and short-lived species, such as *O. minimus*, because reproductive periods are more numerous during the year than in the case of long-lived species (Table 6).

The net reproductive ratio R_0 is calculated by the Southwood (1978) simplified formula. The generation duration t being known, $R_0 = n(x + t)/nx$, with n, the abundance of instar x.

B. Demographic Assessments

The average fecundity of the soil Uropodina during one generation in the Massane conditions reached 6·15; the fecundity was the highest in *N. catalonica* and the lowest in *O. minimus* (Table 6A). The average birth rate

Table 6
Demographic assessments of 7 uropodid populations in the Massane conditions.

(A)	AP	E	R	nxA	F	N	R_0
O. minimus	5·8	15·3	13·7	5·2	2·6	0·46	0·89
N. catalonica	2·6	38·4	28·6	2·2	14·7	0·57	0·84
U. carinatus	0·7	5·8	5·0	0·8	8·2	0·52	1·14
A. coriacea	2·6	8·3	7·4	2·6	3·2	0·55	1·00
T. lamda	0·8	3·5	2·8	0·8	4·3	0·58	1·00
P. quadrangularis	1·6	6·3	5·5	1·4	3·9	0·57	0·87

(B)	AP	E	R	F	N	d_{JUV}	d_A
O. minimus	11·1	37·8	28·3	3·4	17·4	36·0	52·3
N. catalonica	7·1	55·2	50·5	7·7	12·8	5·2	59·0
U. carinatus	1·1	10·5	8·0	9·5	15·7	7·9	67·5
A. coriacea	2·3	11·7	7·5	5·1	9·3	12·7	41·5
T. lamda	1·7	12·4	9·0	7·3	16·6	22·2	73·7
P. quadrangularis	3·6	7·7	6·6	2·1	6·7	10·1	42·8
A. flagelliger	245·8	3687·0	2352·0	15·0	18·5	36·8	88·5

A: assessment of an average generation, B: annual assessments.
AP: number of parent adults ($N/m^2/100$ in soil, N/1000 cc in wood), E: egg number; R: recruitment; nxA: new young adults; F: fecundity; N: birth rate; R_0: net rate of reproduction; d_{JUV}: juvenile death rate for 100 eggs (E–DN/E); d_A: death rate from egg to young adult.

for one generation attained 5·4; this ratio was similar in all the studied species, it was only slightly lower in *O. minimus*. These results were very different for the annual assessments because species with a rapid turn-over, such as *O. minimus* and *T. lamda*, exhibited an annual fecundity lower and a birth rate higher than that of long lived populations (compare Tables 6A and 6B).

The average R_0 ratio reached 0·95 during the studied period (Table 6A); thus, the edaphic population level was rather stable from one generation to another, in the conditions studied, with no major disturbances. The results might be different in the case of populations inhabiting fluctuating micro-habitats. As an example, in the lignicolous phoretic Uropodina *A. flagelliger*, the R_0 value depended on the dynamics of the population, this ratio is the highest in the colonizing period and it decreases in relation to the stage of decomposition of the dead wood (Athias-Binche, 1979b).

C. Production and P/B Ratio

Secondary production is the living material elaborated by a population. It is essentially the sum of production for growth (mass growth, exocuticle elaboration, etc.) and of production for reproduction (production of gametes and mainly egg hatchings).

Usually, production is calculated for a one year period, this method allowing comparison between different populations. Growth production is calculated using the Boysen-Jensen formula (cf Leveque *et al.*, 1977), where $P = n_2(W_2 - W_1) + (n_1 n_2) \cdot (W_2 - W_1)/2$. It can be written more simply: $P = (n_1 + n_2) \cdot (W_2 - W_1)$, where P = the production from time 1 to time 2, n_1 and n_2 = the population density and W_1 and W_2 = the average biomasses. Average biomass is calculated by dividing the total mass of the population by the number of its individuals: the average individual biomass increases when the population lives through a growth period. In the case of the Massane forest, the annual production is the total sum of the productions calculated for all the fortnights in year 1976.

Production for reproduction is estimated following the method used by Lavelle (1978). In our case, the egg mass is estimated as equivalent to the larval biomass (the egg mass is in fact a slight under-estimation because the weight of the chorionic membranes is neglected). The annual production for reproduction is estimated by the total biomass of the eggs produced during one year.

The P/B ratio (production/average biomass) provides an estimation of the turn-over of the population biomass (Leveque *et al.*, 1977; Lavelle, 1978; Lamotte and Meyer, 1978). Annual average biomass equals the cumulative biomass of a population during one year, divided by the number of sampling dates (here $B = B/24$). The P/B ratio allows one to compare various species because it is independent of sample size or collecting method. This ratio decreases with life duration: a long-lived species usually reproducing more slowly than short living forms (Table 7, 8).

Table 7
Production and P/B rate of the Massane Uropodina.

	Pr/♀	Pr/A	Pc	Pr	B	Pc/B	P/B	P%
Om	0·83	0·83	5·87	12·85	12·40	2·21	7·0	68·6
Np	1·46	0·73	11·26	27·06	17·17	1·57	5·3	70·6
Uc	1·07	0·53	17·38	87·50	10·34	1·68	9·5	33·4
Ac	1·33	0·66	2·27	4·20	1·20	1·83	5·2	65·0
Tl	1·67	1·67	4·63	5·22	0·79	5·80	15·0	53·0
Pq	0·88	0·44	1·27	3·38	1·30	0·97	3·5	72·7
Af	1·30	0·65	13·86	22·58	10·42	1·33	3·5	62·0

Pr/♀: ratio production for reproduction/average female biomass; Pr/A: same ratio for all the adults; Pc: annual growth productivity (mg/m^2 for the soil population and mg/1000 cc of dead wood for *A. flagelliger*); Pr: annual productivity allowed to reproduction; B: annual average biomass; Pc/B: P/B ratio for the production allocated to growth; P/B: same ratio for the total production (reproduction + growth); P% = 100 Pr/(Pr + Pc), percentage of the production for reproduction; *Om: O. minimus, Np: N. catalonica, Uc: U. carinatus, Ac: A. coriacea, Tl: T. lamda, Pq: P. quadrangularis, Af: A. flagelliger.*

The results about Uropodina concern the proportion of the biomass devoted to egg production per female and per total number of adults, the production for growth calculated using the Allen equation, the production for reproduction, the annual average biomass, the P/B ratio for growth production and for the total production, and the percentage of production allocated to reproduction (Table 7).

T. lamda was the species producing the highest amount of living matter as eggs with a ratio egg production/adult biomass greater than 1 (Table 7: Pr/A). This species exhibited a low production devoted to sclerotization of the exocuticle, because it is a member of the superfamily Polyaspididoidea (cf Fig. 3). In addition, *T. lamda* is parthenogenetic, and consequently, it is characterized by a high birth rate (Table 7; B). In contrast, *P. quadrangularis*, is also a Polyaspididoidea, but it is a bisexual form with a low birth rate, thus the ratio of the production for growth is more important. In the other species, which all belong to the Uropodidoidea, characterized by a more sclerotized integument, the proportion of production for growth is higher. Among them, *O. minimus*, exhibited the highest rate of production allocated to reproduction (Table 7: Pr/A) because it is a small and parthenogenetic species with a high annual birth rate. On the contrary, in the case of a large bisexual species like *U. carinatus*, with thick, sclerotized and ornamented integuments, the part of production used for gametes is much less. In addition, the shape of the size/mass growth curve is very different in these two species: the growth curve is characterized by extensive production for growth at the moment of the adult appearance in *U. carinatus*.

Growth productivity (Table 7: Pc) was the highest in *N. catalonica* and *U. carinatus*, because the first species presented a density equivalent to that of *O. minimus*, but its individual mass is twice as high (Table 4), and the second of these is a large bodied form.

Production of the soil Uropodid community reached about 43 mg/m^2/year for growth, and 140 mg/m^2/y. for reproduction, that is to say a total production of 183 mg/m^2/y. (1·8 kg/ha/y.).

It will be seen that production for reproduction exceeds production for growth; this result raises the problem of the value of data in the literature for comparative analysis: few authors indicate whether the P/B ratio was calculated from data derived from the growth production only or from the total production (cf Table 8). Whatever the solution of this question, it has to be noticed that the Uropodid productivity in the Massane forest is certainly lower than in temperate forest ecosystems, where the mite biomass is higher because of better microclimatic and edaphic conditions (Athias-Binche, 1981b, 1983a). The P/B ratio of the Massane Uropodid community (with P, the total productivity) reaches an average of 7·6. It is a high value compared with the data of Table 8, but, as already stated, the significance of P (either growth or total productivity) is seldom indicated. Taking into account the total productivity (growth + reproduction), the P/B values observed in the

Table 8
Examples of P/B ratios in various organisms from several authors.

Organisms	P/B	Life duration (years)	Authors
Vertebrates			
Fish	0·3–3·5	1·25–17	Several in Leveque *et al.* (1977)
Perca fluviatilis	0·3	13	*Ibid.*
Reptilia	2·3–3·3	1	Barbault (1974) in Leveque *et al.* (1977)
Bird (*Clethrionomys glareolus*)	3·8	1	Several in Lamotte *et al.* (1977)
Invertebrates			
Marine Nematods	9·4–10·4	0·4	De Bovée (1981)
Arenicola marina	1·14	3	Several in Warwick (1980)
Lumbricus terrestris	0·3–0·5	4–8	Several in Leveque *et al.* (1977)
Oligochaeta	1·1–2·1	2–4	Several in Lamotte *et al.* (1978)
Mollusca	0·13–5·8	1–13	Several in Leveque *et al.* (1977)
Cardium edule	0·69	3·5	Several in Warwick (1980)
Crangon septemspinosa	3.82	3	*Ibid.*
Insecta	2·5–6·3	1–2	Several in Leveque *et al.* (1977)
Heteroptera	9·8–11·8	0·25–0·3	Several in Lamotte *et al.* (1978)
Orthoptera	9·6	0·3	*Ibid.*
Oribatida	1·6–7·5	0·25–2	Several in Luxton (1982)
Collembola	1·4–4·8	—	*Ibid.*
Dipteran larvae	1·2–1·6	—	*Ibid.*

Massane beechwood are close to those of Insects and Oribatid mites (Table 8). Considering the growth production only, the Massane data would be more closely related to the lowest values given for soil Microarthropods. These latter are the most plausible, Uropodina exhibiting a low growth productivity when compared with more abundant groups, like Oribatida or Collembola. These comments underline once more the existence of ambiguity concerning the P/B ratio according to the type of productivity considered.

The P/B ratio is very high in *T. lamda* and *U. carinatus* because the average biomasses of these litter-dwelling species were very low in the Massane forest (Table 7), and because the productivity for reproduction was important in the first species and growth productivity was high in the second one. Apart from these two examples, *O. minimus* exhibited the highest biomass turnover; the other Uropodina; having a P/B ratio close to 5. *A. flagelliger* was characterized by a P/B ratio lower than that of the soil Uropodina, this result being due to the life duration of this species, which is subject to a temporary arrest of its post-embryonic development during the phoretic period (Athias-Binche, 1984a). The Uropodina studied devoted an average of 60% of their productivity to reproduction (Table 7: P%): in spite of hatching a low number of eggs, these are large-sized, thus the female diverts a large part of its productivity to egg formation. The percentage P% was the lowest in *U. carinatus*, which allocated most of its productivity to the composition of a thick, ornamented cuticle.

D. Demographic Strategies

One can observe two kinds of edaphic populations in the Massane forest, on one hand species like *O. minimus* or *T. lamda*, characterized by high birth rate, high reproductive effort and high immature death rate, and on the other hand, populations with lower birth-rate and juvenile mortality. The demographic features of these different populations are compared by means of 6 parameters, selected in order to avoid data based on volume or area scales (*A. flagelliger* not being sampled in the same way as for soil Uropodina). The 6 chosen parameters were: fecundity, annual birth rate, average life duration, juvenile mortality, P/B ratio and percentage of the productivity devoted to reproduction. The species were compared 2 by 2, using the correlation Bravais-Pearson coefficient.

The correlation matrix clearly selects two groups of species (Table 9). One observes a first cluster with the pair constituted by *A. coriacea* and *P. quadrangularis*, then *N. catalonica* and *U. carinatus* are linked to this first group with a slightly lower correlation. The second group is constituted by the pair *O. minimus*/*A. flagelliger*, then *T. lamda* forming the connection between this group and the first one. The first group contains species

Table 9

Interspecific correlation matrix calculated from 6 demographic parameters: birth rate, fecundity, average life duration, immature death rate, P/B ratio, and percentage of production allocated to reproduction. Bold types: correlation coefficient significant at the 99·9% level, italics: id. at the 99% level.

Ac	**0·999**					
Np	**0·992**	**0·991**				
Uc	*0·944*	0·974	*0·969*			
Tl	*0·961*	*0·968*	*0·941*	*0·931*		
Af	0·904	*0·931*	0·866	0·958	**0·974**	
Om	0·914	*0·920*	0·873	0·843	**0·978**	**0·974**
	Pq	*Ac*	*Np*	*Uc*	*Tl*	*Af*

Ac: A. coriacea, Np: N. catalonica, Uc: U. carinatus, Tl: T. lamda, Af: A. flagelliger, Om: O. minimus, Pq: P. quadrangularis.

exhibiting K-strategist trends, with low birth rate, and the second group is more representative of rather *r*-selected species. Between these two sets, *T. lamda* appeared to be an interlinking species.

It has been demonstrated in previous papers that structure and diversity of communities evolve in relation to the stage of ecological maturity of the soil sub-system (Athias-Binche, 1981b, 1983a). In the same way, demographic strategies can be linked with the type of biotope.

One can distinguish on one hand populations characterized by a high growth potential and which are able to quickly disperse when the environmental conditions become less favourable. These are like the "wandering" bird species of Blondel (1975): they spend a lot of energy for reproduction but have a high juvenile death rate, they are characterized by a high ability to colonize ecologically new biotopes and their densities often fluctuate. The mode of scattering and dispersal are well developed. These ideas could be easily applied to the Uropodina and many other phoretic Acari (Gamasina, Tarsonemina, Acaridida) using the example of *A. flagelliger*: it is an *r*-selected species, with a high colonizing capacity being obligatory phoretic, with a high birth rate and exhibiting marked annual and seasonal fluctuations in number (Athias-Binche, 1984a,b). This kind of demographic strategy may be observed in uropodid species colonizing unpredictable habitats or specialized microenvironments: e.g. dead wood, guano, carcasses, dung. One can find also *r*-strategists in soil, able to colonize either degraded, burnt or juvenile soils poor in organic matter such as rankers. *O. minimus* is a typical example of such opportunist and euryecic species. This species, is spread widely over Europe and North Africa, becomes hyperdominant in all ecological niches left empty by other species requiring less severe conditions (Athias-Binche, 1983a). This pioneer species can adapt to life in poor soils, and it is able to colonize special microhabitats, such as ant-nests, and even artificial substrates such as pure cellulose put in litter bags *in situ* (Athias and Mignolet, 1979). In addition, it has high dispersal potentialities because it is a facultative phoretic (Athias-Binche, 1984a). At the opposite extreme, in more stable habitats or more evolved soils, the populations do not migrate very much and possess a low birth rate; they are the "stay-at-home" species of Blondel. In the uropodid community of the Massane forest, this is the case in *P. quadrangularis*, *A. coriacea*, *U. carinatus* or *N. catalonica*, which are mostly found in the rich and biologically active acid mull (Athias-Binche, 1983b). Facultative phoresy is less frequent in such speices than in more *r*-selected populations.

In fact, the *r*-*K* gradient is clearly distinguishable only in some extreme examples, and one can observe a series of intermediate cases. In order to illustrate these transitions, one can make a scatter diagram showing the relative situation of the species in relation to their juvenile mortalities and their birth rates (Fig. 10). The typical *r*-selected species (*A. flagelliger* and *O. minimus*) are situated in the upper right corner of the diagram. Then, one can find transitional species on the left part of the *r*-*K* gradient: *T. lamda*, a strict litter dwelling mite, then *U. carinatus*, less dependent on the litter, and finally *N. catalonica*, indifferent toward the layer. A group of two species are isolated from the previous ones, *A. coriacea* and *P. quadrangularis*, which are strict humicolous forms, non phoretic and characterized by a very low birth rate. These results probably relate to the fact that the litter is more unstable than the deeper layers of the soil: hydric and thermic fluctuations are more marked, the cycle of the litter fall and the seasonal variations of decomposition activity modify the conditions in the different layers of the litter. Furthermore, in the Massane forest, structure and thickness of the litter are frequently disturbed by winds, which are very frequent and violent in the Roussillon region (Athias-Binche, 1981d). The true humicolous populations live in a more stable habitat, but with more scarce trophic resources, consequently, they are mainly confined to the most decomposed humus.

One of the consequences of these remarks is that, paradoxically, litter-dwelling Uropodina (like many other soil Arthropoda) are characterized by demographic strategies more closely related to species living in unpredictable or non-edaphic habitats, than to humicolous species, although living in the same soil system.

To conclude, it must be said that the present quantitative results (Figs 4 to 8, 10; Tables 5 to 7, 9) apply only in the Massane forest, a mediterranean beech-wood, therefore an ecosystem situated at its biological limits. In addition, depending on the conditions or the ecological evolutionary stage of the habitat, the same species, or even the same population, could move from a *r*-type pattern to a *K*-strategy. This is clearly the case in *A. flagelliger*: during the first period of colonization, the population exhibited a very marked expanding *r*-dynamics. Then, in connection with dead wood decom-

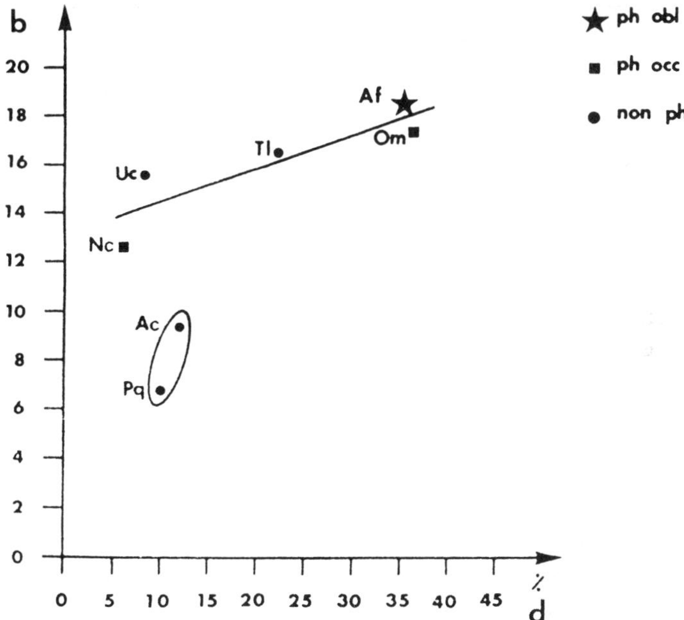

Fig. 10. Massane Uropodina; scatter diagram of juvenile death rate (d) plotted against birth rate (b). ph. obl.: obligatory phoretic, ph. occ.: occasional or facultative phoretic, non ph.: non phoretic, for species names, see Fig. 9.

position, birth rate decreased, emigration dominated over immigration, the sedentary deutonymph proportions increased and, finally, the dynamics became more closely related to a *K*-strategy (Athias-Binche, 1979b). Some disturbances or unpredictable events could unmask *r*-strategies previously not expressed. As an example, during a litter bag experiment in Belguim, *O. minimus* was shown to actively colonize an artificial cellulose substrate, and thus exhibited a pioneer behaviour, whereas this species was a member of the equilibrium community in the control soil (Athias and Mignolet, 1979).

VI. CONCLUSIONS

Despite the fact that Uropodina do not constitute a very numerous group in the Acarina, they give an interesting model for the study both of their ecological and morphological differences and for the wide range of habitats which they colonize. So, the methods used for the present study may be easily applied to many other Acari and soil Arthropoda.

These various biological types are associated with a large spectrum of ecological variables and, thus, with rather numerous types of demographic strategies. Even considering only edaphic communities, one can find pioneer *r*-selected species characteristic of immature-type soils with a low amount of necromass, less prolific species in more evolved soils and litter, and then more *K*-selected species in deeper layers of soil.

Some of the ecological and demographic parameters of the *r* and *K* strategies, as defined by Pianka (1970) and reviewed later by many authors (see for instance Krebs, 1978 or Barbault, 1981), cannot be strictly applied to Acari and small Arthropoda. In actual fact, these ideas have been proposed from studies of Vertebrates, mainly Lizards, and, in addition, Uropodina and many other Acarina exhibit specific adaptations, for instance parthenogenesis and phoresy may play an important role in demographic patterns. The presence of an external inert cuticle, of which the relative mass can be important in some species, may have also an influence on production. Body size can also have a different significance than in Vertebrates, in soil mites, small dimensions could be an advantage in some conditions. For instance, only the smallest mites could migrate down during dry or hot periods, while large species are not able to reach the deepest layers (which might be the reason why large Uropodid species, usually litter-living, are more scarce in mediterranean soils than in deciduous temperate forest ecosystems, Athias-Binche, 1978, 1982b).

When comparing the demographic parameters in Uropodina, it may be better to take into account species living in a similar habitat. For instance, *O. minimus* and *A. flagelliger* are both *r*-selected species, but the main parameters of their dynamics are not the same, *O. minimus* is a small, short-lived species with a high P/B ratio, while *A. flagelliger* is larger, long-lived, with a low P/B and of which the dynamics are especially dependent on the migratory balance. Consequently, it could be better to select two edaphic species, *O. minimus* and *U. carinatus*, for summing up the main demographic strategies. These species were chosen because they contrast in size, demography and production; they represent two extreme cases, and many Uropodina exhibit intermediate features.

These two species differ in usual parameters, such as size, life duration, birth rate or juvenile mortality (Table 10). One of the basic differences concerns the productivity management, *O. minimus* allocates much energy for reproduction and lack of males leads to a relative increase in the part of productivity devoted to reproduction. In contrast, in *U. carinatus*, a large part of production is devoted to growth, it goes towards male production, body size, and development of a thick sclerotized cuticle, with a low metabolic demand, but which could be an advantage in water balance or for

Table 10

Comparison of the main features involved in the demographic strategies in *O. minimus* and *U. carinatus*.

Parameters	*O. minimus*	*U. carinatus*	References
habitat	less specialized	litter	tab. 1
climate	cold to mediterranean	mesomediterranean	chapter A.
population density	very variable	medium	chapter A.
community diversity	very variable	high	Athias-Binche (1981a)
stenotopy	low	medium-high	tab. 1
dominance	high in poor soils	never dominant	Athias-Binche (1981a)
good competitor	yes	no	Athias-Binche (1983a)
body size	small	large	tab. 2
relative duration of development	fast	slower	tab. 5
life duration	short	medium	tab. 5
parthenogenesis	yes	no	chapter A.
% growth production	low	high	tab. 7
% reproduction production	high	low	tab. 7
fecundity	low	high	tab. 6
birth rate	high	medium	tab. 6
juvenile mortality	high	low	tab. 6
survivorship curve	close to type II	close to type I	fig. 10
P/B	high	high, but large species	tab. 7
growth in mass	sub-linear	exponential-like	fig. 9
phoresy	facultative	absent	chapter A.

predation defence. Parthenogenesis may constitute an advantage in *r*-colonizing species because reproduction may occur even in poor habitats with a low population density. In bisexual forms, the probability of meeting between the sexes is, on the contrary, lower in case of scattered individuals.

The two species studied give an example of two different kinds of management for productivity, reproduction and space occupation. Some components of this management are intrinsic and invariant (apart from the effect of genetical factors) e.g. relative size of eggs, cuticle sclerotization, growth model, parthenogenesis or capacity for phoresy. On the contrary, some other parameters may vary, depending on environmental factors, i.e. death and birth rate, density and phenology; this is notably the case in *O. minimus* which possesses a certain ecological plasticity.

Whatever we think (anthropomorphically?) about such or such an advantage of the different types of demographic strategies, it appears to be obvious that a species possessing demographic characteristics closely related to those of *U. carinatus* is not able to withstand a major disturbance of the biotope. On the contrary, *O. minimus* is able to develop more varied behaviour, moving in deep layers, higher reproductive capacities, ability to regulate deutonymphal moult and thus, initiation of adulthood, occurrence of facultative phoresy. In addition, it has a low level of dependence on biological soil and litter activity, unlike most of the other soil Uropodina (Athias-Binche, 1982b). These different features allow this species to withstand severe disturbances, such as wild fire; after a fire or in very poor soils, *O. minimus* is practically the only species to survive (Athias-Binche, 1987b).

To conclude, I would like to mention the role of parthenogenesis in dynamics and adaptative strategies of Invertebrate populations. The problem of parthenogenesis is often approached from an evolutionary, biological or genetical point of view; its significance for ecology or demography is less frequently analysed. One of the reasons could be the fact that the concepts of population dynamics have been mostly studied in vertebrate populations. Parthenogenesis constitutes nevertheless a frequent phenomenon in invertebrates, especially in mites. As suggested by the present results, thelytoky may play an important role in opportunist and *r*-selected soil Uropodina, and it is possible that arrhenotoky exists in obligatory phoretic Uropodina, as in some Gamasid mites like phoretic Macrochelidae (Athias-Binche, 1987a). Thus, parthenogenesis has to be considered as a significant parameter of the demography of Invertebrates; it is not the only factor occurring, but it has to be considered. In actual fact, parthenogenesis appears to be an advantage in the case of low population density or colonizating dynamics, especially when it is associated with phoresy.

ACKNOWLEDGEMENT

I am very grateful to Prof. A. Macfadyen for his kind criticism of the manuscript and his helpful comments.

REFERENCES

Anderson, J. M. and Healy, I. N. (1970). Improvements in the gelatine-embedding technique for woodland soil and litter sample. *Pedobiologia*, **10**, 108–120.

Athias-Binche, F. (1978). Observations morphologiques sur *Allodinychus flagelliger* (Berlese 1910). Au cours du développement postembryonnaire (Acariens, Uropodides). *Acarologia*, **20**, 44–57.

Athias-Binche, F. (1979a). Etude quantitative des Uropodides (Acariens, Anactinotriches) d'un arbre mort de la hêtraie de la Massane. 1. Caractères généraux du peuplement. *Vie Milieu*, **27**, 157–175.

Athias-Binche, F. (1979b). Etude quantitative des Uropodides (Acariens, Anactinotriches) d'un arbre mort de la hêtraie de la Massane. 2. Elèments démographiques d'une population d'*Allodinychus flagelliger* (Berlese 1910). *Vie Milieu*, **28**, 35–60.

Athias-Binche, F. (1981a). Différents types de structures des peuplements d'Uropodides édaphiques de trois écosystèmes forestiers (Arachnides, Anactinotriches). *Acta Oedologica, Oeco. gener.*, **2(2)**, 153–169.

Athias-Binche, F. (1981b). Ecologie des Uropodides édaphiques (Arachnides, Parasitifirmes) de trois écosystèmes forestiers. 1. Introduction, matériel, biologie. *Vie Milieu*, **31(2)**, 137–147.

Athias-Binche, F. (1981c). Ecologie des Uropodides édaphiques (Arachnides, Parasitiformes) de trois écosystèmes forestiers. 2. Stations d'étude, méthodes et techniques, facteurs du milieu. *Vie Milieu*, **31**, 221–241.

Athias-Binche, F. (1982a). Ecologie des Uropodides édaphiques (Arachnides, Parasitiformes) de trois écosystèmes forestiers. 3. Abondance et biomasse des microarthropodes du sol, facteurs du milieu, abondance et distribution spatiale des Uropodides. *Ibid.*, **32**, 47–60.

Athias-Binche, F. (1982b). Ecologie des Uropodides édaphiques (Arachnides, Parasitiformes) de troid écosystèmes forestiers. 4. Abondance, biomasse, distribution verticale, sténo- et eurytopie. *Ibid.*, **32**, 159–170.

Athias-Binche, F. (1983a). Ecologie des Uropodides édaphiques (Arachnides, Parasitiformes) de trois écosystèmes forestiers. 5. Affinités interspécifiques, diversité, structures écologiques et quantitatives des peuplements. *Ibid.*, **33**, 25–34.

Athias-Binche, F. (1983b). Ecologie des Uropodides édaphiques (Arachnides, Parasitiformes).6. Similarités interstationnelles. Conclusions générales. *Ibid.*, **33**, 93–109.

Athias-Binche, F. (1984a). La phorésie chez les Acariens Uropodides (Anactinotriches), une stratégie écologique originale. *Acta Oecologica, Oeco. Gener.*, **5**, 119–133.

Athias-Binche, F. (1984b). Phoresy in Uropodina (Anactinotrichida), occurrence, demographic involvements and ecological significance. In, *Acarology, 1*. Proc VI th Int.Congr.Acarology, Edinburgh 1982, D. A. Griffiths and C. E. Bowman (eds), Ellis Horwood. Chichester, 276–285.

Athias-Binche, F. (1985). Analyses démographiques des populations d'Uropodides

(Arachnides, Anactinotriches) de la hêtraie de la Massane, France. 1. Méthodes, matériel, biologie des cohortes, longévité, croissance linéaire et pondérale. 2. Bilans démographiques, production, rapport P/B et stratégies démographiques. *Pedobiologia*, **28**, 225–253.

Athias-Binche, F. (1987a). Signification adaptative des différents types de développements postembryonnaires chez les Gamasides (Acariens, Anactinotriches). *Can. J. Zool.* (Review), **65**, 1299–1310.

Athias-Binche, F. (1987b). Modalités de cicatrisation des écosystèmes méditerranéens après incendie. 3. Les Acariens Uropodides. *Vie Milieu*, **31**, 39–52.

Athias-Binche, F. and Evans, G. O. (1981). Observations on the genus *Protodinychus* Evans 1957 (Acari, Mesostigmata) with description of the male and phoretic deuteronymph. *Proc.Roy.Irish.Acad.*, **81**, 25–36.

Athias, F. and Mignolet, R. (1979). Colonisation de litière monospécifiques par les Uropodides (Acariens, Anactinotriches) d'une forêt belge. In, *Proc. 4th Int. Congress Acarology*, Saalfenden 1974, E. Piffl (ed.), Akademiai Kiado, Budapest, 101–110.

Athias-Binche, F. and Saulnier, L. (1986). Modalités de cicatrisation des écosystèmes méditerranéens après incendie, cas de certains Arthropodes du sol. 1. Introduction, stations d'étude. *Vie Milieu*, **36**, 117–124.

Athias-Henriot, C. (1976). Sur la bioécologie en France tempérée de *Leitneria granulata*, gamaside édaphique (Arachnides, Parasitiformes). *Pedabiologia*, **16**, 151–160.

Barbault, R. (1981). Ecologie des populations et des peuplements. Des théories aux faits. Masson, Paris, 200 pp.

Bieri, M; Delucchi, V. and Lienhard, C. (1978). Ein abageänderter Macfadyen-Apparat für die dynamisches Extraktion von Bodenarthropoden. *Bull. Soc. Ent. Suisse*, **51**, 119–132.

Block, W. (1967). Recovery of mites from peat and mineral soils using a new flotation method. *J. Anim. Ecol.*, **36**, 323–327.

Blondel, J. (1975). Stratégies écologiques et développement de l'ecosystème. In, J. C. Ruwet (ed.), *Problèmes liès à la gestion des Hautes Fagnes et de la Haute Ardenne.* CR Coll. Univ. Liège, Belgium, 71–99.

Blower, J. G. and Miller, P. F. (1977). The life history of the Iulid millipede *Cylindroiulus nitidus* in a Derbyshire wood. *J. Zool. Lond.*, **183**, 339–351.

Bovée, F. de (1981). Ecologie et dynamique des Nématodes d'une vase sublittorale (Banyuls sur Mer). Thèse d'Etat, Paris VI, 194 pp.

Deleporte, S. (1986). Biologie et écologie du Diptères Sciaridae *Bradysia confinis* (Winn.; Frey) d'une litière de feuillus (Bretagne Intérieure). *Rev. Ecol. Biol. Sol*, **23**, 39–76.

Fernandez, N. A. and Athias-Binche, F. (1986). Analyse démographique d'une population d' *Hydrozetes lemnae* Coggi, Acarien Oribate inféodé à la lentille d'eau *Lemna gibba* L. en Argentine. 1. Méthodes et techniques, analyse de la démographie d'*H. lemnae* et comparaisons avec d'autres Oribates. *Zool. Jb. Syst.*, **113**, 213–228.

Geoffroy, J. J. (1981). Modalités de la coexistence de deux Diplopodes, *Cylindroiulus punctatus* (Seach) et *Cylindroiulus nitidus* (Verhoeff) dans un écosystème forestier du bassin parisien. *Acta Oecologica, Oeco Gene.*, **2**, 227–243.

Gregoire-Wibo, C. (1980). Cycle phénologique de *Folsomia quadrioculata* en forêt (Insecte, Collemboles). *Ann. Soc. R. Belg.*, **109**, 43–65.

Gregoire-Wibo, C. and Snider, R. M. (1978). The intrinsic rate of natural increase, its interest to ecology and its application to various species of Collembola. *Ecol. Bull.* (*Stockholm*), **25**, 442–448.

Harding, J. P. (1949). The use of probability paper for a graphical analysis of polymodal frequency distribution. *J. mar. Biol. Assoc. U.K.*, **28**, 141–153.

Hihm, J. A. and Chang, H. C. (1976). Laboratory studies of the life cycle and reproduction of some soil and manure-inhabiting predatory mite (Acarina, Laelapidae). *Pedobiologia*, **16**, 353–365.

Hutson, B. R. (1981). Age distribution and the annual reproductive cycle of some Collembola colonizing reclaimed land in Northumberland, England. *Pedobiologia*, **21**, 410–416.

Kaliszewski, M., Błoszyk, J. and Sell, D. (1984). Phenology of *Tarsonemus lucifer* (Schaarschmidt 1959) (Acari, Heterostigmata). In, *Acarology 2*. Proc. VIth Int. Congr. Acarology, D. A. Griffiths and C. E. Bowman (eds), Ellis Horwood, Chichester, 929–931.

Krasinskaya, A. L. (1961). Morphological and biological characteristics of the postembryonic development of Uropodina from the Leningrad region (in Russian). *Parasit. Sbornik.*, **20**, 108–147.

Krebs, C. J. (1978). Ecology. The experimental analysis of distribution and abundance. 2nd edn, Sharper International, NY, 678 pp.

Lamotte, M. and Meyer, J. A. (1978). Utilisation du taux de renouvellement P/B dans l'analyse du fonctionnement énergétique des écosystèmes. *C. R. Acad. Sci. Paris*, **286(D) 19**, 1387–1390.

Lavelle, P. (1978). Les vers de terre des la savane de Lamto (Côte d'Ivoire), peuplement, populations et fonction dans l'écosystème. Thèse d'Etat, Paris VI, 301 pp.

Lasebikan, B. A. (1975). Characteristics and efficiency of four dynamic-type extractors. *Pedobiologia*, **15**, 29–39.

Lasebikan, B. A.; Belfield, W. and Gibson, N. H. E. (1978). Comparison of relative efficiency of methods for the extraction of soil microarthropods. *Rev. Ecol. Biol. Sol.*, **15**, 39–65.

Lebrun, P. (1977). Comparaisons des effets des températures constantes ou variables sur la durée de développement de *Dameus onustus* (Acarina, Oribatei). *Acarologia*, **19**, 136–143.

Lebrun, P. (1984). Determination of the dynamics of an edaphic oribatid population (*Nothrus palustris* C. L. Koch 1839). In, *Acarology II*. Proc. VIthe Int. Congr. Acarology, D. A. Griffiths and C. E. Bowman (eds), Ellis Horwood, Chichester, 860–870.

Leveque, C., Durand, J. R. and Ecoutin, J. M. (1977). Relations entre le rapport P/B et la longévité des organismes. *Cah. Orstom* (*Hydrobiol.*), **11**, 17–31.

Lindquist, E. E. (1986). The world genera of Tarsonemidae (Acari, Heterostigmata). *Meme. Ent. Soc. Canada,* **136**, 517 pp.

Longstaff, B. C. (1976). The dynamics of collembolan populations, competitive relationships in an experimental system. *Can. J. Zool.*, **54**, 948–962.

Longstaff, B. C. (1977). The dynamics of collembolan populations, a matrix model of single species population growth. *Can. J. Zool.*, **55**, 314–324.

Luxton, M. (1982). Quantitative utilization of energy by soil fauna. *Oikos*, **39**, 342–354.

Macfadyen, A. (1968). Notes methods for the extraction of small soil arthropods by the high gradient apparatus. *Pedobiologia*, **8**, 401–406.

Mitchell, M. (1977). Population dynamics of oribatid mites (Acari, Cryptostigmata) in an aspen woodland soil. *Pedobiologia*, **17**, 305–319.

Pande, Y. D. and Berthet, P. (1973). Comparison of the Tullgren funnel and soil section methods for surveying oribatid populations. *Oikos*, **24**, 273–277.

Petersen, H. (1980). Population dynamics and metabolic characterization of

Collembola species in a beech forest ecosystem. *In Soil biology as related to land use practices.*, Proc. VII Int. Soil Zool. Coll., Dindal J. L. (ed.), ISSS EPA, Washington, 806–833.

Pianka, E. R. (1970). On r- and K-selection. *Amer. Nat.*, **104**, 592–597.

Radinovsky, S. (1965). The biology and ecology of granary mites of the Pacific Northwest. 3. Life history and development of *Leiodinychus krameri* (Acarina, Uropodidae). *Ann. Ent. Soc. Amer.*, **58**, 259–267.

Schatz, H. (1983). Überlebensrate von *Oromurcia sudetica* Willmann (Acari, Oribatei) von einer alpinen Wiese Tirols (Obergurgl, Central Alps). *Zool. Jb. Syst.*, **110**, 97–109.

Schatz, H. (1985). The life cycle of an alpine oribatid mite, *Oromurcia sudetica* Willmann. *Acarologia*, **26**, 95–100.

Schrenker, R. (1986). Population dynamics of oribatid mites (Acari, Oribatei) in a forest soil ecosystem. *Pedobiologia*, **29**, 239–246.

Southwood, T. R. E. (1978). Ecological methods, with particular reference to the study of insect populations. Chapman & Hall, London, 524 pp.

Stamou, G. P. (1986). A phenological model applied to oribatid mites data. *Rev. Ecol. Biol. Sol*, **23**, 453–460.

Straalen, N. M. van (1985a). Comparative demography of forest floor Collembola population. *Oikos*, **45**, 253–265.

Straalen, N. M. van (1985b). Production and biomass turnover in stationary stage structured populations. *J. theor. Biol.*, **113**, 331–352.

Suomalainen, E., Saura, A. and Lokki, J. (1976). Evolution of parthenogenetic insects. *Evol. Biol.*, **9**, 209–257.

Thiebaut, B. (1982). Existe-t-il une hêtraie méditerranéenne distincte des autres forêts de hêtre en Europe occidentale? *Vegetatio*, **50**, 23–42.

Vallee, A. (1954). Intorno allo sviluppo postembryonale di "*Phaulotrachytes rackei*" (OUDM) (Acari, Phaulodinychidae). *Comm. pontifica Acad. Sci. Vaticano*, **16**, 291–314.

Warwick, R. M. (1980). Population dynamics and production of benthos. In, *Marine benthic dynamics*, K. R. Tendre and B. C. Doull (eds), Univ. South Carolina Press, 1–24.

Williamson, M. (1972). The analysis of biological populations. Edward Arnold, London, 180 pp.

Wood, T. G. and Lawton, J. H. (1973). Experimental studies on the respiratory rates of mites (Acari) from beech-woodland leaf litter. *Oecologia* (*Berl.*), **12**, 169–191.

Index

Advances in Ecological Research Volumes 1–18

Cumulative List of Titles